Fairness and Machine Learning

Fairness and Machine Learning

Limitations and Opportunities

Solon Barocas, Moritz Hardt, and Arvind Narayanan

The MIT Press
Cambridge, Massachusetts
London, England

The MIT Press would like to thank the anonymous peer reviewers who provided comments on drafts of this book. The generous work of academic experts is essential for establishing the authority and quality of our publications. We acknowledge with gratitude the contributions of these otherwise uncredited readers.

This book was set in Times New Roman by Westchester Publishing Services. Printed and bound in the United States of America.

Library of Congress Cataloging-in-Publication Data is available.

ISBN: 978-0-262-04861-3

10 9 8 7 6 5 4 3 2 1

Contents

Contents vii

Preface

Institutions of all kinds, from firms to governments, represent populations as data tables. Rows reference individuals. Columns contain measurements about them. Statistical machinery applied to these tables empowers their owners to mine patterns that fit the aggregate.

Then comes a leap of faith. We have to imagine that unknown outcomes, future or unobserved, in the life trajectory of an individual follow the patterns that have been found. We must accept decisions made as if all individuals were going to follow the rule of the aggregate. We must pretend that to look into the future is to look into the past. It's a leap of faith that has been the basis of consequential decisions for centuries. Fueled by early successes in insurance pricing and financial risk assessment, statistical decision making of this kind has found its way into nearly all aspects of our lives. What has accelerated its adoption in recent years has been the explosive growth of machine learning, often under the name of artificial intelligence.

Machine learning shares long-established decision-theoretic foundations with large parts of statistics, economics, and computer science. What machine learning adds is a rapidly growing repertoire of heuristics that finds decision rules from sufficiently large datasets. These techniques for fitting huge statistical models on large datasets have led to several impressive technological achievements. Image classification, speech recognition, and natural language processing have all made leaps forward. Although these advances often don't directly relate to specific decision-making settings, they shape narratives about the new capabilities of machine learning.

As useful as machine learning is for some positive applications, it is also used to great effect for tracking, surveillance, and warfare. Commercially, its most successful use cases to date are targeted advertising and digital content recommendation, both of questionable value to society. From its roots in World War II–era cybernetics and control theory, machine learning has always been political. Advances in artificial intelligence feed into a global industrial military complex and are funded by it. The success stories told about machine learning also support those who would like to adopt algorithms in domains outside those studied by computer scientists. An opaque marketplace of software vendors renders algorithmic decision-making tools for use in law enforcement, criminal justice, education, and social services. In many cases what are marketed and sold as artificial intelligence are statistical methods that virtually haven't changed in decades.

Many take the leap of faith behind statistical decision making for granted to the extent that it has become difficult to question. Entire disciplines have embraced mathematical models of optimal decision making in their theoretical foundations. Much of economic theory takes optimal decisions as an assumption and an ideal of human behavior. In turn, other disciplines label deviations from mathematical optimality as "bias" that invites elimination. Volumes of academic papers speak to the evident biases of human decision makers.

In this book, we take machine learning as a reason to revisit this leap of faith and to interrogate how institutions make decisions about individuals. Institutional decision making has long been formalized via bureaucratic procedures, and machine learning has much in common with it. In many cases, machine learning is adopted to improve and sometimes automate the high-stakes decisions routinely made by institutions. Thus, we do not compare machine learning models to the subjective judgments of individual humans but instead to institutional decision making. Interrogating machine learning is a way of interrogating institutional decision making in society today and for the foreseeable future.

If machine learning is our way into studying institutional decision making, fairness is the moral lens through which we examine those decisions. Much of our discussion applies to concrete screening, selection, and allocation scenarios. A typical example is that of an employer accepting or rejecting job applicants. One way to construe fairness in such decision-making scenarios is as the absence of discrimination. This perspective is micro insofar as individuals are the unit of analysis. We study how measured characteristics of an individual lead to different outcomes. Individuals are the sociological building block. A population is a collection of individuals. Groups are subsets of the population. A decision maker has the power to accept or reject individuals for an opportunity they seek. Discrimination in this view is about wrongful consideration on the basis of group membership. The problem is as much about what wrongful means as on what basis it relies. Discrimination is also not a general concept. It's domain specific as it relates to opportunities that affect people's lives. It's concerned with socially salient categories that have served as the basis for unjustified and systematically adverse treatment.

The first chapter after the introduction explores the properties that make automated decision making a matter of significant and unique normative concern. In particular, we situate our exploration of machine learning in a longer history of critical reflection on the perils of bureaucratic decision making and its mechanical application of formalized rules. Before we even turn to questions of discrimination, we first ask what makes automated decision making legitimate in the first place. In so doing, we isolate the specific properties of machine learning that distinguish it from other forms of automation along a range of normative dimensions.

Since the 1950s, scholars have developed formal models of discrimination that describe the unequal treatment by a decision maker of multiple different groups in the population. In chapter 3, we dive into statistical decision theory, allowing us to formalize a number of fairness criteria. Statistical fairness criteria express different notions of equality between groups. We boil down the vast space of formal definitions to essentially three different mutually exclusive definitions. Each definition resonates with a different moral intuition. None is sufficient to support conclusive claims of fairness. Nor are these definitions suitable targets to optimize for. Satisfying one of these criteria permits blatantly unfair solutions. Despite their significant limitations, these definitions have been influential in the debate around fairness.

Chapter 4 explores the normative underpinnings of objections to systematic differences in the treatment of different groups and inequalities in the outcomes experienced by these groups. We review the many accounts of the wrongfulness of discrimination and show how these relate to various views of what it would mean to provide equality of opportunity. In doing so, we highlight some tensions between competing visions of equality of opportunity—some quite narrow and others quite sweeping—and the various arguments that have been advanced to help settle these conflicts. With this in place, we then explore how common moral intuitions and established moral theories can help us make sense of the formalisms introduced in chapter 3, with the goal of giving these definitions greater normative substance.

Present in both technical and legal scholarship on discrimination is the idea of assigning normative weight to causal relationships. Was group membership the cause of rejection? Would the applicant have been rejected had they been of a different race? Would they have been accepted but for their gender? To understand these kinds of statements and the role that causality plays in discrimination, chapter 5 of this book is a self-contained introduction to the formal concepts of causality.

Following our formal encounter with fairness definitions, both statistical and causal, we turn to the legal dimensions of discrimination in the United States in chapter 6. The legal situation maps cleanly on to neither the moral foundations nor the formal work, complicating the situation considerably. The two dominant legal doctrines, disparate treatment and disparate impact, appear to create a tension between explicit consideration of group membership and intervening to avoid discrimination.

Expanding on both the causal and legal chapters, chapter 7 goes into detail about the complexities of testing for discrimination in practice through experiments and audits.

Studying discrimination in decision making has been criticized as a narrow perspective on a broader system of injustice for at least two reasons. First, as a notion of discrimination, it neglects powerful structural determinants of discrimination, such as laws and policies, infrastructure, and education. Second, it orients the space of intervention toward solutions that reform existing decision-making systems, in the case of machine learning typically via updates to an algorithm. As such, the perspective can seem to prioritize "tech fixes" over more powerful structural interventions and alternatives to deploying a machine learning system altogether. Rather than predicting failure to appear in court and punishing defendants for it, for example, perhaps the better intervention is to facilitate access to court appointments by providing transportation and child care. Chapter 8 introduces the reader to this broader perspective and its associated space of interventions from an empirical angle.

Recognizing the importance of a broader social and structural perspective, why should we continue to study the notion of discrimination in decision making? One benefit is that it provides a political and legal strategy to put pressure on individual decision makers. We can bring forward claims of discrimination against a specific person, firm, or institution. We can discuss what interventions exist within reasonable proximity to the decision maker that we therefore expect the decision maker to implement. Some such micro interventions may also be more directly feasible than structural interventions.

Taking on a micro perspective decidedly does not mean to ignore context. In fact, allocation rules that avoid explicit consideration of group membership while creating opportunity for a group likely do so by connecting the allocation rule with external social facts. One prominent example is the "Texas 10 percent rule" that guarantees Texas students who graduated in the top 10 percent of their high school class automatic admission to all state-funded universities. The rule wouldn't be effective in promoting racial diversity on public university campuses if high school classes weren't segregated to begin with. This example illustrates that there is no mutual exclusivity between examining specific decision rules in detail and paying attention to broader social context. Rather, these go hand in hand.

A consequential point of contact between the broader social world and the machine learning ecosystem are datasets. The whole of chapter 9 explores the history, significance, and scientific basis of machine learning datasets. Detailed consideration of datasets, the collection and construction of data, as well as the harms associated with data tend to be lacking from machine learning curricula.

Fairness remains an active research area that is far from settled. We wrote this book during a time of explosive research activity. Thousands of related papers have appeared in the last five years of writing. Research on fairness-promoting algorithmic interventions is especially fast-moving, so we do not attempt to survey it.

The book has some serious, perhaps obvious, limitations.

Large parts of our book are specific to the United States. Written by three authors educated and employed at US institutions, the book is based on Western moral tradition, assumes the laws and legal theory of the United States, and references the industrial and political context of the United States throughout. We made no attempt to address this serious limitation within this book. Indeed, it would require an entirely different book to address this limitation.

A second limitation stems from the fact that our primary goal was to develop the moral, normative, and technical foundations necessary to engage with the topic. Due to its focus on foundations, the book will strike some as a step removed from the important experiences of those individuals and communities most seriously wronged and harmed by the use of algorithms. This shortcoming is exacerbated by the fact that the authors of this book lack first-hand experience of the systems of oppression of which algorithms are a part. Consequently, this book is no substitute for the vital work of those activists, journalists, and scholars who have taught us about the dangers of algorithmic decision making in context. We build on these essential contributions in writing this book. We aimed to highlight them throughout, anticipating that we likely fell short in some significant ways.

The book is neither a wholesale endorsement of algorithmic decision making, nor a broad indictment. In writing this book, we attempt what is likely the least popular position on any topic: a balance. We try to work out where algorithmic decision making has merit, while committing significant attention to its harms and limitations. Some will see our balancing act as a lack of political commitments, a sort of bothsidesism.

Despite the urgency of the political situation, our book provides no direct practical guide to fair decisions. As a matter of fact, we wrote this book for the long haul. We're convinced that the debates around algorithmic decision making will persist. Our goal is to strengthen the intellectual foundations of debates to come, which will play out in thousands of specific instances. Anyone hoping to shape this future of algorithmic decision making in society will likely find some worthwhile material in this book.

A few chapters, specifically chapter 3 on classification and chapter 5 on causality, require significant mathematical prerequisites, primarily in undergraduate probability and statistics. However, the other chapters we dedicate to much broader audiences. We hope that students in multiple fields will find this book helpful in preparing for research in related areas. The book does not fit neatly into the disciplinary boundaries of any single department. As a result it gives readers an opportunity to go beyond established curricula in their primary discipline.

Since we started publishing material from this book years ago, instructors have incorporated the material into a variety of courses, at both the undergraduate and the graduate level, in different departments. Hundreds of readers have sent us tremendously helpful feedback for which we are deeply grateful.

And to those lamenting our slow progress in writing this book, we respond empathetically: That's fair.

Online Materials

We have created a website to accompany this textbook: https://fairmlbook.org/. The website includes a freely accessible online version of the textbook based on an earlier draft of the manuscript. It also includes a number of exercises. And it provides links to video recordings of a number of tutorials that we have given on these topics as well as the syllabi for our classes that served as the inspiration for the textbook.

Acknowledgments

This book wouldn't have been possible without the profound contributions of our collaborators and the community at large.

We are grateful to our students for their active participation in pilot courses at Berkeley, Cornell, and Princeton. Thanks in particular to Brian McInnis and Claudia Roberts for lecture notes of the Cornell and Princeton courses, respectively.

Special thanks to Katherine Yen for editorial and technical help with the book.

Moritz Hardt is indebted to Cynthia Dwork for introducing him to the topic of this book during a formative internship in 2010.

We benefited from substantial discussions, feedback, and comments from Rediet Abebe, Andrew Brunskill, Aylin Caliskan, Alex Chouldechova, André Cruz, Frances Ding, Michaela Hardt, Jake Hofman, Lily Hu, Ben Hutchinson, Shan Jiang, Sayash Kapoor, Lauren Kaplan, Niki Kilbertus, Been Kim, Kathy Kleiman, Jon Kleinberg, Issa Kohler-Hausmann, Mihir Kshirsagar, Eric Lawrence, Karen Levy, Zachary Lipton, Lydia T. Liu, John Miller, Smitha Milli, Shira Mitchell, Jared Moore, Robert Netzorg, Helen Nissenbaum, David Parkes, Juan Carlos Perdomo, Eike Willi Petersen, Manish Raghavan, Daniele Regoli, Ofir Reich, Claudia Roberts, Olga Russakovsky, Matthew J. Salganik, Carsten Schwemmer, Ludwig Schmidt, Andrew Selbst, Matthew Sun, Hanna Wallach, Angelina Wang, Christo Wilson, Annette Zimmermann, and Tijana Zrnic.

Solon Barocas is grateful for support from the John D. and Catherine T. MacArthur Foundation and Cornell Center for Social Sciences. Arvind Narayanan is grateful for support from the National Science Foundation under grants IIS-1763642 and CHS-1704444.

1

Introduction

Our success, happiness, and well-being are never fully of our own making. Others' decisions can profoundly affect the course of our lives: whether to admit us to a particular school, offer us a job, or grant us a mortgage. Arbitrary, inconsistent, or faulty decision making thus raises serious concerns because it risks limiting our ability to achieve the goals that we have set for ourselves and access the opportunities for which we are qualified.

So how do we ensure that these decisions are made the right way and for the right reasons? While there is much to value in fixed rules, applied consistently, *good* decisions take available evidence into account. We expect admissions, employment, and lending decisions to rest on factors that are relevant to the outcome of interest.

Identifying details that are relevant to a decision might happen informally and without much thought: employers might observe that people who study math seem to perform particularly well in the financial industry. But they could test these observations against historical evidence by examining the degree to which a person's major correlates with success on the job. This is the traditional work of statistics—and it promises to provide a more reliable basis for decision making by quantifying how much weight to assign certain details in our determinations.

A body of research has compared the accuracy of statistical models to the judgments of humans, even experts with years of experience. In many head-to-head comparisons on fixed tasks, data-driven decisions are more accurate than those based on intuition or expertise. As one example, a 2002 study shows that automated underwriting of loans was both more accurate and less racially disparate [1]. These results have been welcomed as a way to ensure that the high-stakes decisions that shape our life chances are both accurate and fair.

Machine learning promises to bring greater discipline to decision making because it offers to uncover factors relevant to decision making that humans might overlook, given the complexity or subtlety of the relationships in historical evidence. Rather than starting with some intuition about the relationship between certain factors and an outcome of interest, machine learning lets us defer the question of relevance to the data themselves: which factors—among all that we have observed—bear a statistical relationship to the outcome.

Uncovering patterns in historical evidence can be even more powerful than this might seem to suggest. Breakthroughs in computer vision—specifically object recognition—reveal just how much pattern discovery can achieve. In this domain, machine learning has helped to overcome a strange fact of human cognition: while we may be able to effortlessly identify objects in a scene, we are unable to specify the full set of rules that we rely on to make these determinations. We cannot hand-code a program that exhaustively enumerates all the relevant factors that allow us to recognize objects from every possible perspective or in all their potential visual con-figurations. Machine learning aims to solve this problem by abandoning the attempt to teach a computer through explicit instruction in favor of a process of learning by example. By exposing the computer to many examples of images containing preidentified objects, we hope the computer will learn the patterns that reliably dis-tinguish different objects from one another and from the environments in which they appear.

This can feel like a remarkable achievement, not only because computers can now execute complex tasks but also because the rules for deciding what appears in an image seem to emerge from the data themselves.

But there are serious risks in learning from examples. Learning is not a process of simply committing examples to memory. Instead, it involves generalizing from examples: homing in on those details that are characteristic of, say, cats in general, not just the specific cats that happen to appear in the examples. This is the process of induction: drawing general rules from specific examples—rules that effectively account for past cases, but that also apply to future, as yet unseen cases. The hope is that we'll figure out how future cases are likely to be similar to past cases, even if they are not exactly the same.

Reliably generalizing from historical examples to future cases requires that we provide the computer with *good* examples: a sufficiently large number of examples to uncover subtle patterns, a sufficiently diverse set of examples to showcase the many different types of appearances that objects might take, a sufficiently well-annotated set of examples to furnish machine learning with reliable ground truth, and so on. Thus, evidence-based decision making is only as reliable as the evidence on which it is based, and high-quality examples are critically important to machine learning. The fact that machine learning is "evidence-based" by no means ensures that it will lead to accurate, reliable, or fair decisions.

This is especially true when using machine learning to model human behavior and characteristics. Our historical examples of the relevant outcomes will almost always reflect historical prejudices, prevailing cultural stereotypes, and existing demographic inequalities. And finding patterns in these data will often mean replicating these very same dynamics.

Something else is lost in moving to automated, predictive decision making. Human decision makers rarely try to maximize predictive accuracy at all costs; frequently, they might consider factors such as whether the attributes used for prediction are morally relevant. For example, although younger defendants are statistically more likely to reoffend, judges are loath to take this into account in deciding sentence lengths, viewing younger defendants as less morally culpable. This is one reason to be cautious of comparisons that seem to show the superiority of statistical decision making [2]. Humans are also unlikely to make decisions that are obviously absurd, but this could happen with automated decision making, perhaps due to erroneous data. These and many other differences between human and automated decision making are reasons why decision-making systems that rely on machine learning might be unjust.

We write this book as machine learning begins to play a role in especially consequential decision making. In the criminal justice system, as alluded to above, defendants are assigned statistical scores that are intended to predict the risk of committing future crimes, and these scores inform decisions about bail, sentencing, and parole. In the commercial sphere, firms use machine learning to analyze and filter résumés of job applicants. And statistical methods are of course the bread and butter of lending, credit, and insurance underwriting.

We now begin to survey the risks in these and many other applications of machine learning, and provide a critical review of an emerging set of proposed solutions. We will see how even well-intentioned applications of machine learning might give rise to objectionable results.

Demographic Disparities

Amazon uses a data-driven system to determine the neighborhoods in which to offer free same-day delivery. A 2016 investigation found stark disparities in the demographic makeup of these neighborhoods: in many US cities, White residents were more than twice as likely as Black residents to live in one of the qualifying neighborhoods [3].

Now, we don't know the details of how Amazon's system works, and in particular we don't know to what extent it uses machine learning. The same is true of many other systems reported in the press. Nonetheless, we'll use these as motivating examples when a machine learning system for the task at hand would plausibly show the same behavior.

In chapter 3 we will see how to make our intuition about demographic disparities mathematically precise, and we will see that there are many possible ways of measuring these inequalities. The pervasiveness of such disparities in machine learning applications is a key concern of this book.

Disparities don't imply that the designer of the system intended for such inequalities to arise. Looking beyond intent, it is important to understand when observed disparities can be considered to be due to discrimination. In turn, two key questions to ask are whether the disparities are justified and whether they are harmful. These questions rarely have simple answers, but the extensive literature on discrimination in philosophy and sociology can help us reason about them.

To understand why the racial disparities in Amazon's system might be harmful, we must keep in mind the history of racial prejudice in the United States, its relationship to geographic segregation and inequalities, and the perpetuation of those inequalities over time. Amazon argued that its system was justified because it was designed based on efficiency and cost considerations and that race wasn't an explicit factor. Nonetheless, it has the effect of providing different opportunities to consumers at racially disparate rates. The concern is that this might contribute to the perpetuation of long-lasting cycles of inequality. If, instead, the system had been found to be partial to ZIP codes ending in an odd digit, it would not have triggered a similar outcry.

The term *bias* is often used to refer to demographic disparities in algorithmic systems that are objectionable for societal reasons. We'll minimize the use of this sense of the word "bias" in this book, since different disciplines and communities understand the term differently, and this can lead to confusion. There's a more traditional use of the term in statistics and machine learning. Suppose that Amazon's estimates of delivery dates or times were consistently too early by a few hours. This would be a case of *statistical bias*. A statistical estimator is said to be biased if its expected or average value differs from the true value that it aims to estimate. Statistical bias is a fundamental concept in statistics, and there is a rich set of established techniques for analyzing and avoiding it.

There are many other measures that quantify desirable statistical properties of a predictor or an estimator, such as precision, recall, and calibration. These are similarly well understood; none of them requires any knowledge of social groups and they are relatively straightforward to measure. The attention to demographic criteria in statistics and machine learning is a relatively new direction. It reflects a change in how we conceptualize machine learning systems and the responsibilities of those building them. Is our goal to faithfully reflect the data? Or do we have an obligation to question the data and to design our systems to conform to some notion of equitable behavior, regardless of whether or not that's supported by the data currently available to us? These perspectives are often in tension, and the difference between them will become clearer when we delve into the stages of machine learning.

The Machine Learning Loop

Let's study the pipeline of machine learning and understand how demographic disparities propagate through it. This approach lets us glimpse into the black box of machine learning and will prepare us for the more detailed analyses in later chapters. Studying the stages of machine learning is crucial if we want to intervene to minimize disparities.

Figure 1.1 shows the stages of a typical system that produces outputs using machine learning. Like any such diagram, it is a simplification, but it is useful for our purposes.

The first stage is measurement, which is the process by which the state of the world is reduced to a set of rows, columns, and values in a dataset. It's a messy process, because the real world is messy. The term "measurement" is misleading, evoking an image of a dispassionate scientist recording what they observe, whereas we'll see that it requires subjective human decisions.

The "learning" in machine learning refers to the next stage, which is to turn that data into a model. A model summarizes the patterns in the training data; it makes generalizations. A model could be trained using supervised learning via an algorithm such as Support Vector Machines, or using unsupervised learning via an algorithm such as k-means clustering. It could take many forms: a hyperplane or a set of regions in n-dimensional space, or a set of distributions. It is typically represented as a set of weights or parameters.

The next stage is the action we take based on the model's *predictions*, which are applications of the model to new, unseen inputs. By the way, "prediction" is another misleading term—while it does sometimes involve trying to predict the future ("Is this patient at high risk for cancer?"), sometimes it doesn't ("Is this social media account a bot?").

Prediction can take the form of classification (determine whether a piece of email is spam), regression (assigning risk scores to defendants), or information retrieval (finding documents that best match a search query). The actions in these three

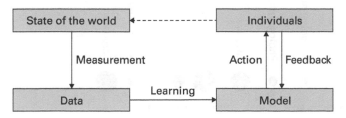

Figure 1.1
The machine learning loop.

applications might be: depositing the email in the user's inbox or spam folder, deciding whether to set bail for the defendant's pretrial release, and displaying the retrieved search results to the user. They may differ greatly in their significance to the individual, but they have in common that the collective responses of individuals to these decisions alter the state of the world—that is, the underlying patterns that the system aims to model.

Some machine learning systems record feedback from users (how users react to actions) and use them to refine their models. For example, search engines track what users click on as an implicit signal of relevance or quality. Feedback can also occur unintentionally, or even adversarially; these are more problematic, as we explore later in this chapter.

The State of Society

In this book, we're concerned with applications of machine learning that involve data about *people*. In these applications, the available training data will likely encode the demographic disparities that exist in our society. For example, figure 1.2 shows the gender breakdown of a sample of occupations in the United States, based on data released by the Bureau of Labor Statistics for the year 2017.

Unsurprisingly, many occupations have stark gender imbalances. If we're building a machine learning system that screens job candidates, we should be keenly aware that this is the baseline we're starting from. It doesn't necessarily mean that the outputs of our system will be inaccurate or discriminatory, but throughout this chapter we'll see how it complicates things.

Why do these disparities exist? There are many potentially contributing factors, including a history of explicit discrimination, implicit attitudes and stereotypes

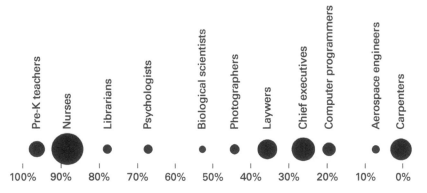

Figure 1.2
A sample of occupations in the United States in decreasing order of the percentage of women. The area of the bubble represents the number of workers.

about gender, and differences in the distribution of certain characteristics by gender. We'll see that even in the absence of explicit discrimination, stereotypes can be self-fulfilling and persist for a long time in society. As we integrate machine learning into decision making, we should be careful to ensure that it doesn't become a part of this feedback loop.

What about applications that aren't about people? Consider "Street Bump," a project by the city of Boston to crowdsource data on potholes. The smartphone app automatically detects potholes using data from the smartphone's sensors and sends the data to the city. Infrastructure seems like a comfortably boring application of data-driven decision making, far removed from the ethical quandaries we've been discussing. And yet! Kate Crawford points out that the data reflect the patterns of smartphone ownership, which are higher in wealthier parts of the city compared to lower-income areas and areas with large elderly populations [4]. The lesson here is that it's rare for machine learning applications not to be about people. In the case of Street Bump, the data are collected by people, and hence reflect demographic disparities; besides, the reason we're interested in improving infrastructure in the first place is its effect on people's lives.

To drive home the point that most machine learning applications involve people, we analyzed Kaggle, a well-known platform for data science competitions. We focused on the top thirty competitions sorted by prize amount. In fourteen of these competitions, we observed that the task is to make decisions about individuals. In most of these cases, there exist societal stereotypes or disparities that may be perpetuated by the application of machine learning. For example, the Automated Essay Scoring [5] task seeks algorithms that attempt to match the scores of human graders of student essays. Students' linguistic choices are signifiers of social group membership, and human graders are known to sometimes have prejudices based on such factors [6, 7]. Thus, because human graders must provide the original labels, automated grading systems risk enshrining any such discriminatory patterns that are captured in the training data.

In a further five of the thirty competitions, the task did not call for making decisions about people, but decisions made using the model would nevertheless directly impact people. For example, one competition sponsored by real-estate company Zillow calls for improving the company's "Zestimate" algorithm for predicting home sale prices. Any system that predicts a home's future sale price (and publicizes these predictions) is likely to create a self-fulfilling feedback loop in which homes predicted to have lower sale prices deter future buyers, suppressing demand and lowering the final sale price.

In nine of the thirty competitions, we did not find an obvious, direct impact on people. One competition, for example, predicted ocean health (of course, even such competitions have indirect impacts on people, due to actions that we might take on the basis of the knowledge gained). In two cases, we didn't have enough information to make a determination.

To summarize, human society is full of demographic disparities, and training data will likely reflect these. We will now turn to the process by which training data are constructed, and see that things are even trickier.

The Trouble with Measurement

The term "measurement" suggests a straightforward process, calling to mind a camera objectively recording a scene. In fact, measurement is fraught with subjective decisions and technical difficulties.

Consider a seemingly straightforward task: measuring the demographic diversity of college campuses. A 2017 *New York Times* article aimed to do just this, and was titled "Even with Affirmative Action, Blacks and Hispanics Are More Underrepresented at Top Colleges than 35 Years Ago" [8]. The authors argue that the gap between enrolled Black and Hispanic freshmen and the Black and Hispanic college-age population has grown over the past thirty-five years. To support their claim, they present demographic information for more than 100 American universities and colleges from the year 1980 to 2015, and show how the percentages of Black, Hispanic, Asian, White, and multiracial students have changed over the years. Interestingly, the multiracial category was introduced only in 2008, but the comparisons in the article ignore the introduction of this new category. How many students who might have checked the "White" or "Black" box checked the "multiracial" box instead? How might this have affected the percentages of "White" and "Black" students at these universities? Furthermore, individuals' and society's conception of race changes over time. Would a person with Black and Latino parents be more inclined to self-identify as Black in 2015 than in the 1980s? The point is that even a seemingly straightforward question about trends in demographic diversity is impossible to answer without making some assumptions, and it illustrates the difficulties of measurement in a world that resists falling neatly into a set of checkboxes. Race is not a stable category; how we measure race often changes how we conceive of it, and changing conceptions of race may force us to alter what we measure.

To be clear, this situation is typical: measuring almost any attribute about people is similarly subjective and challenging. If anything, things are more chaotic when machine learning researchers have to create categories, as is often the case.

One area where machine learning practitioners often have to define new categories is in defining the target variable [9]. This is the outcome that we're trying to predict: Will the defendant recidivate if released on bail? Will the candidate be a good employee if hired? And so on.

Biases in the definition of the target variable are especially critical, because they are guaranteed to bias the predictions relative to the actual construct we intended to predict, as is the case when we use arrests as a measure of crime, or sales as a measure of job performance, or GPA as a measure of academic success. This is not necessarily so with other attributes. But the target variable is arguably the hardest from a measurement standpoint, because it is often a construct that is made up for the purposes of the problem at hand rather than one that is widely understood and measured. For example, "creditworthiness" is a construct that was created in the context of the problem of how to successfully extend credit to consumers [9]; it is not an intrinsic property that people either possess or lack.

If our target variable is the idea of a "good employee," we might use performance review scores to quantify it. This means that our data inherit any biases present in managers' evaluations of their reports. Another example is the use of computer vision to automatically rank people's physical attractiveness [10, 11]. The training data consist of human evaluation of attractiveness, and, unsurprisingly, all these classifiers showed a preference for lighter skin.

In some cases we might be able to get closer to a more objective definition for a target variable, at least in principle. For example, in criminal risk assessment, the training data are not judges' decisions about bail, but they are rather based on who actually went on to commit a crime. But there's at least one big caveat—we can't really measure who committed a crime, so we use arrests as a proxy. This means that the training data contain distortions not due to the prejudices of judges but due to discriminatory policing. On the other hand, if our target variable is whether the defendant appears or fails to appear in court for trial, we would be able to measure it directly with perfect accuracy. That said, we may still have concerns about a system that treats defendants differently based on predicted probability of appearance, given that some reasons for failing to appear are less objectionable than others (trying to hold down a job that would not allow for time off versus trying to avoid prosecution) [12].

In hiring, instead of relying on performance reviews for, say, a sales job, we might rely on the number of sales closed. But is that an objective measurement or is it subject to the prejudices of the potential customers (who might respond more positively to certain salespeople than to others) and workplace conditions (which might be a hostile environment for some but not others)?

In some applications, researchers repurpose an existing scheme of classification to define the target variable rather than creating one from scratch. For example, an object recognition system can be created by training a classifier on ImageNet, a database of images organized in a hierarchy of concepts [13]. ImageNet's hierarchy comes from Wordnet, a database of words, categories, and the relationships among them [14]. Wordnet's authors in turn imported the word lists from a number of older sources, such as thesauri. As a result, WordNet (and ImageNet) categories contain numerous outmoded words and associations, such as occupations that no longer exist and stereotyped gender associations [15].

We think of technology changing rapidly and society being slow to adapt but, at least in this instance, the categorization scheme at the heart of much of today's machine learning technology has been frozen in time while social norms have changed.

Our favorite example of measurement bias has to do with cameras, which we referenced at the beginning of the section as the exemplar of dispassionate observation and recording. But are they?

The visual world has an essentially infinite bandwidth compared to what can be captured by cameras, whether film or digital, which means that photography technology involves a series of choices about what is relevant and what isn't, and transforms the captured data based on those choices. Both film and digital cameras have historically been more adept at photographing lighter-skinned individuals [16]. One reason is the default settings, such as color balance, which were optimized for lighter skin tones. Another, deeper reason is the limited "dynamic range" of cameras, which makes it hard to capture brighter and darker tones in the same image. This started changing in the 1970s, in part due to complaints from furniture companies and chocolate companies about the difficulty of photographically capturing the details of furniture and chocolate, respectively! Another impetus came from the increasing diversity of television subjects at this time.

When we go from individual images to datasets of images, we introduce another layer of potential biases. Consider the image datasets that are used to train today's computer vision systems for tasks such as object recognition. If these datasets were representative samples of an underlying visual world, we might expect that a computer vision system trained on one such dataset would do well on another dataset. But in reality, we observe a big drop in accuracy when we train and test on different datasets [17]. This shows that these datasets are biased relative to each other in a statistical sense, and this is a good starting point for investigating whether these biases include cultural stereotypes.

It's not all bad news: machine learning can in fact help mitigate measurement biases. Returning to the issue of dynamic range in cameras, computational techniques, including machine learning, are making it possible to improve the representation of tones in images [18, 19, 20]. Another example comes from medicine: diagnoses and treatments are sometimes personalized by race. But it turns out that race is used as a crude proxy for ancestry and genetics, and sometimes environmental and behavioral factors [21, 22]. If we can measure the factors that are medically relevant and incorporate them—instead of race—into statistical models of disease and drug response, we can increase the accuracy of diagnoses and treatments while mitigating racial disparities.

To summarize, measurement involves defining variables of interest, developing a process for turning observations into numbers, and then actually collecting the data. Often machine learning practitioners don't think about these steps, because someone else has already done those things. And yet it is crucial to understand the provenance of the data. Even if someone else has collected the data, they are almost always too messy for algorithms to handle, hence the dreaded "data cleaning" step. But the messiness of the real world isn't just an annoyance to be dealt with by cleaning. It is a manifestation of a diverse world in which people don't fit neatly into categories. Being inattentive to these nuances can particularly hurt marginalized populations.

From Data to Models

We've seen that training data reflects the disparities, distortions, and biases from the real world and the measurement process. This leads to an obvious question: When we learn a model from such data, are these disparities preserved, mitigated, or exacerbated?

Predictive models trained with supervised learning methods are often good at calibration, ensuring that the model subsumes all features in the data for the purpose of predicting the outcome. But calibration also means that by default, we should expect our models to faithfully reflect disparities found in the input data.

Here's another way to think about it. Some patterns in the training data (smoking is associated with cancer) represent knowledge that we wish to mine using machine learning, while other patterns (girls like pink and boys like blue) represent stereotypes that we might wish to avoid learning. But learning algorithms have no general way to distinguish between these two types of patterns, because the patterns are the result of social norms and moral judgments. Absent specific intervention, machine learning will extract stereotypes, including incorrect and harmful ones, in the same way that it extracts knowledge.

A telling example of this comes from machine translation. Figure 1.3 shows the result of translating sentences from English to Turkish and back [23]. The same stereotyped translations result for many pairs of languages and other occupation words in all translation engines we've tested. It's easy to see why. Turkish has gender neutral pronouns, and when translating such a pronoun to English, the system picks the sentence that best matches the statistics of the training set (which is typically a large, minimally curated corpus of historical text and text found on the web).

When we build a statistical model of language from such text, we should expect the gender associations of occupation words to roughly mirror real-world labor statistics. In addition, because of the male-as-norm bias [24] (the use of male pronouns when the gender is unknown), we should expect translations to favor male

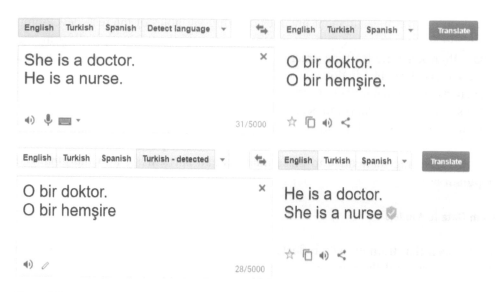

Figure 1.3
Translating from English to Turkish then back to English injects gender stereotypes. The translation was performed in 2017.

pronouns. It turns out that when we repeat the experiment with dozens of occupation words, these two factors—labor statistics and the male-as-norm bias—together almost perfectly predict which pronoun will be returned [23].

Here's a tempting response to the observation that models reflect data biases. Suppose we're building a model for scoring résumés for a programming job. What if we simply withhold gender from the data? Is that a sufficient response to concerns about gender discrimination? Unfortunately, it's not that simple, because of the problem of proxies [9] or redundant encodings [25], as we discuss in chapter 3. There are any number of other attributes in the data that might correlate with gender. For example, in our society, the age at which someone starts programming is correlated with gender. This illustrates why we can't just get rid of proxies: they may be genuinely relevant to the decision at hand. How long someone has been programming is a factor that gives us valuable information about their suitability for a programming job, but it also reflects the reality of gender stereotyping.

Another common reason why machine learning might perform worse for some groups than others is sample size disparity. If we construct our training set by sampling uniformly from the training data, then by definition we'll have fewer data points about minorities. Of course, machine learning works better when there are more data, so it will work less well for members of minority groups, assuming that members of the majority and minority groups are systematically different in terms of the prediction task [25].

Worse, in many settings minority groups are underrepresented relative to population statistics. For example, minority groups are underrepresented in the tech industry. Different groups might also adopt technology at different rates, which might skew datasets assembled from social media. If training sets are drawn from these unrepresentative contexts, there will be even fewer training points from minority individuals.

When we develop machine learning models, we typically test only their overall accuracy; so a "5 percent error" statistic might hide the fact that a model performs terribly for a minority group. Reporting accuracy rates by group will help alert us to problems like the above example. In chapter 3, we look at metrics that quantify the error-rate disparity between groups.

There's one application of machine learning where we find especially high error rates for minority groups: anomaly detection. This is the idea of detecting behavior that deviates from the norm . This technique is often used to identify abuse against a system. A good example is the *Nymwars* controversy, where Google, Facebook, and other tech companies aimed to block users who used uncommon (hence, presumably fake) names. To understand how this approach could cause problems, consider the fact that in some cultures, most people receive names from a small set of names, whereas in other cultures, names might be more diverse, and it might be common for names to be unique. For users in the latter culture, a popular name would be more likely to be fake. In other words, the same feature that constitutes evidence towards a prediction in one group might constitute evidence against the prediction for another group [25].

If we're not careful, learning algorithms will generalize based on the majority culture, leading to a high error rate for minority groups. Attempting to avoid this by making the model more complex runs into a different problem: overfitting to the training data, that is, picking up patterns that arise due to random noise rather than true differences. One way to avoid this is to explicitly model the differences between groups, although there are both technical and ethical challenges associated with this.

The Pitfalls of Action

Any real machine learning system seeks to make some change in the world. To understand its effects, then, we have to consider it in the context of the larger sociotechnical system in which it is embedded.

In chapter 3, we see that if a model is calibrated—it faithfully captures the patterns in the underlying data—predictions made using that model will inevitably have disparate error rates for different groups, if those groups have different *base rates*, that is, rates of positive or negative outcomes. In other words, understanding the properties of a prediction requires understanding not just the model but also the population differences between the groups on which the predictions are applied.

Further, population characteristics can shift over time; this is a well-known machine learning phenomenon known as drift. If subpopulations change differently over time, but the model isn't retrained, that can introduce disparities. An additional wrinkle: whether or not disparities are objectionable may differ between cultures and may change over time as social norms evolve.

When people are subject to automated decisions, their perception of those decisions depends not only on the outcomes but also on the process of decision making. An ethical decision-making process might require, among other things, the ability to explain a prediction or decision, which might not be feasible with black box models.

A major limitation of machine learning is that it reveals only correlations, but we often use its predictions as if they reveal causation. This is a persistent source of problems. For example, an early machine learning system in health care famously learned the seemingly nonsensical rule that patients with asthma had lower risk of developing pneumonia. This was a true pattern in the data, but the likely reason was that asthmatic patients were more likely to receive in-patient care [26]. So it's not valid to use the prediction to decide whether or not to admit a patient. We discuss causality in chapter 5.

Another way to view this example is that the prediction affects the outcome (because of the actions taken on the basis of the prediction), and thus invalidates itself. The same principle is also seen in the use of machine learning for predicting traffic congestion: if sufficiently many people choose their routes based on the prediction, then the route predicted to be clear will in fact be congested. The effect can also work in the opposite direction: the prediction might reinforce the outcome, resulting in feedback loops. To better understand how, let's talk about the final stage in our loop: feedback.

Feedback and Feedback Loops

Many systems receive feedback when they make predictions. When a search engine serves results, it typically records the links that the user clicks on and how long the user spends on those pages and treats these as implicit signals about which results were found to be most relevant. When a video-sharing website recommends a video, it uses the thumbs up/down feedback as an explicit signal. Such feedback is used to refine the model.

But feedback is tricky to interpret correctly. If a user clicked on the first link on a page of search results, is that simply because it was first, or because it was in fact the most relevant? This is again a case of the action (the ordering of search results) affecting the outcome (the link(s) the user clicks on). This is an active area of research; there are techniques that aim to learn accurately from this kind of biased feedback [27].

Bias in feedback might also reflect cultural prejudices, which are of course much harder to characterize than the effects of the ordering of search results. For example, the clicks on the targeted ads that appear alongside search results might reflect gender and racial stereotypes. There's a well-known study by Latanya Sweeney that hints at this: Google searches for Black-sounding names such as "Latanya Farrell" were much more likely to result in ads for arrest records ("Latanya Farrell, Arrested?") than searches for White-sounding names ("Kristen Haring") [28]. One potential explanation is that users are more likely to click on ads that conform to stereotypes, and the advertising system is optimized for maximizing clicks.

In other words, even feedback that is designed into systems can lead to unexpected or undesirable biases. But on top of that, there are many unintended ways in which feedback might arise, and these are more pernicious and harder to control. Let's look at three.

Self-Fulfilling Predictions

Suppose a predictive policing system determines certain areas of a city to be at high risk for crime. More police officers might be deployed to such areas. Alternatively, officers in areas predicted to be high risk might be subtly lowering their threshold for stopping, searching, or arresting people—perhaps even unconsciously. Either way, the prediction will appear to be validated, even if it had been made purely based on data biases.

Here's another example of how acting on a prediction can change the outcome. In the United States, some criminal defendants are released prior to trial, whereas for others, a bail amount is set as a precondition of release. Many defendants are unable to post bail. Does the release or detention affect the outcome of the case? Perhaps defendants who are detained face greater pressure to plead guilty. At any rate, how could one possibly test the causal impact of detention without doing an experiment? Intriguingly, we can take advantage of a pseudo-experiment, namely that defendants are assigned bail judges quasi-randomly, and some judges are stricter than others. Thus, pretrial detention is partially random, in a quantifiable way. Studies using this technique have confirmed that detention indeed causes an increase in the likelihood of a conviction [29]. If bail were set based on risk predictions, whether human or algorithmic, and we evaluated its efficacy by examining case outcomes, we would see a self-fulfilling effect.

Predictions That Affect the Training Set

Continuing this example, predictive policing activity will lead to arrests, records of which might be added to the algorithm's training set. These areas, and perhaps also other areas with a similar demographic composition, might then continue to appear to be at high risk of crime, depending on the feature set used for predictions. The disparities might even compound over time.

A 2016 paper by Kristian Lum and William Isaac analyzed a predictive policing algorithm by PredPol. This is one of the few predictive policing algorithms to be published in a peer-reviewed journal, for which the company deserves praise. By applying the algorithm to data derived from Oakland police records, the authors found that Black people would be targeted for predictive policing of drug crimes at roughly twice the rate of White people, even though the two groups have roughly equal rates of drug use [30]. Their simulation showed that this initial bias would be amplified by a feedback loop, with policing increasingly concentrated on targeted areas. This is despite the fact that the PredPol algorithm does not explicitly take demographics into account.

A follow-up paper built on this idea and showed mathematically how feedback loops occur when data discovered on the basis of predictions are used to update the model [31]. The paper also shows how to tweak the model to avoid feedback loops in a simulated setting, by quantifying how surprising an observation of crime is given the predictions, and updating the model only in response to surprising events.

Predictions That Affect the Phenomenon and Society at Large

Prejudicial policing on a large scale, algorithmic or not, will affect society over time, contributing to the cycle of poverty and crime. This is a well-established thesis, and we briefly review the sociological literature on durable inequality and the persistence of stereotypes in chapter 8.

Let us remind ourselves that we deploy machine learning so that we can act on its predictions. It is hard even conceptually to eliminate the effects of predictions on outcomes, future training sets, the phenomena themselves, or society at large. The more central machine learning becomes in our lives, the stronger this effect.

Returning to the example of a search engine, in the short term it might be possible to extract an unbiased signal from user clicks but in the long run results that are returned more often will be linked to and thus will rank more highly. As a side effect of fulfilling its purpose of retrieving relevant information, a search engine will necessarily change the very thing that it aims to measure, sort, and rank. Similarly, most machine learning systems will affect the phenomena that they predict. This is why we've depicted the machine learning process as a loop.

Throughout this book we'll learn methods for mitigating societal biases in machine learning, but we should keep in mind that there are fundamental limits to what we can achieve, especially when we consider machine learning as a sociotechnical system instead of a mathematical abstraction. The textbook model of training and test data being independent and identically distributed is a simplification, and might be unachievable in practice.

Getting Concrete with a Toy Example

Now let's look at a concrete setting, albeit a toy problem, to illustrate many of the ideas discussed so far, and some new ones.

Let's say you're on a hiring committee, making decisions based on just two attributes of each applicant: their college GPA and their interview score (we did say it's a toy problem!). We formulate this as a machine learning problem: the task is to use these two variables to predict some measure of the "quality" of an applicant. For example, quality could be based on the average performance review score after two years at the company. We'll assume we have data from past candidates that allows us to train a model to predict performance scores based on GPA and interview score.

Obviously, this is a reductive formulation—we're assuming that an applicant's worth can be reduced to a single number and that we know how to measure that number. This is a valid criticism, and applies to most applications of data-driven decision making today. But it has one big advantage: once we do formulate the decision as a prediction problem, statistical methods tend to do better than humans, even domain experts with years of training, in making decisions based on noisy predictors.

Given this formulation, the simplest thing we can do is to use linear regression to predict the average job performance rating from the two observed variables, and then use a cutoff based on the number of candidates we want to hire. Figure 1.4 shows what this might look like. In reality, the variables under consideration need not satisfy a linear relationship, thus suggesting the use of a nonlinear model, which we avoid for simplicity.

As you can see in figure 1.4, our candidates fall into two demographic groups, represented by triangles and squares. This binary categorization is a simplification for the purposes of our thought experiment. But when building real systems, enforcing rigid categories of people can be ethically questionable.

Note that the classifier didn't take into account which group a candidate belonged to. Does this mean that the classifier is fair? We might hope that it is, based on the fairness-as-blindness idea, symbolized by the icon of Lady Justice wearing a blindfold. In this view, an impartial model—one that doesn't use group membership in the regression—is fair; a model that gives different scores to otherwise-identical members of different groups is discriminatory.

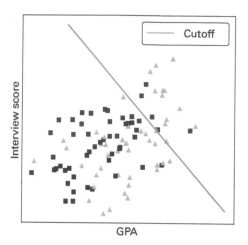

Figure 1.4
Toy example: a hiring classifier that predicts job performance (not shown) based on GPA and interview score, and then applies a cutoff.

We'll defer a richer understanding of what fairness means to later chapters, so let's ask a simpler question: are candidates from the two groups equally likely to be positively classified? The answer is no: the triangles are more likely to be selected than the squares. That's because a dataset is a social mirror; the "ground truth" labels that we're predicting—job performance ratings—are systematically lower for the squares than the triangles.

There are many possible reasons for this disparity. First, the managers who score the employees' performance might discriminate against one group. Or the overall workplace might be less welcoming to one group, preventing them from reaching their potential and leading to lower performance. Alternately, the disparity might originate from before the candidates were hired. For example, it might arise from disparities in educational institutions attended by the two groups. Or there might be intrinsic differences between them. Of course, it might be a combination of these factors. We can't tell from our data how much of the disparity is attributable to these different factors. In general, such a determination is methodologically hard, and requires causal reasoning [32].

For now, let's assume that we have evidence that the level of demographic disparity produced by our selection procedure is unjustified, and we're interested in intervening to decrease it. How could we do it? We observe that GPA is correlated with the demographic attribute—it's a proxy. Perhaps we could simply omit that variable as a predictor? Unfortunately, we would also hobble the accuracy of our model. In real datasets, most attributes tend to be proxies for demographic variables, and dropping them may not be a reasonable option.

Another crude approach would be to pick different cutoffs so that candidates from both groups have the same probability of being hired. Or we could mitigate the demographic disparity instead of eliminating it, by decreasing the difference in the cutoffs.

Given the available data, there is no mathematically principled way to know which cutoffs to pick. In some situations there is a legal baseline: for example, guidelines from the US Equal Employment Opportunity Commission state that if the probability of selection for two groups differs by more than 20 percent, it might constitute a sufficient disparate impact to initiate a lawsuit. But a disparate impact alone is not illegal; the disparity needs to be unjustified or avoidable for courts to find liability. Even these quantitative guidelines do not provide easy answers or bright lines.

At any rate, the pick-different-thresholds approach to mitigating disparities seems unsatisfying, because it is crude and uses the group attribute as the sole criterion for redistribution. It does not account for the underlying reasons why two candidates with the same observable attributes (except for group membership) may be deserving of different treatment.

But there are other possible interventions, and we'll discuss one. To motivate it, let's take a step back and ask why the company wants to decrease the demographic disparity in hiring.

One answer is rooted in justice to individuals and the specific social groups to which they belong. But a different answer comes from the firm's selfish interests: diverse teams work better [33, 34]. From this perspective, increasing the diversity of the cohort that is hired would benefit the firm and everyone in the cohort. As an analogy, picking eleven goalkeepers, even if individually excellent, would make for a poor soccer team.

How do we operationalize diversity in a selection task? If we had a distance function between pairs of candidates, we could measure the average distance between selected candidates. As a strawman, let's say we use the Euclidean distance based on the GPA and interview score. If we incorporated such a diversity criterion into the objective function, it would result in a model where the GPA is weighted less. This technique doesn't explicitly consider group membership. Rather, as a side effect of insisting on diversity of the other observable attributes, it also improves demographic diversity. However, a careless application of such an intervention can easily go wrong: for example, the model might give weight to attributes that are completely irrelevant to the task.

More generally, there are many possible algorithmic interventions beyond picking different thresholds for different groups. In particular, the idea of a similarity function between pairs of individuals is a powerful one, and we'll see other interventions that make use of it. But coming up with a suitable similarity function in practice isn't easy: it may not be clear which attributes are relevant, how to weight them, and how to deal with correlations between attributes.

Justice beyond Fair Decision Making

The core concern of this book is group disparities in decision making. But ethical obligations don't end with addressing those disparities. Fairly rendered decisions under unfair circumstances may do little to improve people's lives. In many cases, we cannot achieve any reasonable notion of fairness through changes to decision making alone; we need to change the conditions under which these decisions are made. In other cases, the very purpose of the system might be oppressive, and we should ask whether it should be deployed at all.

Further, decision making systems aren't the only places where machine learning is used that can harm people: for example, online search and recommendation algorithms are also of concern, even though they don't make decisions about people. Let's briefly discuss these broader questions.

Interventions That Target Underlying Inequities

Let's return to the hiring example above. When using machine learning to make predictions about how someone might fare in a specific workplace or occupation, we tend to treat the environment that people will confront in these roles as a constant and ask how people's performance will vary according to their observable characteristics. In other words, we treat the current state of the world as a given, leaving us to select the person who will do best under these circumstances. This approach risks overlooking more fundamental changes that we could make to the workplace (culture, family-friendly policies, on-the-job training) that might make it a more welcoming and productive environment for people who have not flourished under previous conditions [35].

The tendency with work on fairness in machine learning is to ask whether an employer is using a fair selection process, even though we might have the opportunity to intervene in the workplace dynamics that actually account for differences in predicted outcomes along the lines of race, gender, disability, and other characteristics [36].

We can learn a lot from the so-called social model of disability, which views a predicted difference in a disabled person's ability to excel on the job as the result of a lack of appropriate accommodations (an accessible workplace, necessary equipment, flexible working arrangements) rather than of any inherent capacity of the person. A person is disabled only in the sense that we have not built physical environments or adopted appropriate policies to ensure their equal participation.

The same might be true of people with other characteristics, and changes to the selection process alone will not help us address the fundamental injustice of conditions that keep certain people from contributing as effectively as others. We examine these questions in chapter 8.

It may not be ethical to deploy an automated decision-making system at all if the underlying conditions are unjust and the automated system would serve only to reify it. Or a system may be ill-conceived, and its intended purpose may be unjust, even if it were to work flawlessly and perform equally well for everyone. The question of which automated systems should be deployed shouldn't be left to the logic (and whims) of the marketplace. For example, we may want to regulate the police's access to facial recognition. Our civil rights—freedom or movement and association—are threatened by these technologies both when they fail and when they work well. These are concerns about the *legitimacy* of an automated decision-making system, and we explore them in chapter 2.

The Harms of Information Systems

When a defendant is unjustly detained pretrial, the harm is clear. But beyond algorithmic decision making, information systems such as search and recommendation algorithms can also have negative effects, but here the harm is indirect and harder to define.

Here's one example. Image search results for occupation terms such as CEO or software developer reflect (and arguably exaggerate) the prevailing gender composition and stereotypes about those occupations [37]. Another example that we encountered earlier is the gender stereotyping in online translation. These and other examples that are disturbing to varying degrees—such as Google's app labeling photos of Black Americans as "gorillas," or offensive results in autocomplete—seem to fall into a different moral category than, say, a discriminatory system used in criminal justice, which has immediate and tangible consequences.

A talk by Kate Crawford lays out the differences [38]. When decision-making systems in criminal justice, health care, and so on are discriminatory, they create *allocative harms*, which are caused when a system withholds from certain groups an opportunity or a resource. In contrast, the other examples—stereotype perpetuation and cultural denigration—are examples of *representational harms*, which occur when systems reinforce the subordination of some groups along the lines of identity—race, class, gender, and so on.

Allocative harms have received much attention both because their effects are immediate and because they are easier to formalize and study in computer science and in economics. Representational harms have long-term effects, and resist formal characterization. But as machine learning has become a part of how we make sense of the world—through technologies such as search, translation, voice assistants, and image labeling—representational harms will leave an imprint on our culture and influence identity formation and stereotype perpetuation. Thus, these are critical concerns for the fields of natural language processing and computer vision. Although this book is primarily about allocative harms, we briefly discuss representational harms in chapters 7 and 9.

The majority of content consumed online is mediated by recommendation algorithms that influence which users see which content. Thus, these algorithms influence which messages are amplified. Social media algorithms have been blamed for a litany of ills: being echo chambers in which users are exposed to content that conforms to their prior beliefs, exacerbating political polarization, radicalizing some users into fringe beliefs, stoking ethnic resentment and violence, causing a deterioration of mental health, and so on. Research on these questions is nascent and establishing causality is hard, and it remains unclear how much of these effects are due to the design of the algorithm rather than user behavior. But there is little doubt that algorithms have some role. Twitter experimentally compared a nonalgorithmic (reverse chronological) content feed to an algorithmic feed and found that content from the mainstream political right was consistently favored in the algorithmic setting over content from the mainstream political left [39]. While important, this topic is out of scope for us. However, we briefly touch on discrimination in ad targeting and in online marketplaces in chapter 7.

Our Outlook: Limitations and Opportunities

We've seen how machine learning propagates inequalities in the state of the world through the stages of measurement, learning, action, and feedback. Machine learning systems that affect people are best thought of as closed loops, since the actions we take based on predictions in turn affect the state of the world. One major goal of fair machine learning is to develop an understanding of when these disparities are harmful, unjustified, or otherwise unacceptable, and to develop interventions to mitigate such disparities.

There are fundamental challenges and limitations to this goal. Unbiased measurement might be infeasible even in principle, such as when the construct itself (e.g., race) is unstable. There are additional practical limitations arising from the fact that the decision maker is typically not involved in the measurement stage. Further, observational data can be insufficient to identify the causes of disparities, which is needed in the design of meaningful interventions and in order to understand the effects of intervention. Most attempts to "debias" machine learning in the current research literature assume simplistic mathematical systems, often ignoring the effect of algorithmic interventions on individuals and on the long-term state of society.

Despite these important limitations, there are reasons to be cautiously optimistic about fairness and machine learning. First, data-driven decision making has the potential to be more transparent compared to human decision making. It forces us to articulate our decision-making objectives and enables us to clearly understand the trade-offs between desiderata. However, there are challenges to overcome to achieve this potential for transparency. One challenge is improving the interpretability and explainability of modern machine learning methods, which is a topic of vigorous ongoing research. Another challenge arises from the proprietary nature of datasets and systems that are crucial to an informed public debate on this topic. Many commentators have called for a change in the status quo [40].

Second, effective interventions do exist in many machine learning applications, especially in natural-language processing and computer vision. Tasks in these domains (say, transcribing speech) are subject to less inherent uncertainty than traditional decision making (say, predicting if a loan applicant will repay), removing some of the statistical constraints that we study in chapter 3.

Our final and most important reason for optimism is that the turn to automated decision making and machine learning offers an opportunity to reconnect with the moral foundations of fairness. Algorithms force us to be explicit about what we want to achieve with decision making. And it's far more difficult to paper over our poorly specified or true intentions when we have to state these objectives formally. In this way, machine learning has the potential to help us debate the fairness of different policies and decision-making procedures more effectively.

We should not expect work on fairness in machine learning to deliver easy answers. And we should be suspicious of efforts that treat fairness as something that can be reduced to an algorithmic stamp of approval. We must try to confront, not avoid, the hard questions when it comes to debating and defining fairness. We may even need to reevaluate the meaningfulness and enforceability of existing approaches to discrimination in law and policy [9], expanding the tools at our disposal to reason about fairness and seek out justice.

We hope that this book can play a small role in stimulating this interdisciplinary inquiry.

Bibliographic Notes and Further Reading

This chapter draws from several taxonomies of biases in machine learning and data-driven decision making: a blog post by Moritz Hardt [25], a paper by Solon Barocas and Andrew Selbst [9], and a 2016 report by the White House Office of Science and Technology Policy [41]. For a broad survey of challenges raised by AI, machine learning, and algorithmic systems, see the AI Now report [42].

An early work that investigated fairness in algorithmic systems is by Batya Friedman and Helen Nissenbaum in 1996 [43]. Papers studying demographic disparities in classification began appearing regularly starting in 2008 [44]; the locus of this research was in Europe, and in the data-mining research community. With the establishment of the FAT/ML workshop in 2014, a new community emerged, and the topic has since grown in popularity. Several books for a popular audience have delivered critiques of algorithmic systems in modern society: *The Black Box Society* by Frank Pasquale [45], *Weapons of Math Destruction* by Cathy O'Neil [46], *Automating Inequality* by Virginia Eubanks [47], and *Algorithms of Oppression* by Safiya Noble [48].

2

When Is Automated Decision Making Legitimate?

These three scenarios have something in common:

- A student is proud of the creative essay she wrote for a standardized test. She receives a perfect score, but is disappointed to learn that the test had in fact been graded by a computer.
- A defendant finds that a criminal risk prediction system categorized him as high risk for failure to appear in court, based on the behavior of others like him, despite his having every intention of appearing in court on the appointed date.
- An automated system locked out a social media user for violating the platform's policy on acceptable behavior. The user insists that they did nothing wrong, but the platform won't provide further details nor any appeal process.

All of these are automated decision-making or decision support systems that likely feel unfair or unjust. Yet this is a sense of unfairness that is distinct from what we talked about in the first chapter (and which we will return to in the next chapter). It is not about the relative treatment of different groups. Instead, what these questions are about is *legitimacy*—whether it is fair to deploy such a system at all in a given scenario. And this, in turn, affects the legitimacy of the organization deploying it.

Most institutions need legitimacy to be able to function effectively. People have to believe that the institution is broadly aligned with social values. The reason for this is relatively clear in the case of public institutions such as the government, or schools, which are directly or indirectly accountable to the public. It is less clear why private firms need legitimacy.

One answer is that the more power a firm has over individuals, the more the exercise of that power needs to be perceived as legitimate. And decision making about people involves exercising power over them, so it is important to ensure legitimacy. Otherwise, people will find various ways to resist, notably through law. A loss of legitimacy might also hurt a firm's ability to compete in the market.

Questions about firms' legitimacy have repeatedly come up in the digital technology industry. For example, ride-sharing firms have faced such questions, leading to activism, litigation, and regulation. Firms whose business models rely on personal data, especially covertly collected data, have also undergone crises of perception. In addition to legal responses, such firms have seen competitors capitalize on their lax privacy practices. For instance, Apple made it harder for Facebook to track users on iOS, putting a dent in its revenue [49]. This move enjoyed public support despite Facebook's vociferous protests, arguably because the underlying business model had lost legitimacy.

For these reasons, a book on fairness is incomplete without a discussion of legitimacy. Moreover, the legitimacy question should precede other fairness questions. Debating distributive justice in the context of a fundamentally unjust institution at best is a waste of time and at worst helps prop up the institution's legitimacy, and is thus counterproductive. For example, improving facial analysis technology to decrease the disparity in error rates between racial groups is not a useful response to concerns about the use of such technologies for oppressive purposes [50].

In the discourse on fairness in machine learning, discussions of legitimacy have been overshadowed by discussions of bias and discrimination. Often, advocates have chosen to focus on distributional considerations as a way of attacking legitimacy, since it tends to be an easier argument to make. But this can backfire, as many firms have co-opted fairness discourse, and find it relatively easy to ensure parity in the decisions between demographic groups without addressing the legitimacy concerns [51].

This chapter is all about legitimacy: whether it is morally justifiable to use machine learning or automated methods at all in certain scenarios.

Although we have stressed the overriding importance of legitimacy, readers interested in distributive questions may skip to chapter 3 for a technical treatment or to chapter 4 for a normative account; those chapters, chapter 3 in particular, do not directly build on this one.

Machine Learning Is Not a Replacement for Human Decision Making

Machine learning plays an important role in decisions that allocate resources and opportunities that are critical to people's life chances. The stakes are clearly high. But people have been making high-stakes decisions about each other for a long time, and those decisions seem to be subject to far less critical examination. Here's a strawman view: decisions based on machine learning are analogous to decision making by humans, and so machine learning doesn't warrant special concern. While it's true that machine learning models might be difficult for people to understand, humans are black boxes, too. And while there can be systematic bias in machine learning models, they are often demonstrably less biased than humans.

We reject this analogy of machine learning to human decision making. By understanding why it fails and which analogies are more appropriate, we will develop a better appreciation for what makes machine learning uniquely dangerous as a way of making high-stakes decisions.

While machine learning is sometimes used to automate the tasks performed inside a human's head, many of the high-stakes decisions that are the focus of work on fairness and machine learning are those that have been traditionally performed by *bureaucracies*. For example, hiring, credit, and admissions decisions are rarely left to one person to make on their own as they see fit. Instead, these decisions are guided by formal rules and procedures, involving many actors with prescribed roles and responsibilities. Bureaucracy arose in part as a response to the subjectivity, arbitrariness, and inconsistency of human decision making; its institutionalized rules and procedures aim to minimize the effects of humans' frailties as individual decision makers [52].

Of course, bureaucracies aren't perfect. The very term bureaucracy tends to have a negative connotation—a needlessly convoluted process that is difficult or impossible to navigate. And despite their overly formalistic (one might say cold) approach to decision making, bureaucracies rarely succeed in fully disciplining the individual decision makers that occupy their ranks. Bureaucracies risk being as capricious and inscrutable as humans, but far more dehumanizing [52].

That's why bureaucracies often incorporate procedural protections: mechanisms that ensure that decisions are made transparently, on the basis of the right and relevant information, and with the opportunity for challenge and correction. Once we realize that machine learning is being used to automate bureaucratic rather than individual decisions, asserting that humans don't need to—or simply cannot—account for their everyday decisions does not excuse machine learning from these expectations. As Katherine Strandburg has argued, "[r]eason giving is a core requirement in conventional decision systems *precisely because* human decision makers are inscrutable and prone to bias and error, not because of any expectation that they will, or even can, provide accurate and detailed descriptions of their thought processes" [53].

In analogizing machine learning to bureaucratic—rather than individual—decision making, we can better appreciate the source of some of the concerns about machine learning. When it is used in high-stakes domains, it undermines the kinds of protections that we often put in place to ensure that bureaucracies are engaged in well-executed and well-justified decision making.

Bureaucracy as a Bulwark against Arbitrary Decision Making

The kind of problematic decision making that bureaucracies protect against can be called *arbitrary* decision making. Kathleen Creel and Deborah Hellman have usefully distinguished betweeen two flavors of arbitrariness [54]. First, arbitrariness might refer to decisions made on an inconsistent or ad hoc basis. Second, arbitrariness might refer to the basis for decision making lacking reasoning, even if the decisions are made consistently on that basis. This first view of arbitrariness is principally concerned with procedural regularity [55]: whether a decision making scheme is executed consistently and correctly. Worries about arbitrariness, in this case, are really worries about whether the rules governing important decisions are fixed in advance and applied appropriately, with the goal of reducing decision makers' capacity to make decisions in a haphazard manner.

When decision making is arbitrary in this sense of the term, individuals may find that they are subject to different decision-making schemes and receive different decisions simply because they happen to go through the decision-making process at different times. Not only might the decision-making scheme change over time, human decision makers might be inconsistent in how they apply these schemes as they make their way through different cases. The latter could be true of one individual decision maker whose behavior is inconsistent over time, but it could also be true if the decision-making process allocates cases to different individuals who are individually consistent but differ from one another. Thus, even two people who are identical when it comes to the decision criteria may receive different decisions, violating the expectation that similar people should be treated similarly when it comes to high-stakes decisions.

This principle is premised on the belief that people are entitled to similar decisions unless there are reasons to treat them differently (we'll soon address what determines if these are *good* reasons). For especially consequential decisions, people may have good reason to wonder why someone who resembles them received the desired outcome from the decision-making process while they did not.

Inconsistency is also problematic when it prevents people from developing effective life plans based on expectations about the decision-making systems they must navigate in order to obtain desirable resources and opportunities [54]. Thus, inconsistent decision making is unjust both because it might result in unjustified differential treatment of similar individuals and also because it threatens individual autonomy by preventing people from making effective decisions about how best to pursue their life goals.

The second view of arbitrariness is getting at a deeper concern: Are there good reasons—or any reasons—why the decision-making scheme looks the way that it does? For example, if a coach picks a track team based on the color of runners' sneakers, but does so consistently, it is still arbitrary because the criterion lacks a valid basis. It does not help advance the decision maker's goals (e.g., assembling a team of runners that will win the upcoming meet).

Arbitrariness, from this perspective, is problematic because it undermines a bedrock justification for the chosen decision-making scheme: that it actually helps to advance the goals of the decision maker. If the decision-making scheme does nothing to serve these goals, then there is no justified reason to have settled on that decision-making scheme, and to treat people accordingly. When desirable resources and opportunities are allocated arbitrarily, it needlessly subjects individuals to different decisions, despite the fact that all individuals may have equal interest in these resources and opportunities.

In the context of government decision making, there is often a legal requirement that there be a *rational* basis for decision making—that is, that there be good reasons for making decisions in the way that they are [54]. Rules that do not help the government achieve its stated policy goals run afoul of the principles of due process. This could be either because the rules were chosen arbitrarily or because of some evident fault with the reasoning behind these rules. These requirements stem from the fact that the government has a monopoly over certain highly consequential decisions, leaving people with no opportunity to seek recourse by trying their case with another decision maker.

There is no corresponding legal obligation when the decision makers are private actors, as Kathleen Creel and Deborah Hellman point out. Companies are often free to make poorly reasoned—even completely arbitrary—decisions. In theory, decision-making schemes that seem to do nothing to advance private actors' goals should be pushed out of the market by competing schemes that are more effective [54].

Despite this, we often expect that decisions of major consequence, even when they are made by private actors, are made for good reasons. We are not likely to tolerate employers, lenders, or admission officers who make decisions about applicants by flipping a coin or according to the color of applicants' sneakers. Why might this be?

Arbitrary decision making fails to respect the gravity of these decisions and shows a lack of respect for the people subject to them. Even if we accept that we cannot dictate the goals of institutions, we still object to giving them complete freedom to treat people however they like. When the stakes are sufficiently high, decision makers bear some burden of justifying their decision-making schemes out of respect for the interests of people affected by these decisions. The fact that people might try their luck with other decision makers in the same domain (e.g., another employer, lender, or admission officer) may do little to lessen these expectations.

Three Forms of Automation

To recap our earlier discussion, automation might undermine important procedural protections in bureaucratic decision making. But what, exactly, does machine learning help to automate? It turns out that there are three different types of automation.

The first kind of automation involves taking decision-making rules that have been set down by hand (e.g., worked out through a traditional policy-making process) and translating these into software, with the goal of automating their application to particular cases [56]. For example, many government agencies follow this approach when they adopt software to automate benefits eligibility determinations in accordance with preexisting policies. Likewise, employers follow this approach when they identify certain minimum qualifications for a job and develop software to automatically reject applicants who do not possess them. In both of these cases, the rules are still set by humans, but their application is automated by a computer; machine learning has no obvious role here.

But what about cases where human decision makers have primarily relied on informal judgment rather than formally specified rules? This is where the second kind of automation comes in. It uses machine learning to figure out how to replicate the informal judgements of humans. Having automatically discovered a decision-making scheme that produces the same decisions as humans have made in the past, it then implements this scheme in software to replace the humans who had been making these decisions. The student whose creative essay was subject to computerized assessment, described in the opening of this chapter, is an example of just such an approach: the software in this case seeks to replicate the subjective evaluations of human graders.

The final kind of automation is quite different from the first two. It does not rely on an existing bureaucratic decision-making scheme or human judgment. Instead, it involves learning decision-making rules from data. It uses a computer to uncover patterns in a dataset that predict an outcome or property of policy interest—and then bases decisions on those predictions. Note that such rules could be applied either manually (by humans) or automatically (through software). The relevant point of automation, in this case, is in the process of developing the rules, not necessarily applying them. For example, these could be rules that instruct police to patrol certain areas, given predictions about the likely incidence of crime based on past observations of crime. Or they could be rules that suggest that lenders grant credit to certain applicants, given the repayment histories of past recipients like them. Machine learning and other statistical techniques are crucial to this form of automation.

As we will see over the next three sections, each type of automation raises its own unique concerns.

Automating Preexisting Decision-Making Rules

In many respects, the first form of automation—translating preexisting rules into software so that decisions can be executed automatically—is a direct response to arbitrariness as inconsistency. Automation helps ensure consistency in decision making because it requires that the scheme for making decisions be fixed. It also means that the scheme is applied the same way every time.

And yet, many things can go wrong. Danielle Citron offers a compelling account of the dangers of automating decision-making rules established via a deliberative policy-making or rule-making process [56]. Automating the execution of a preexisting decision-making scheme requires translating such a scheme into code. Programmers might make errors in that process, leading to automated decisions that diverge from the policy that the software is meant to execute. Another problem is that the policy that programmers are tasked with automating may be insufficiently explicit or precise; in the face of such ambiguity, programmers might take it upon themselves to make their own judgment calls, effectively usurping the authority to define policy. And at the most basic level, software may be buggy. For example, hundreds of British postmasters were convicted for theft or fraud over a twenty-year period based on flawed software in what has been called the biggest miscarriage of justice in British history [57].

Automating decision making can also be problematic when it completely stamps out any room for discretion. While human discretion presents its own issues, as described above, it can be useful when it is difficult or impossible to fully specify how decisions should be made in accordance with the goals and principles of the institution [58]. Automation requires that an institution determine in advance all of the criteria that a decision-making scheme will take into account; there is no room to consider the relevance of additional details that might not have been contemplated or anticipated at the time that the software was developed.

Automated decision making is thus likely to be much more brittle than decision making that involves manual review because it limits the opportunity for decision subjects to introduce information into the decision-making process. People are confined to providing evidence that corresponds to a preestablished field in the software. Such constraints can result in absurd situations in which the strict application of decision-making rules leads to outcomes that are directly counter to the goals behind these rules. New evidence that would immediately reverse the assessment of a human decision maker may have no place in automated decision making [59]. For example, in an automated system developed to assess whether people with illnesses were eligible for a state-provided caregiver, one field asked if there were any foot problems. An assessor visited a certain person and filled out the field to indicate that they didn't have any problems—because they were an amputee [60].

Discretion is valuable in these cases because humans are often able to reflect on the relevance of additional information to the decision at hand and the underlying goal that such decisions are meant to serve. In effect, human review leaves room to expand the criteria under consideration and to reflect on when the mechanical application of the rules fails to serve their intended purpose [59, 61].

These same constraints can also restrict people's ability to point out errors or to challenge the ultimate decision [62]. When interacting with a loan officer, a person could point out that their credit file contains erroneous information. When applying for a loan via an automated process, they may have no equivalent opportunity. Or perhaps a person recognizes that the rules dictating their eligibility for government benefits have been applied incorrectly. When caseworkers are replaced by software, people subject to these decisions may have no means to raise justified objections [63].

Finally, automation runs the serious risk of limiting accountability and exacerbating the dehumanizing effects of dealing with bureaucracies. Automation can make it difficult to identify the agent responsible for a decision; software often has the effect of dispersing the locus of accountability because the decision seems to be made by no one [64]. People may have more effective means of disputing decisions and contesting the decision-making scheme when decision making is vested in identifiable people. Likewise, automation's ability to remove humans from the decision-making process may contribute to people's sense that an institution does not view them as worthy of the respect that would grant them an opportunity to make legitimate corrections, introduce additional relevant information, or describe mitigating circumstances [65]. This is precisely the problem highlighted by the opening example of a social media user who had been kicked off a platform without explanation or opportunity for appeal.

We have highlighted many normative concerns that arise from simply automating the application of a preexisting decision-making scheme. While many of these issues are commonly attributed to the adoption of machine learning, none of them originates from the use of machine learning specifically. Long-standing efforts to automate decision making with traditional software pose many dangers of their own. The fact that machine learning is not the exclusive cause of these types of problems is no reason to take them any less seriously, but effective responses to these problems require that we be clear about their origins.

Learning Decision-Making Rules from Data on Past Decisions in Order to Automate Them

Decision makers may have a preexisting but informal process for making decisions which they may like to automate. In this case, machine learning (or other statistical techniques) might be employed to "predict" how a human would make a decision, given certain criteria. The goal isn't necessarily to perfectly recover the specific weight that past decision makers had implicitly assigned to different criteria, but rather to ensure that the model produces a similar set of decisions as humans. To return to one of our recurring examples, an educational institution may want to automate the process of grading essays, and it may attempt to do that by relying on machine learning to learn to mimic the grades teachers have assigned to similar work in the past.

This form of automation might help to address concerns with arbitrariness in human decision making by formalizing and fixing a decision-making scheme in ways similar to what humans might have been employing in the past. In this respect, machine learning might be desirable because it can help to smooth out any inconsistencies in the human decisions from which it has inferred some decision-making rule. For example, the essay grading model described above might reduce some of the variance observed in the grading of teachers whose subjective evaluations the model is learning to replicate. Automation can once again help to address concerns with arbitrariness understood as inconsistency, even when it is subjective judgments that are being automated.

A few decades ago, there was a popular approach to automation that relied on explicitly encoding the reasoning that humans relied on to make decisions [66]. This approach, called "expert systems," failed for many reasons, including the fact that people aren't always able to explain their own reasoning [67]. Expert systems eventually gave way to the approach of simply asking people to label examples and having learning algorithms discover how to best predict the label that humans would assign. While this approach has proved powerful, it has its dangers.

First, it may give the veneer of objective assessment to decision-making schemes that simply automate the subjective judgment of humans. As a result, people may be more likely to view its decisions as less worthy of critical investigation. This is particularly worrisome because learning decision-making rules from the previous decisions made by humans runs the obvious risk of replicating and exaggerating any objectionable qualities of human decision making by learning from the bad examples set by humans. (In fact, many attempts to learn a rule to predict some seemingly objective target of interest—the form of automation that we will discuss in the next section—are really just a version of replicating human judgment in disguise. If we can't obtain objective ground truth for the chosen target of prediction, there is no way to escape human judgment [68]).

Second, such decision-making schemes may be regarded as equivalent to those employed by humans and thus likely to operate in the same way, even though the model might reach its decisions differently and produce quite different error patterns [69]. Even when the model is able to predict with a high degree of accuracy the decisions that humans would make given any particular input, there is no guarantee that the model will have inherited all of the nuance and considerations that go into human decision making. Worse, models might also learn to rely on criteria in ways that humans would find worrisome or objectionable, even if doing so still produces a set of decisions similar to those humans would make [70]. For example, a model that automates essay grading by assigning higher scores to papers that employ sophisticated vocabulary may do a reasonably good job replicating the judgments of human graders (likely because higher-quality writing tends to rely on more sophisticated vocabulary), but checking for the presense of certain words is unlikely to be a reliable substitute for assessing an essay for logical coherence and factual correctness [71].

In short, the use of machine learning to automate decisions previously performed by humans can be problematic because machine learning models can end up being both too much like human decision makers and too different from them.

Deriving Decision-Making Rules by Learning to Predict a Target

The final form of automation is one in which decision makers rely on machine learning to learn a decision-making rule or policy from data. This form of automation, which we'll call predictive optimization, speaks directly to concerns with reasoned decision making. Note that neither of the first two forms of automation does so. Consistently executing a preexisting policy via automation does not ensure that the policy itself is a reasoned one. Nor does relying on past human decisions to induce a decision-making rule guarantee that the basis for automated decision making will reflect reasoned judgments. In both cases, the decision-making scheme will be only as reasoned as the formal policy or informal judgments whose execution is being automated.

In contrast, predictive optimization tries to provide a more rigorous foundation for decision making by relying on criteria only to the extent that they demonstrably predict the outcome or quality of interest. When employed in this manner, machine learning seems to ensure reasoned decisions because the criteria that have been incorporated into the decision making scheme—and their particular weighing—are dictated by how well they predict the target. And so long as the chosen target is a good proxy for decision makers' goals, relying on criteria that predict this target to make decisions would seem well reasoned because doing so will help to achieve decision makers' goals.

Unlike the first two forms of automation, predictive optimization is a radical departure from the traditional approach to decision making. In the traditional approach, a set of decision makers has some goal—even if this goal is amorphous and hard to specify—and would like to develop an explicit decision-making scheme to help realize their goal. They engage in discussion and deliberation to try to come to some agreement about the criteria that are relevant to the decision and the weight to assign to each criterion in the decision-making scheme. Relying on

Table 2.1
Comparison of traditional decision making to predictive optimization

	Traditional approach	Predictive optimization approach
Example: college admissions	Holistic approach that takes into account achievements, character, special circumstances, and other factors	Train a model based on past students' data to predict applicants' GPA if admitted; admit highest-scoring applicants
Goal and target	No explicit target; goal is implicit (and there are usually multiple goals)	Define an explicit target; assume it is a good proxy for the goal
Focus of deliberation	Debate is about how the criteria should affect the decision	Debate is largely about the choice of target
Effectiveness	May fail to produce rules that meet their putative objectives	Predictive accuracy can be quantified
Range of normative considerations	Easier to incorporate multiple normative principles such as need	Harder to incorporate multiple normative principles
Justification	Can be difficult to divine rule makers' reasons for choosing a certain decision making scheme	Reasons for the chosen decision making scheme are made explicit in choice of target

intuition, prior evidence, and normative reasoning, decision makers will choose and combine features in ways that are thought to help realize their goals.

The statistical or machine learning approach works differently. First, the decision makers try to identify an explicit target for prediction which they view as synonymous with their goal—or a reasonable proxy for it. In a college admissions scenario, one goal might be scholastic achievement in college, and college grade point average (GPA) might be a proxy for it. Once this is settled, the decision makers use data to discover which criteria to use and how to weight them in order to best predict the target. While they might exercise discretion in choosing the criteria to use, the weighting of these criteria would be dictated entirely by the goal of maximizing the accuracy of the resulting prediction of the chosen target. In other words, the decision-making rule would, in large part, be learned from data, rather than set down according to decision makers' subjective intuitions, expectations, and normative commitments (table 2.1).

Each approach has pluses and minuses from a normative perspective. The traditional approach makes it possible to express multiple goals and normative values through the choice of criteria and the weight assigned to them.

In the machine learning approach, multiple goals and normative considerations need to be packed into the choice of target. In college admissions, those goals and considerations might include—in addition to scholastic potential—athletic and leadership potential, the extent to which the applicant would contribute to campus life, whether the applicant brings unusual life experiences, their degree of need, and many others. The most common approach is to define a composite target variable that linearly combines multiple components, but this quickly becomes unwieldy and is rarely subject to robust debate. There is also some room to exercise normative judgment about the choice to include or exclude certain decision criteria, but is a far cry from deliberative policy making.

On the other hand, if we believe that a target does, in fact, capture the full range of goals that decision makers have in mind, machine learning models might be able to serve these goals more effectively. For example, in a paper that compares the two approaches to policy making, Rebecca Johnson and Simone Zhang show that the traditional approach (i.e., manually crafting rules via a process of deliberation and debate) often fails to produce rules that meet their putative objectives [72]. In examining rules for allocating housing assistance, they find that housing authorities prioritize veterans above particularly rent-burdened households, despite the fact that supporting such households would seem to be more in line with the policy's most basic goal. Johnson and Zhang assert that while this prioritization might be the actual intent of the policy makers setting the rules, the reasons for this prioritization are rarely made explicit in the process of deliberation and are especially difficult to discern after the fact. Were these rules developed instead using machine learning, policy makers would need to agree on an explicit target of prediction, which would leave much less room for confusion about policy makers' intent. And it would ensure that the resulting rules are *only* designed to predict that target [72]. As Rediet Abebe, Solon Barocas, Jon Kleinberg, and colleagues have argued, "[t]he nature of computing is such that it requires explicit choices about inputs, objectives, constraints, and assumptions in a system" [73]—and this may be a good thing if it forces certain policy considerations and normative judgements into the open.

The machine learning approach nevertheless runs the serious risk of focusing narrowly on the accuracy of predictions. In other words, "good" decisions are those that accurately predict the target. But decision making might be "good" for other reasons: focusing on the right qualities or outcomes (in other words, the target is a good proxy for the goal), considering only relevant factors, considering the full set of relevant factors, incorporating other normative principles (e.g., need, desert, etc.), or allowing people to understand and potentially contest the policy. Even a decision-making process that is not terribly accurate might be seen as good if it has some of these other properties [74]. In the next few sections, we explore how each of these concerns might apply to machine learning.

Mismatch between Target and Goal

Identifying a target of prediction that is a good match for the goals of the decision maker is rarely straightforward. Decision makers often don't have a preexisting, clear, and discrete goal in mind [75]. When they do, the goal can be far more complex and multifaceted than one discrete and easily measurable outcome [76]. In fact, decision makers can have multiple conflicting goals, perhaps involving some trade-offs between them. For example, the decision-making schemes adopted by college admission officers often encode a range of different goals. They do not simply rank applicants by their predicted GPA and then admit the top candidates to fill the physical capacity of the school. Aside from the obvious fact that this would favor candidates who take "easy" classes, admissions officers aim to recruit a student body with diverse interests and a capacity to contribute to the broader community.

Besides, there might be serious practical challenges in measuring the true outcome of interest, leaving decision makers to find alternatives that might serve as reasonable proxies for it. In most cases, decision makers settle on a target of convenience—that is, on a target for which there is easily accessible data [56, 77]. For example, arrest data (i.e., whether someone has been arrested) is often adopted as a proxy for crime data (i.e., whether someone has committed a crime), even though many crimes are never observed and thus never result in arrest and even though the police might be quite selective in choosing to arrest someone for an observed crime [78]. Without condoning the decision to adopt this target, we might still recognize the practical challenges that would encourage the police to rely on arrests. It is simply impossible to observe all crime and so decision makers might feel justified in settling on arrests as a substitute.

Even if decision makers had some way of obtaining information on crime, it is *still* not obvious how well this chosen target would match the underlying goals of the police. Accurately predicting the occurrence of future crimes is not the same thing as helping to reduce crime; in fact, accurate predictions of crime might simply cause the police to observe more crimes and generate more arrests rather than preventing those crimes from happening in the first place [79]. If the police's actual goal is to reduce crime and not simply to ensure that all crimes result in arrests, then even using crime as the target of prediction might not help the police to realize these goals. The police might be better off estimating the deterrent effect of police intervention, but this is a far more complicated task than making predictions on the basis of observational data; answering these questions requires experimentation. (Of course, even this formulation of the problem should be subject to further critical analysis because it fails to consider the many other kinds of interventions that might help to reduce crime beyond improving the deterrent effect of police presence.) Yet even when there are good reasons to favor a more nuanced approach along these lines, decision makers may favor imperfect simplifications of the problem because they are less costly or more tractable [56, 61].

Finally, decision makers and decision subjects might have very different ideas about what would constitute the right target of prediction. Much of the discussion in this chapter has so far been premised on the idea that decision makers' goals are widely perceived as desirable in the first place, and thus defensible. But there are many times when the normative issue is not with the way decisions are being made but with the goal of the decision-making process itself [77]. In some cases, we may disagree with the goals of a decision maker because we don't think that they are what is in the best interest of the decision maker themself. More often, we might disagree with these goals because they are at odds with the interests of other people who will be negatively impacted by decision makers' pursuit of these goals. As Oscar Gandy has argued, "certain kind[s] of bias are inherent in the selection of the goals or objective functions that automated systems will [be] designed to support" [80].

To appreciate how this is different from a target-goal mismatch, consider a well-known study by Ziad Obermeyer, Brian Powers, Christine Vogeli et al. on bias in an algorithm employed by a health care system to predict which patients would benefit most from a "high-risk care management" program [81]. They found that the algorithm exhibited racial bias—specifically, that it underestimated the degree to which Black patients' health would benefit from enrollment in the program. That's because the developers adopted health care costs as the target of prediction, on the apparent belief that it would serve as a reasonable proxy for health care needs. The common recounting of this story suggests that decision makers simply failed to recognize the fact that there are racial disparities in both care seeking and the provision of health care that cause Black patients of equivalently poor health to be less expensive than non-Black patients. On this account, fixing the problem would only require adopting a target that better reflected the health care system's goals: maximizing the overall health benefits of the program. Yet it is entirely possible that the original target of prediction reflected the health care system's true goals, which might have been to simply reduce costs without any regard for whose health would benefit most from these interventions. If that were the case, then the choice of target was not simply a poor match for decision makers' goals; the goals themselves were problematic. We must be careful not to confuse cases where we object to the goals for cases where we object to the particular choice of target.

Failing to Consider Relevant Information

Bureaucracies are often criticized for not being sufficiently individualized or particularized in their assessments, lumping people into needlessly coarse groups. Had decision makers only considered some additional detail, they would have realized that the person in question is actually unlike the rest of the people with whom they had been grouped.

Supervised machine learning is a form of inductive reasoning. It aims to draw general rules from a set of specific examples, identifying the features and feature values that reliably co-occur with an outcome of interest. As it turns out, the limitation of being insufficiently individualized is an unavoidable part of inductive reasoning.

Imagine a car insurance company that is trying to predict the likelihood that a person applying for an insurance policy will get into a costly accident. The insurer will try to answer this question by looking at the frequency of past accidents that involved other people similar to the applicant. This is inductive reasoning: the applicant is likely to exhibit similar behavior or experience similar outcomes as previous policyholders because the applicant possesses many other qualities in common with these policyholders. Perhaps the person is applying for insurance to cover their bright red sports car—a type of car that is involved in accidents far more frequently than other types of cars. Noting this historical pattern, the insurer might therefore conclude that there is a heightened chance that the applicant will need to make a claim against their policy—and offer to insure the applicant only at an elevated price. Having received the offer, the applicant, who is, in fact, a highly skilled driver with an excellent command of the car, might balk at the price, objecting to the idea that they present a risk anything like the other policyholders with the same car.

What has happened here? The insurer has made its prediction on the basis of rather coarse details (in this case, on the basis of only the model and color of the car), treating the rate at which accidents happen among previous policyholders with such a car as a reliable indicator of the probability of the applicant having an accident of their own. Frederick Schauer refers to this as the problem of "statistically sound but nonuniversal generalizations": when an individual fulfills all the criteria for inclusion in a particular group but fails to possess the quality that these criteria are expected to predict [82].

Situations of this sort can give rise to claims of stereotyping or profiling and to demands that decision makers assess people as individuals, not merely as members of a group. Yet, as Schauer has explained, it can be difficult to specify what it means to treat someone as an individual or to make individualized decisions. It is unclear, for example, how an insurer could make predictions about an individual's likelihood of getting into a car accident without comparing the applicant to other people that resemble them. At issue in these cases is not the failure to treat someone as an individual, but the failure to take additional relevant criteria into account that would distinguish a person from the other people with whom they would otherwise be lumped in [82]. If the insurer had access to additional details (in particular, details about the applicant's and past policyholders' driving skills), the insurer might have made a more discerning judgment about the applicant. This is exactly what is going on when insurers agree to offer lower prices to applicants who voluntarily install black boxes in their cars and who demonstrate themselves to be careful drivers. It is easy to misinterpret this trend as a move toward individualized assessment, as if insurers are judging each individual person on their unique merits as a driver. The correct interpretation requires that we recognize that insurers are only able to make use of the data from a specific driver's black box by comparing it to the data from the black boxes of other drivers whose driving records are being used to make a prediction about the driver in question. Even if we accept that decisions can never be fully individualized, we might still expect that decision makers take into account the full range of relevant information at their potential disposal. To carry forward the example above, we might say that the car insurance company had an obligation to consider the applicants' driving skills, not just the model and color of their car, even if doing so still meant that they were being assessed according to how often other people with similar driving skills and similar cars had gotten into accidents in the past.

But how far should this expectation extend? What obligations do decision makers have to seek out every last bit of conceivable information that might enable more accurate predictions? Well, at some point, additional information ceases to be helpful because there aren't enough training data. For example, people who live near a specific intersection may be more likely to get into accidents because the intersection is poorly designed and thus dangerous. But the insurer can learn this only if it has enough data from enough people who live near this intersection.

There is also a very practical reason why we might not hold decision makers to a standard by which they are required to consider all information that might be conceivably relevant. Collecting and considering all of this information can be expensive, intrusive, and impractical. In fact, the cost of doing so could easily outweigh the perceived benefits that come from more granular decision making—not just to the decision maker but to the subjects of the decisions as well. While black boxes can help to achieve far more granularity in insurance pricing, they are also controversial because they are quite intrusive and pose a threat to drivers' privacy. For reasons along these lines, Frederick Schauer and others have suggested that decision makers are justified in making decisions on the basis of a limited set of information, even when additional relevant information might exist, if the cost of obtaining that information exceeds its benefits [82, 83].

There are three things to note about these arguments. First, they are not arguments about automated decision making specifically; they are general statements about any form of decision making, whether automated or not. Yet, as we discussed earlier in the chapter, automated decision making often limits the opportunity to introduce additional relevant information into the decision-making process. The cost savings that might be achieved by automating certain decisions (often by way of replacing human workers with software) come at the cost of depriving people of the chance to highlight relevant information that has no place in the automated process. Given that people might be both very willing and perfectly able to volunteer this information (i.e., able to do so at little cost), automated decision making that simply denies people the opportunity to do so might fail the cost-benefit analysis. Second, the cost-benefit analysis that undergirds Schauer's arguments does not take into account any distributional considerations, like which groups might be enjoying more of the benefits or experiencing more of the costs. In chapter 4, we return to this question, asking whether decision makers are justified in subjecting certain groups to less granular and thus less accurate decisions simply because there is less information about them. Finally, these arguments don't grapple with the fact that decision makers and decision subjects might arrive at quite different conclusions when performing a cost-benefit analysis if they are performing this analysis from their own perspectives. A decision maker might find that the cost of collecting more information does not generate a sufficiently large corresponding benefit *for them as the decision maker*, despite the fact that certain decision subjects would surely benefit from such an investment. It is not obvious why the cost-benefit analysis of decision makers alone should be allowed to determine the level of granularity that is acceptable. One possible explanation might be that increasing the costs of making decisions (by, for example, seeking out and taking more information into account) will encourage decision makers to simply pass these costs onto decision subjects. For instance, if developing a much more detailed assessment of applicants for car insurance increases the operating costs of the insurer, the insurer is likely to charge applicants a higher price to offset these additional costs. From this perspective, the costs to the decision maker are really just costs to decision subjects. Of course, this perspective doesn't contemplate the possibility of the insurer simply assuming these costs and accepting less profit.

The Limits of Induction

Beyond cost considerations, there are other limits to inductive reasoning. Suppose the coach of a track team assesses potential members of the team according to the color of their sneakers rather than the speed at which they can run. Imagine that just by coincidence, slower runners in the pool happen to prefer red sneakers and faster runners happen to prefer blue sneakers—but that no such relationship obtains in other pools of runners. Thus, any lessons the coach might draw from these particular runners about the relationship between sneaker colors and speed would be unreliable when applied to other runners. This is the problem of overfitting [84]. It is a form of arbitrary decision making because the predictive validity that serves as its justification is an illusion.

Overfitting is a well-understood problem in machine learning and there are many ways to counteract it. Since the spurious relationship occurs due to coincidence, the bigger the sample, the less likely it is to occur. Further, one can penalize models that are overly complicated to make it less likely that they pick up on chance patterns in the data. And most importantly, it is standard practice to separate the examples that are used to train and test machine learning models. This allows a realistic assessment of how well the relationships observed in the training data carry over to unseen examples. For these reasons, unless dealing with small sample sizes, overfitting is generally not a serious problem in practice.

But variants of the overfitting problem can be much more severe and thorny. It is common practice in machine learning to take one existing dataset—in which all the data have been gathered in a similar way—and simply split it into training and test sets. The small differences between these sets will help to avoid overfitting and may give some sense for performance on unseen data. But these split sets are still much more similar to each other than the future population to which the model might be applied [85, 86]. This is the problem of "distribution shifts," of which there are many different kinds. They are common in practice and they present a foundational problem for the machine learning paradigm.

Returning to our earlier example, imagine that runners are able to buy sneakers from only one supplier and that the supplier sells only one type of sneaker, but varies the color of the sneaker by size (all sizes below 8 are red, while all sizes 8 and above are blue). Further, assume that runners with larger feet are faster than those with smaller feet and that there is a large step change in runners' speed once their foot size exceeds 7. Under these circumstances, selecting runners according to the color of their sneakers will reliably result in a team composed of faster runners, but it will do so for reasons that we still might find foolish or even objectionable. Why? The relationship between the color of a runner's sneakers and running speed is obviously spurious in the sense that we know that the color of a runner's sneakers has no causal effect on speed. But is this relationship truly spurious? It is not just an artifact of the particular set of examples from which a general rule has been induced; it's a stable relationship in the real world. So long as there remains only one supplier and the supplier only offers different colors in these specific sizes, sneaker color will reliably distinguish faster runners from slower runners. So what's the problem with making decisions on this basis? Well, we might not always have a way to determine whether we are operating under the conditions described. Generalizing from specific examples always admits the possibility of drawing lessons that do not apply to the situation that decision makers will confront in the future.

One response to these concerns is to assert that there is a normative obligation that decision criteria bear a causal relationship to the outcome that they are being used to predict. The problem with using sneaker color as a criterion is obvious to us because we can recognize the complete absence of any plausible causal influence on running speed. When machine learning is used, the resulting models, unconcerned with causality, may seize upon unstable correlations [87]. This gives rise to demands that no one should be subject to decision-making schemes that are based on findings that lack scientific merit—that is, on findings that are spurious and thus invalid. They likely account for concerns of scholars like Frank Pasquale, who talks about cases where machine learning is "facially invalid" [88], and Pauline Kim and Erika Hanson, who have argued that "because data mining uncovers statistical relationships that may not be causal, relying on those correlations to make predictions about future cases may result in arbitrary treatment of individuals" [89]. Asserting that decision-making schemes should be based only on criteria that have a causal relationship to the outcome of interest is likely perceived as a way to avoid these situations—that is, as a way to ensure that the basis for decision making is well reasoned, not arbitrary.

A Right to Accurate Predictions?

In the previous two sections, we discussed several reasons why predictions using inductive reasoning may be inaccurate, including failing to consider relevant information and distribution shift. But even if we set aside those reasons—and assume that the decision maker considers all available information, there is no distribution shift, and so on—there might be insurmountable limits to the accuracy of predicting future outcomes. These limits might persist whether or not inductive reasoning is employed [90]. For example, at least some cases of recidivism are due to spur-of-the-moment crimes committed when opportunities fortuitously presented themselves, and these might not be predictable in advance. (We will review some of the empirical evidence of limits to prediction in later chapters.)

What are the implications of these limits to prediction? From the decision maker's perspective, even a small increase in predictive accuracy compared to a baseline (human judgment or rule-based policy) can be valuable. Consider a child protection agency employing a predictive screening tool to determine which children are at risk of child abuse. Increased accuracy may mean fewer children placed in foster care. It might also result in substantial cost savings, with fewer case workers required to make visits to homes.

A typical model deployed in practice may have an accuracy (more precisely, AUC) of between 0.7 and 0.8 [91]. That's better than a coin toss but still results in a substantial number of false positives and false negatives. A claim that the system makes the most accurate decision possible at the time of screening is cold comfort to families where children are separated from their parents due to the model's prediction of future abuse, or cases of abuse that the model predicted to be low risk. If the model's outputs were random, we would clearly consider it arbitrary and illegitimate (and even cruel). But what is the accuracy threshold for legitimacy? In other words, how high must accuracy be in order to be able to justify the use of a predictive system at all [92]?

Low accuracy becomes even more problematic when we consider that it is measured with respect to a prediction target that typically requires sacrificing some of the multifaceted goals that decision makers might have. For example, a child welfare risk prediction model might not be able to reason about the differential effects that an intervention such as foster care might have on different children and families. How much of an increase in predictive accuracy is needed to justify the mismatch between the actual goals of the system and those realized by the model?

Obviously, these questions don't have easy answers, but they represent important and underappreciated threats to the legitimacy of predictive decision making.

Agency, Recourse, and Culpability

Let's now consider a very different concern: Could criteria that exhibit statistical relevance and enable accurate predictions still be normatively inappropriate as the basis for decision making?

Perhaps the criterion in question is an immutable characteristic. Perhaps it is a mutable characteristic, but not one that the specific person in question has any capacity to change. Or perhaps the characteristic has been affected by the actions of others, and is not the result of the person's own actions. Each of these reasons, in slightly different ways, concerns the degree of control that a person is understood to have over the characteristic in question—and each provides some normative justification for either ignoring or discounting the characteristic even when it might be demonstrably predictive of the outcome of interest. Let's dig into each of these concerns further.

Decisions based on immutable characteristics can be cause for concern because they threaten people's agency. By definition, there is nothing anyone can do to change immutable characteristics (such as one's country of birth). By extension, there is nothing anyone can do to change decisions made on the basis of immutable characteristics. Under these circumstances, people are condemned to their fates and are no longer agents of their own lives. There is something disquieting about the idea of depriving people of the capacity to make changes that would result in a different outcome from the decision-making process, especially when these decisions might significantly affect a person's life chances and life course. This might be viewed as especially problematic when there seem to be alternative ways for a decision maker to render effective judgment about a person without relying on immutable characteristics. In this view, if it is possible to develop decision-making schemes that are equally accurate, but that still leave room for decision subjects to adapt their behavior so as to improve their chances of a favorable decision, then decision makers have a moral obligation to adopt such a scheme out of respect for people's agency.

Recourse is a related but more general idea about the degree to which people have the capacity to make changes that result in different decisions [93]. While there is nothing anyone might do to change an immutable characteristic, people might be more or less capable of changing those characteristics that are, in principle, mutable [94, 95]. Some people might need to expend far more resources than others to obtain the outcome that they want from the decision-making process. Choosing certain criteria to serve as the basis for decision making is also a choice about the kinds of actions that will be available for people to undertake in seeking a different decision. And people in different circumstances will have different abilities to successfully do so. In some cases, people may never have sufficient resources to achieve this—bringing us back to the same situation discussed above. For example, one applicant for credit might be well positioned to move to a new neighborhood so as to make herself a more attractive candidate for a new loan, assuming that the decision-making scheme uses location as an important criterion. But another applicant might not be able to do so, for financial, cultural, or many other reasons.

Research on recourse in machine learning has largely focused on ensuring that people receive *explanations* of ways to achieve a different decision from a model that people can actually execute in reality [96]. Given that there are many possible ways to explain the decisions of a machine learning model, the goal of this work is to ensure that the proffered explanations direct people to take viable actions rather than suggest that the only way to get the desired outcome is to do something beyond their capacity. Even when developing a decision-making scheme that relies on only mutable characteristics, decision makers can do more to preserve recourse by adapting their explanation of a model's decisions to focus on those actions that are easiest for people to change. On this account, the better able people are to make changes that give them the desired outcome, the better the decision-making scheme and the better the explanation.

Finally, as mentioned earlier in this section, we might view certain decision-making schemes as unfair if they hold people accountable for characteristics outside their control. Basic ideas about moral culpability almost always rest on some understanding of the actions that brought about the outcomes of concern. For example, we might be upset with a person who has bumped into us and caused us to drop and break some precious item. Upon discovering that they have been pushed by somebody else, we are likely to hold them blameless and redirect our disapprobation to the person who pushed them. This same reasoning often carries over to the way that we think about the fairness of relying on certain criteria when making decisions that allocate desirable resources and opportunities. Unless we know *why* certain outcomes come to pass, we cannot judge whether decision makers are normatively justified in relying on criteria that accurately predict if that outcome will come to pass. We need to understand the cause of the outcome of interest so that we might reflect on whether the subject of the decision bears moral responsibility for the outcome, given its cause.

For example, as Barbara Kiviat has explored, laws in many US states limit the degree to which car insurance providers can take into account "extraordinary life circumstances" when making underwriting or pricing decisions, including such events as the death of a spouse, child, or parent [97]. These laws forbid insurers from considering a range of factors over which people cannot exercise any control—like a death in the family—but which may nevertheless contribute to someone experiencing financial hardship and thus to increasing the likelihood of making a claim against their car insurance policy in the event of even a minor accident. These prohibitions reflect an underlying belief that people should not be subject to adverse decisions if they were not responsible for whatever it is that makes them appear less deserving of more favorable treatment. Or to put it another way: people should be judged only on the basis of factors for which they are morally culpable. Fully implementing this principle is impractical, since most attributes that the decision maker might use (say, income) are partly but not fully the result of the individual's choices. However, attributes like a death in the family seem to fall fairly clearly on one side of the line.

Of course, there is a flip side to all of this. If people can easily change the features that are used to make decisions about them, they might "game" the decision-making process. By gaming we mean changing the value of features in order to change the decision without changing the expected outcome that the features are meant to predict [98]. "Teaching to the test" is a familiar scenario that is an example of gaming. Here, the test score is a feature that predicts future performance (say, at a job). Assume that the test, in fact, has predictive value, because people who do well at the test tend to have mastered some underlying body of knowledge, and such mastery improves their job performance. Teaching to the test refers to methods of preparation that increase the test score without correspondingly increasing the underlying ability that the score is meant to reflect. For example, teachers might help students prepare for the test by exploiting the fact that the test assesses very specific knowledge or competencies—not the full range of knowledge or competencies that the test purports to measure—and focus preparation on only those parts that will be assessed [99]. Jane Bambauer and Tal Zarsky give many examples of gaming decision making systems [100].

Gaming is a common problem because most models do not discover the causal mechanism that accounts for the outcome. Thus, preventing gaming requires causal modeling [98]. Furthermore, a gameable scheme becomes less effective over time and may undermine the goals of the decision maker and the proper allocation of the resource. In fact, gaming can be a problem even when decision subjects are not act-ing adversarially. Job seekers may expend considerable effort and money to obtain meaningless credentials that they are told matter in their industry, only to find that while this helps them land a job, it does not make them any better prepared to actu-ally perform it [101]. Under such circumstances, strategic behavior may represent wasteful investment of effort on the part of well-intentioned actors.

Concluding Thoughts

In this chapter, we teased apart three forms of automation. We discussed how each responds to concerns about arbitrary decision making in some ways, while at the same time opening up new concerns about legitimacy. We then delved deep into the third type of automation, predictive optimization, which is what we will be concerned with in most of this book.

To be clear, we make no blanket claims about the legitimacy of automated deci-sion making. In applications that aren't consequential to people's life chances, questions of legitimacy are less salient. For example, in credit card fraud detection, statistical models are used to find patterns in transaction data, such as a sudden change in location, that might indicate fraud resulting from stolen credit card infor-mation. The stakes to individuals tend to be quite low. For example, in the United States, individual liability is capped at $50 provided certain conditions are met. Thus, while errors are costly, the cost is primarily borne by the decision maker (in this example, the bank). So banks tend to deploy such models based on cost con-siderations without worrying about (for instance) providing a way for customers to contest the model.

In consequential applications, however, to establish legitimacy, decision makers must be able to affirmatively justify their scheme along the dimensions we've laid out: explain how the target relates to goals that all stakeholders can agree on, vali-date the accuracy of the deployed system, allow methods for recourse, and so forth. In many cases, it is possible to put procedural protections around automated sys-tems to achieve this justification, yet decision makers are loath to do so because it undercuts the cost savings that automation is meant to achieve.

3

Classification

The goal of classification is to leverage patterns in natural and social processes to conjecture about uncertain outcomes. An outcome may be uncertain because it lies in the future. This is the case when we try to predict whether a loan applicant will pay back a loan by looking at various characteristics such as credit history and income. Classification also applies to situations where the outcome has already occurred, but we are unsure about it. For example, we might try to classify whether financial fraud has occurred by looking at financial transactions.

What makes classification possible is the existence of patterns that connect the outcome of interest in a population to pieces of information that we can observe. Classification is specific to a population and the patterns prevalent in the population. Risky loan applicants might have a track record of high credit utilization. Financial fraud often coincides with irregularities in the distribution of digits in financial statements. These patterns might exist in some contexts but not others. As a result, the degree to which classification works varies.

We formalize classification in two steps. The first is to represent a population as a probability distribution. While often taken for granted in quantitative work today, the act of representing a dynamic population of individuals as a probability distribution is a significant assumption. The second step is to apply statistics, specifically statistical decision theory, to the probability distribution that represents the population. Statistical decision theory formalizes the classification objective, allowing us to talk about the quality of different classifiers.

The statistical decision-theoretic treatment of classification forms the foundation of supervised machine learning. Supervised learning makes classification algorithmic in how it provides heuristics to turn samples from a population into good classification rules.

Age. Curt.	Per-ſons.	Age. Curt.	Per-ſons	Age. Curt.	Per-ſons	Age. Curt.	Per-ſons	Age. Curt.	Per-ſons	Age. Curt.	Per-ſons
1	1000	8	680	15	628	22	586	29	539	36	481
2	855	9	670	16	622	23	579	30	531	37	472
3	798	10	661	17	616	24	573	31	523	38	463
4	760	11	653	18	610	25	567	32	515	39	454
5	732	12	646	19	604	26	560	33	507	40	445
6	710	13	640	20	598	27	553	34	499	41	436
7	692	14	634	21	592	28	546	35	490	42	427
Age. Curt	Per-ſons.	Age. Curt.	Per-ſons	Age. Curt.	Per-ſons	Age Curt	Per-ſons	Age. Curt.	Per-ſons	Age. Curt.	Per-ſons
43	417	50	346	57	272	64	202	71	131	78	58
44	407	51	335	58	262	65	192	72	120	79	49
45	397	52	324	59	252	66	182	73	109	80	41
46	387	53	313	60	242	67	172	74	98	81	34
47	377	54	302	61	232	68	162	75	88	82	28
48	367	55	292	62	222	69	152	76	78	83	23
49	357	56	282	63	212	70	142	77	68	84	20

Age.	Perſons.
7	5547
14	4584
21	4270
28	3964
35	3604
42	3178
49	2709
56	2194
63	1694
70	1204
77	692
84	253
100	107
Sum Total.	34000

Figure 3.1
Halley's life table (1693).

Modeling Populations as Probability Distributions

One of the earliest applications of probability to the study of human populations is Halley's *life table* from 1693. Halley tabulated births and deaths in a small town in order to estimate life expectancy in the population (figure 3.1). Estimates of life expectancy, then as novel as probability theory itself, found use in accurately pricing investments that paid an amount of money annually for the remainder of a person's life.

However, for centuries that followed, the use of probability to model human populations remained contentious, both scientifically and politically [102, 103, 104]. Among the first to apply statistics to the social sciences was the nineteenth-century astronomer and sociologist Adolphe Quetelet. In a scientific program he called "social physics," Quetelet sought to demonstrate the existence of statistical *laws* in human populations. He introduced the concept of the "average man," characterized by the mean values of measured variables, such as height, that followed a normal distribution. Averages were as much a descriptive as a normative proposal. Quetelet regarded them as an ideal to be pursued. Among others, his work influenced Francis Galton in the development of eugenics.

The success of statistics throughout the twentieth century cemented in the use of probability to model human populations. Few raise an eyebrow today if we talk about a survey as sampling responses from a distribution. It seems obvious now that we would like to estimate parameters such as mean and standard deviation from distributions of incomes, household sizes, or other such attributes. Statistics is so deeply embedded in the social sciences that we rarely revisit the premise that we can represent a human population as a probability distribution.

The differences between a human population and a distribution are stark. Human populations change over time, sometimes rapidly, due to different actions, mechanisms, and interactions among individuals. A distribution, in contrast, can be thought of as a static array where rows correspond to individuals and columns correspond to measured covariates of an individual. The mathematical abstraction for such an array is a set of nonnegative numbers, called *probabilities*, that sum up to 1 and give us for each row the relative weight of this setting of covariates in the population. To sample from such a distribution corresponds to picking one of the rows in the table at random in proportion to its weight. We can repeat this process without change or deterioration. In this view, the distribution is immutable. Nothing we do can change the population.

Much of statistics deals with samples and the question of how we can relate quantities computed on a sample, such as the sample average, to corresponding parameters of a distribution, such as the population mean. The focus in our chapter is different. We use statistics to talk about properties of populations as distributions and by extension classification rules applied to a population. While sampling introduces many additional issues, the questions we raise in this chapter come out most clearly at the population level.

Formalizing Classification

The goal of classification is to determine a plausible value for an unknown *target* Y given observed *covariates* X. Typically, the covariates are represented as an an array of continuous or discrete variables, while the target is a discrete, often binary, value. Formally, the covariates X and target Y are jointly distributed random variables. This means that there is one probability distribution over pairs of values (x, y) that the random variables (X, Y) might take on. This probability distribution models a population of instances of the classification problem. In most of our examples, we think of each instance as the covariates and target of one individual.

At the time of classification, the value of the target variable is not known to us, but we observe the covariates X and make a guess $\widehat{Y} = f(X)$ based on what we observed. The function f that maps our covariates into our guess $\widehat{Y}$ is called a

Table 3.1
Common classification criteria

Event	Condition	Resulting notion ($\mathbb{P}\{\text{event}\mid\text{condition}\}$)
$\widehat{Y}=1$	$Y=1$	True positive rate, recall
$\widehat{Y}=0$	$Y=1$	False negative rate
$\widehat{Y}=1$	$Y=0$	False positive rate
$\widehat{Y}=0$	$Y=0$	True negative rate

classifier, or *predictor*. The output of the classifier is called *label* or *prediction*. Throughout this chapter we are primarily interested with the random variable $\widehat{Y}$ and how it relates to other random variables. The function that defines this random variable is secondary. For this reason, we stretch the terminology slightly and refer to $\widehat{Y}$ itself as the classifier.

Implicit in this formal setup of classification is a major assumption. Whatever we do on the basis of the covariates X cannot influence the outcome Y. After all, our distribution assigns a fixed weight to each pair (x, y). In particular, our prediction $\widehat{Y}$ cannot influence the outcome Y. This assumption is often violated when predictions motivate actions that influence the outcome. For example, the prediction that a student is at risk of dropout might be followed with educational interventions that make dropout less likely.

To be able to choose a classifier out of many possibilities, we need to formalize what makes a classifier *good*. This question often does not have a fully satisfying answer, but statistical decision theory provides criteria that can help highlight different qualities of a classifier that can inform our choice.

Perhaps the best-known property of a classifier $\widehat{Y}$ is its *classification accuracy*, or *accuracy* for short, defined as $\mathbb{P}\{Y = \widehat{Y}\}$, the probability of correctly predicting the target variable. We define *classification error* as $\mathbb{P}\{Y \neq \widehat{Y}\}$. Accuracy is easy to define, but misses some important aspects when evaluating a classifier. A classifier that always predicts *no traffic fatality in the next year* might have high accuracy on any given individual, simply because fatal accidents are unlikely. However, it's a constant function that has no value in assessing the risk of a traffic fatality.

Other decision-theoretic criteria highlight different aspects of a classifier. We can define the most common ones by considering the conditional probability $\mathbb{P}\{\text{event}\mid\text{condition}\}$ for various different settings (table 3.1).

The true positive rate corresponds to the frequency with which the classifier correctly assigns a positive label when the outcome is positive. We call this a *true positive*. The other terms *false positive*, *false negative*, and *true negative* derive analogously from the respective definitions. It is not important to memorize all these terms. They do, however, come up regularly in the classification settings.

Table 3.2
Additional classification criteria

Event	Condition	Resulting notion ($\mathbb{P}\{\text{event} \mid \text{condition}\}$)
$Y = 1$	$\widehat{Y} = 1$	Positive predictive value, precision
$Y = 0$	$\widehat{Y} = 0$	Negative predictive value

Another family of classification criteria arises from swapping event and condition. We'll highlight only two of the four possible notions (table 3.2).

Optimal Classification

Suppose we assign a quantified cost (or negative reward) to each of the four possible classification outcomes: true positive, false positive, true negative, false negative. The problem of optimal classification is to find a classifier that minimizes cost in expectation over a population. We can write the cost as a real number $\ell(\widehat{y}, y)$, called *loss*, that we experience when we classify a target value y with a label $\widehat{y}$. An *optimal classifier* is any classifier that minimizes the expected loss:

$$\mathbb{E}[\ell(\widehat{Y}, Y)].$$

This objective is called classification *risk* and *risk minimization* refers to the optimization problem of finding a classifier that minimizes risk.

As an example, choose the losses $\ell(0, 1) = \ell(1, 0) = 1$ and $\ell(1, 1) = \ell(0, 0) = 0$. For this choice of loss function, the optimal classifier is the one that minimizes classification error. The resulting optimal classifier has an intuitive solution.

Fact 1   *The optimal predictor minimizing classification error satisfies*

$$\widehat{Y} = f(X), \quad \text{where} \quad f(x) = \begin{cases} 1 & \text{if } \mathbb{P}\{Y = 1 \mid X = x\} > 1/2. \\ 0 & \text{otherwise.} \end{cases}$$

The optimal classifier checks whether the propensity of positive outcomes given the observed covariates X is greater than $\frac{1}{2}$. If it is, it makes the guess that the outcome is 1. Otherwise, it guesses that the outcome is 0. The optimal predictor above is specific to classification error. If our loss function were different, the threshold $\frac{1}{2}$ in the definition above would need to change. This makes intuitive sense. If our cost for false positives was much higher than our cost for false negatives, we had better err on the side of not declaring a positive.

The optimal predictor is a theoretical construction that we may not be able to build from data. For example, when the vector of covariates X is high-dimensional, a finite sample is likely going to miss out on some settings $X = x$ that the covariates might take on. In this case, it's not clear how to get at the probability

$\mathbb{P}\{Y=1 \mid X=x\}$. There is a vast technical repertoire in statistics and machine learning for finding good predictors from finite samples. Throughout this chapter we focus on problems that persist even if we had access to the optimal predictor for a given population.

Risk Scores

The optimal classifier we just saw has an important property. We were able to write it as a threshold applied to the function

$$r(x) = \mathbb{P}\{Y=1 \mid X=x\} = \mathbb{E}[Y \mid X=x].$$

This function is an example of a *risk score*. Statistical decision theory tells us that optimal classifiers can generally be written as a threshold applied to this risk score. The risk score we see here is a particularly important and natural one. We can think of it as taking the available evidence $X = x$ and calculating the expected outcome given the observed information. This is called the *posterior probability* of the outcome Y given X. In an intuitive sense, the conditional expectation is a statistical *lookup table* that gives us for each setting of features the frequency of positive outcomes given these features. The risk score is sometimes called *Bayes optimal*. It minimizes the *squared loss*

$$\mathbb{E}(Y - r(X))^2$$

among all possible real-valued risk scores $r(X)$. Minimization problems where we try to approximate the target variable Y with a real-valued risk score are called *regression* problems. In this context, risk scores are often called *regressors*. Although our loss function was specific, there is a general lesson. Classification is often attacked by first solving a regression problem to summarize the data in a single real-valued risk score. We then turn the risk score into a classifier by thresholding.

Risk scores need not be optimal or learned from data. For an illustrative example consider the well-known body mass index, which we owe to Quetelet by the way, which summarizes the *weight* and *height* of a person into a single real number (figure 3.2). In our formal notation, the features are $X = (H, W)$ where H denotes height in meters and W denotes weight in kilograms. The body mass index corresponds to the score function $R = W/H^2$.

We could interpret the body mass index as measuring risk of, say, diabetes. Taking a thresholding of value 30, we might decide that individuals with a body mass index above this are at risk of developing diabetes while others are not. It does not take a medical degree to worry that the resulting classifier may not be very accurate. The body mass index has a number of known issues leading to errors when used for

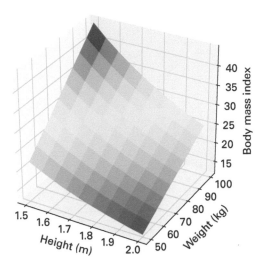

Figure 3.2
Plot of the body mass index.

classification. We won't go into detail, but it's worth noting that these classification errors can systematically align with certain groups in the population. For instance, the body mass index tends to be inflated as a risk measure for taller people due to scaling issues.

A more refined approach to finding a risk score for diabetes would be to solve a regression problem involving the available covariates and the outcome variable. Solved optimally, the resulting risk score would tell us, for every setting of weight (say, rounded to the nearest kg unit) and every physical height (rounded to the nearest cm unit), the incidence rate of diabetes among individuals with these values. The target variable in this case is a binary indicator of diabetes. So, $r((176, 68))$ would be the incidence rate of diabetes among individuals who are 1.76 m tall and weigh 68 kg. The conditional expectation is likely more useful as a risk measure of diabetes than the body mass index we saw earlier. After all, the conditional expectation directly reflects the incidence rate of diabetes given the observed characteristics, while the body mass index didn't solve this specific regression problem.

Varying Thresholds and ROC Curves
In the optimal predictor for classification error we chose a threshold of $\frac{1}{2}$. This exact number was a consequence of the equal cost for false positives and false negatives. If a false positive was significantly more costly, we might wish to choose a higher threshold for declaring a positive. Each choice of a threshold results in a specific

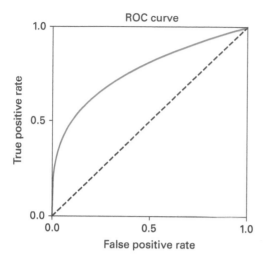

Figure 3.3
Example of an ROC curve.

trade-off between true positive rate and false positive rate. By varying the threshold from 0 to 1, we can trace a curve in a two-dimensional space where the axes correspond to true positive rate and false positive rate. This curve is called an *ROC curve* (figure 3.3). ROC stands for receiver operator characteristic, a name pointing at the roots of the concept in signal processing.

In statistical decision theory, the ROC curve is a property of a distribution (X, Y). It gives us for each setting of false positive rate, the optimal true positive rate that can be achieved for the given false positive rate on the distribution (X, Y). This leads to several nice theoretical properties of the ROC curve. In the machine learning context, ROC curves are computed more liberally for any given risk score, even if it isn't optimal. The ROC curve is often used to eyeball how predictive our score is of the target variable. A common measure of predictiveness is the area under the curve (AUC), which equals the probability that a random positive instance gets a score higher than a random negative instance. An area of $\frac{1}{2}$ corresponds to random guessing, and an area of 1 corresponds to perfect classification.

Supervised Learning

Supervised learning is what makes classification algorithmic. It's about how to construct good classifiers from samples drawn from a population. The details of supervised learning won't matter for this chapter, but it is still worthwhile to have a working understanding of the basic idea.

Suppose we have labeled data, also called *training examples*, of the form $(x_1, y_1), \ldots, (x_n, y_n)$, where each *example* is a pair (x_i, y_i) of an *instance* x_i and a *label* y_i. We typically assume that these examples were drawn independently and repeatedly from the same distribution (X, Y). A supervised learning algorithm takes in training examples and returns a classifier, typically a threshold of a score: $f(x) = \mathbb{1}\{r(x) > t\}$. A simple example of a learning algorithm is the familiar least squares method that attempts to minimize the objective function

$$\sum_{i=1}^{n} (r(x_i) - y_i)^2 .$$

We saw earlier that at the population level, the optimal score is the conditional expectation $r(x) = \mathbb{E}\left[Y \mid X = x\right]$. The problem is that we don't necessarily have enough data to estimate each of the conditional probabilities required to construct this score. After all, the number of possible values that x can assume is exponential in the number of covariates.

The whole trick in supervised learning is to approximate this optimal solution with algorithmically feasible solutions. In doing so, supervised learning must negotiate a balance along three axes:

- **Representation**: Choose a family of functions that the score r comes from. A common choice is linear functions $r(x) = \langle w, x \rangle$ that take the inner product of the covariates x with some vector of coefficients w. More complex representations involve nonlinear functions, such as *artificial neural networks*. This function family is often called the *model class* and the coefficients w are called *model parameters*.

- **Optimization**: Solve the resulting optimization problem by finding model parameters that minimize the loss function on the training examples.

- **Generalization**: Ensure that small loss on the training examples implies small loss on the population that we drew the training examples from.

The three goals of supervised learning are entangled. A powerful representation might make it easier to express complicated patterns, but it might also burden optimization and generalization. Likewise, there are tricks to make optimization feasible at the expense of representation or generalization.

For the remainder of this chapter, we can think of supervised learning as a black box that provides us with classifiers when given labeled training data. What matters is which properties these classifiers have at the population level. At the population level, we interpret a classifier as a random variable by considering $\widehat{Y} = f(X)$. We ignore how $\widehat{Y}$ was learned from a finite sample, what the functional form of the classifier is, and how we estimate various statistical quantities from finite samples.

While finite sample considerations are fundamental to machine learning, they are not central to the conceptual and technical questions around fairness that we will discuss in this chapter.

Groups in the Population

Chapter 2 introduced some of the reasons why individuals might want to object to the use of statistical classification rules in consequential decisions. We now turn to one specific concern, namely, *discrimination on the basis of membership in specific groups of the population*. Discrimination is not a general concept. It is concerned with socially salient categories that have served as the basis for unjustified and systematically adverse treatment in the past. United States law recognizes certain *protected categories*, including race, sex (which extends to sexual orientation), religion, disability status, and place of birth.

In many classification tasks, the features X implicitly or explicitly encode an individual's status in a protected category. We will set aside the letter A to designate a discrete random variable that captures one or multiple sensitive characteristics. Different settings of the random variable A correspond to different mutually disjoint groups of the population. The random variable A is often called a *sensitive attribute* in the technical literature.

Note that formally we can always represent any number of discrete protected categories as a single discrete attribute whose support corresponds to each of the possible settings of the original attributes. Consequently, our formal treatment in this chapter does apply to the case of multiple protected categories. This formal maneuver, however, does not address the important concept of *intersectionality*, which refers to the particular forms of disadvantage that members of multiple protected categories may experience [105].

The fact that we allocate a special random variable for group membership does not mean that we can cleanly partition the set of features into two independent categories such as "neutral" and "sensitive." In fact, we will see shortly that sufficiently many seemingly neutral features can often give highly accurate predictions of group membership. This should not be surprising. After all, if we think of A as the target variable in a classification problem, there is reason to believe that the remaining features would give a nontrivial classifier for A.

The choice of sensitive attributes will generally have profound consequences as it determines which groups of the population we highlight and what conclusions we draw from our investigation. The taxonomy induced by discretization can on its own be a source of harm if it is too coarse, too granular, misleading, or inaccurate. The act of classifying status in protected categories, and collecting associated data, can on its own can be problematic. We revisit this important discussion in the next chapter.

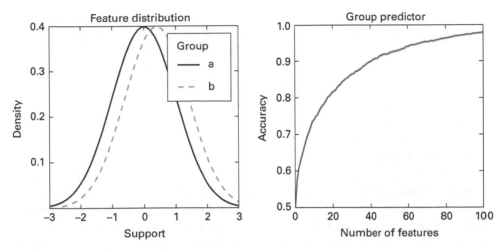

Figure 3.4
On the left, we see the distribution of a single feature that differs only very slightly between the two groups. In both groups the feature follows a normal distribution. Only the means are slightly different in each group. Multiple features like this can be used to build a highly accurate group membership classifier. On the right, we see how the accuracy grows as more and more features become available.

No Fairness through Unawareness

Some have hoped that removing or ignoring sensitive attributes would somehow ensure the impartiality of the resulting classifier. Unfortunately, this practice can be ineffective and even harmful.

In a typical dataset, we have many features that are slightly correlated with the sensitive attribute. At the time of writing, visiting the website `pinterest.com` in the United States, for example, had a small statistical correlation with being female. The correlation on its own is too small to classify someone's gender with high accuracy. However, if numerous such features are available, as is the case in a typical browsing history, the task of classifying gender becomes feasible at higher accuracy levels (figure 3.4).

Several features that are slightly predictive of the sensitive attribute can be used to build high-accuracy classifiers for that attribute. In large feature spaces sensitive attributes are generally *redundant* given the other features. If a classifier trained on the original data uses the sensitive attribute and we remove the attribute, the classifier will then find a redundant encoding in terms of the other features. This results in an essentially equivalent classifier, in the sense that it implements the same function.

To further illustrate the issue, consider a fictitious start-up that sets out to predict your income from your genome. At first, this task might seem impossible. How could someone's DNA reveal their income? However, we know that DNA encodes information about ancestry, which in turn correlates with income in some countries such as the United States. Hence, DNA can likely be used to predict income better than random guessing. The resulting classifier uses ancestry in an entirely implicit manner. Removing redundant encodings of ancestry from the genome is a difficult task that cannot be accomplished by removing a few individual genetic markers. What we learn from this is that machine learning can wind up building classifiers for sensitive attributes without explicitly being asked to, simply because it is an available route to improving accuracy.

Redundant encodings typically abound in large feature spaces. For example, gender can be predicted from retinal photographs with very high accuracy [106]. What about few hand-crafted features? In some studies, features are chosen carefully so as to be roughly statistically independent of each other. In such cases, the sensitive attribute may not have good redundant encodings. That does not mean that removing it is a good idea. Medication, for example, sometimes depends on race in legitimate ways if these correlate with underlying causal factors [21]. Forcing medications to be uncorrelated with race in such cases can harm the individual.

Statistical Nondiscrimination Criteria

Statistical nondiscrimination criteria aim to define the absence of discrimination in terms of statistical expressions involving random variables describing a classification or decision-making scenario.

Formally, statistical nondiscrimination criteria are properties of the joint distribution of the sensitive attribute A, the target variable Y, the classifier $\widehat{Y}$ or score R, and in some cases also features X. This means that we can unambiguously decide whether or not a criterion is satisfied by looking at the joint distribution of these random variables.

Broadly speaking, different statistical fairness criteria all equalize some group-dependent statistical quantity across groups defined by the different settings of A. For example, we could ask to equalize acceptance rates across all groups. This corresponds to imposing the constraint for all groups a and b:

$$\mathbb{P}\{\widehat{Y} = 1 \mid A = a\} = \mathbb{P}\{\widehat{Y} = 1 \mid A = b\}.$$

In the case where $\widehat{Y} \in \{0, 1\}$ is a binary classifier and we have two groups a and b, we can determine if acceptance rates are equal in the two groups by knowing the three probabilities $\mathbb{P}\{\widehat{Y} = 1, A = a\}$, $\mathbb{P}\{\widehat{Y} = 1, A = b\}$, and $\mathbb{P}\{A = a\}$ that fully specify the joint distribution of $\widehat{Y}$ and A. We can also estimate the relevant probabilities given random samples from the joint distribution using standard statistical arguments that are not the focus of this chapter.

Table 3.3
Nondiscrimination criteria

Independence	Separation	Sufficiency
$R \perp A$	$R \perp A \mid Y$	$Y \perp A \mid R$

Researchers have proposed dozens of different criteria, each trying to capture different intuitions about what is *fair*. Simplifying the landscape of fairness criteria, we can say that there are essentially three fundamentally different ones. Each of these equalizes one of the following three statistics across all groups:

- Acceptance rate $\mathbb{P}\{\widehat{Y} = 1\}$ of a classifier $\widehat{Y}$;
- Error rates $\mathbb{P}\{\widehat{Y} = 0 \mid Y = 1\}$ and $\mathbb{P}\{\widehat{Y} = 1 \mid Y = 0\}$ of a classifier $\widehat{Y}$;
- Outcome frequency given score value $\mathbb{P}\{Y = 1 \mid R = r\}$ of a score R.

The three criteria can be generalized to score functions using simple (conditional) independence statements. We use the notation $U \perp V \mid W$ to denote that random variables U and V are conditionally independent given W. This means that conditional on any setting $W = w$, the random variables U and V are independent (table 3.3).

Below we introduce and discuss each of these conditions in detail. This chapter focuses on the mathematical properties of and relationships between these different criteria. Once we have acquired familiarity with the technical matter, we will have a broader debate around the moral and normative content of these definitions in chapter 4.

Independence

Our first formal criterion requires the sensitive characteristic to be statistically independent of the score.

Definition 1   *Random variables (A, R) satisfy* independence *if $A \perp R$.*

If R is a score function that satisfies independence, then any classifier $\widehat{Y} = \mathbb{1}\{R > t\}$ that thresholds the score at value t also satisfies independence. This is true so long as the threshold is independent of group membership. Group-specific thresholds may not preserve independence.

Independence has been explored through many equivalent and related definitions. When applied to a binary classifier $\widehat{Y}$, independence is often referred to as *demographic parity, statistical parity, group fairness, disparate impact*, among other terms. In this case, independence corresponds to the condition

$$\mathbb{P}\{\widehat{Y} = 1 \mid A = a\} = \mathbb{P}\{\widehat{Y} = 1 \mid A = b\},$$

for all groups a, b. Thinking of the event $\widehat{Y} = 1$ as "acceptance," the condition requires the acceptance rate to be the same in all groups. A relaxation of the constraint introduces a positive amount of slack $\epsilon > 0$ and requires that

$$\mathbb{P}\{\widehat{Y} = 1 \mid A = a\} \geq \mathbb{P}\{\widehat{Y} = 1 \mid A = b\} - \epsilon.$$

Note that we can swap a and b to get an inequality in the other direction. An alternative relaxation is to consider a ratio condition, such as

$$\frac{\mathbb{P}\{\widehat{Y} = 1 \mid A = a\}}{\mathbb{P}\{\widehat{Y} = 1 \mid A = b\}} \geq 1 - \epsilon.$$

Some have argued that, for $\epsilon = 0.2$, this condition relates to the *80 percent rule* that appears in discussions around disparate impact law [107].

Yet another way to state the independence condition in full generality is to require that A and R must have zero mutual information $I(A; R) = 0$. Mutual information quantifies the amount of information that one random variable reveals about the other. We can define it in terms of the more standard entropy function as $I(A; R) = H(A) + H(R) - H(A, R)$. The characterization in terms of mutual information leads to useful relaxations of the constraint. For example, we could require $I(A; R) \leq \epsilon$.

Limitations of Independence

Independence is pursued as a criterion in many papers, for multiple reasons. Some argue that the condition reflects an assumption of equality: All groups have an equal claim to acceptance and resources should therefore be allocated proportionally. What we encounter here is a question about the *normative* significance of independence, which we expand on in chapter 4. But there is a more mundane reason for the prevalence of this criterion, too. Independence has convenient technical properties, which makes the criterion appealing to machine learning researchers. It is often the easiest one to work with mathematically and algorithmically.

However, decisions based on a classifier that satisfies independence can have undesirable properties (and similar arguments apply to other statistical criteria). We can easily illustrate one way in which this can happen by imagining a callous or ill-intentioned decision maker. Imagine a company that in group a hires diligently selected applicants at some rate $p > 0$. In group b, the company hires carelessly selected applicants at the same rate p. Even though the acceptance rates in both groups are identical, it is far more likely that unqualified applicants are selected in one group than in the other. As a result, it will appear in hindsight that members of group b performed worse than members of group a, thus establishing a negative track record for group b.

A real-world phenomenon similar to this hypothetical example is termed the *glass cliff*: women and people of color are more likely to be appointed CEO when a firm is struggling. When the firm performs poorly during their tenure, they are likely to be replaced by White men [108, 109].

This situation might arise without our having to posit malice: the company might have historically hired employees primarily from group a, giving them a better understanding of this group. As a technical matter, the company might have substantially more training data in group a, thus potentially leading to lower error rates of a learned classifier within that group. The last point is a bit subtle. After all, if both groups were entirely homogeneous in all ways relevant to the classification task, more training data in one group would equally benefit both. Then again, the mere fact that we chose to distinguish these two groups indicates that we believe they might be heterogeneous in relevant aspects.

Separation

Our next criterion engages with the limitation of independence that we described. In a typical classification problem, there is a difference between accepting a positive instance or accepting a negative instance. The target variable Y suggests one way of partitioning the population into strata of equal claim to acceptance. Viewed this way, the target variable gives us a sense of *merit*. A particular demographic group $(A = a)$ may be more or less well represented in these different strata defined by the target variable. A decision maker might argue that in such cases it is justified to accept more or fewer individuals from group a.

These considerations motivate a criterion that demands independence within each stratum of the population defined by target variable. We can formalize this requirement using a conditional independence statement.

Definition 2 *Random variables* (R, A, Y) *satisfy* separation *if* $R \perp A \mid Y$.

The conditional independence statement applies even if the variables take on more than two values each. For example, the target variable might partition the population into many different types of individuals.

We can display separation as a graphical model in which R is separated from A by the target variable Y (figure 3.5).

If you haven't seen graphical models before, don't worry. All this says is that R is conditionally independent of A given Y.

Figure 3.5
Graphical model representation of separation.

In the case of a binary classifier, separation is equivalent to requiring for all groups a, b the two constraints

$$\mathbb{P}\{\widehat{Y} = 1 \mid Y = 1, A = a\} = \mathbb{P}\{\widehat{Y} = 1 \mid Y = 1, A = b\}$$

$$\mathbb{P}\{\widehat{Y} = 1 \mid Y = 0, A = a\} = \mathbb{P}\{\widehat{Y} = 1 \mid Y = 0, A = b\}.$$

Recall that $\mathbb{P}\{\widehat{Y} = 1 \mid Y = 1\}$ is called the *true positive rate* of the classifier. It is the rate at which the classifier correctly recognizes positive instances. The *false positive rate* $\mathbb{P}\{\widehat{Y} = 1 \mid Y = 0\}$ highlights the rate at which the classifier mistakenly assigns positive outcomes to negative instances. Recall that the true positive rate equals 1 minus the false negative rate. What separation therefore requires is that all groups experience the same false negative rate and the same false positive rate. Consequently, the definition asks for *error rate parity*.

This interpretation in terms of equality of error rates leads to natural relaxations. For example, we could require equality of only false negative rates. A false negative, intuitively speaking, corresponds to a denied opportunity in scenarios where acceptance is desirable, such as in hiring. In contrast, when the task is to identify high-risk individuals, as in the case of loan default prediction, it is common to denote the undesirable outcome as the "positive" class. This inverts the meaning of false positives and false negatives, and is a frequent source of terminological confusion.

Why Equalize Error Rates?

The idea of equalizing error rates across has been subject to critique. Much of the debate has to do with the fact that an optimal predictor need not have equal error rates in all groups. Specifically, when the propensity of positive outcomes ($\mathbb{P}\{Y = 1\}$) differs between groups, an optimal predictor will generally have different error rates. In such cases, enforcing equality of error rates leads to a predictor that performs worse in some groups than an unconstrained predictor. How is that *fair*?

One response is that separation puts emphasis on the question: Who bears the cost of misclassification? A violation of separation highlights the fact that different groups experience different costs of misclassification. There is concern that higher error rates coincide with historically marginalized and disadvantaged groups, thus inflicting additional harm on these groups.

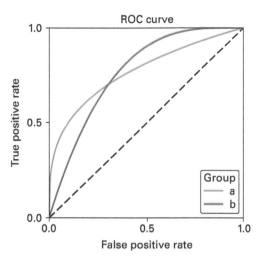

Figure 3.6
ROC curve by group.

The act of measuring and reporting group-specific error rates can create an incentive for decision makers to work toward improving error rates through collecting better datasets and building better models. If there is no way to improve error rates in some group relative to others, this raises questions about the legitimate use of machine learning in such cases. We return to this normative question in later chapters.

A second line of concern with the separation criterion relates to the use of the target variable as a stand-in for merit. Researchers have rightfully pointed out that in many cases machine learning practitioners use target variables that reflect existing inequality and injustice. In such cases, satisfying separation with respect to an inadequate target variable does no good. This valid concern, however, applies equally to the use of supervised learning at large in such cases. If we cannot agree on an adequate target variable, the right action may be to suspend the use of supervised learning.

These observations hint at the subtle role that nondiscrimination criteria play. Rather than presenting constraints that we can optimize for without further thought, they can help surface issues with the use of machine learning in specific scenarios.

Visualizing Separation
A binary classifier that satisfies separation must achieve the same true positive rates and the same false positive rates in all groups. We can visualize this condition by plotting group-specific ROC curves (figure 3.6).

We see the ROC curves of a score displayed for each group separately. The two groups have different curves indicating that not all trade-offs between true and false

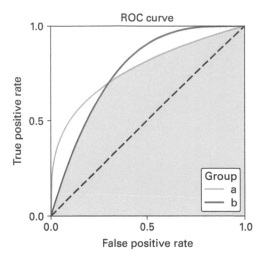

Figure 3.7
Intersection of area under the curves.

positive rate are achievable in both groups. The trade-offs that are achievable in both groups are precisely those that lie under both curves, corresponding to the intersection of the regions enclosed by the curves (figure 3.7).

The highlighted region in figure 3.7 is the *feasible region* of trade-offs that we can achieve in all groups. However, the thresholds that achieve these trade-offs are in general also group-specific. In other words, the bar for acceptance varies by group. Trade-offs that are not exactly on the curves, but rather in the interior of the region, require randomization. To understand this point, think about how we can realize trade-offs on the the dashed line in the plot. Take one classifier that accepts everyone. This corresponds to true and false positive rate 1, hence achieving the upper right corner of the plot. Take another classifier that accepts no one, resulting in true and false positive rate 0, the lower left corner of the plot. Now, construct a third classifier that given an instance randomly picks and applies the first classifier with probability $1 - p$, and the second with probability p. This classifier achieves true and false positive rate p, thus giving us one point on the dashed line in the plot. In the same manner, we could have picked any other pair of classifiers and randomized between them. This way we can realize the entire area under the ROC curve.

Conditional Acceptance Rates
A relative of the independence and separation criteria is common in debates around discrimination. Here, we designate a random variable W and ask for conditional independence of the decision $\widehat{Y}$ and group status A conditional on the variable W.

Figure 3.8
Graphical model representation of sufficiency.

That is, for all values w that W could take on, and all groups a and b we demand:

$$\mathbb{P}\{\widehat{Y}=1 \mid W=w, A=a\} = \mathbb{P}\{\widehat{Y}=1 \mid W=w, A=b\}.$$

Formally, this is equivalent to replacing Y with W in our definition of separation. Often W corresponds to a subset of the covariates of X. For example, we might demand that independence holds among all individuals of equal *educational attainment*. In this case, we would choose W to reflect educational attainment. In doing so, we license the decision maker to distinguish between individuals with different educational backgrounds. When we apply this criterion, the burden falls on the proper choice of what to condition on, which determines whether we detect discrimination or not. In particular, we must be careful not to condition on the mechanism by which the decision maker discriminates. For example, an ill-intentioned decision maker might discriminate by imposing excessive educational requirements for a specific job, exploiting the fact that this level of education is distributed unevenly among different groups. We will be able to return to the question of what to condition on with significantly more substance once we reach familiarity with causality in chapter 5.

Sufficiency

Our third criterion formalizes that the score already subsumes the sensitive characteristic for the purpose of predicting the target. This idea again boils down to a conditional independence statement.

Definition 3 *We say the random variables (R, A, Y) satisfy* sufficiency *if $Y \perp A \mid R$.*

We can display sufficiency as a graphical model as we did with separation before (figure 3.8).

Let us write out the definition more explicitly in the binary case where $Y \in \{0, 1\}$. In this case, a random variable R is sufficient for A if and only if for all groups a, b and all values r in the support of R, we have

$$\mathbb{P}\{Y=1 \mid R=r, A=a\} = \mathbb{P}\{Y=1 \mid R=r, A=b\}.$$

If we replace R by a binary predictor $\widehat{Y}$, we recognize this condition as requiring a parity of positive/negative predictive values across all groups.

Calibration and Sufficiency

Sufficiency is closely related to an important notion called *calibration*. In some applications it is desirable to be able to interpret the values of the score functions as if they were probabilities. The notion of calibration allows us to move in this direction. Restricting our attention to binary outcome variables, we say that a score R is *calibrated* with respect to an outcome variable Y if for all score values $r \in [0, 1]$, we have

$$\mathbb{P}\{Y = 1 \mid R = r\} = r.$$

This condition means that the set of all instances assigned a score value r has an r fraction of positive instances among them. The condition refers to the group of all individuals receiving a particular score value. Calibration need not hold in subgroups of the population. In particular, it's important not to interpret the score as an *individual probability*. Calibration does not tell us anything about the outcome of a specific individual that receives a particular value.

From the definition, we can see that sufficiency is closely related to the idea of calibration. To formalize the connection we say that the score R satisfies *calibration by group* if it satisfies

$$\mathbb{P}\{Y = 1 \mid R = r, A = a\} = r$$

for all score values r and groups a. Observe that calibration is the same requirement at the population level without the conditioning on A.

Fact 2 *Calibration by group implies sufficiency.*

Conversely, sufficiency is only slightly weaker than calibration by group in the sense that a simple renaming of score values goes from one property to the other.

Proposition 1 *If a score R satisfies sufficiency, then there exists a function $\ell \colon [0, 1] \to [0, 1]$ so that $\ell(R)$ satisfies calibration by group.*

Proof. Fix any group a and put $\ell(r) = \mathbb{P}\{Y = 1 \mid R = r, A = a\}$. Since R satisfies sufficiency, this probability is the same for all groups a and hence this map ℓ is the same regardless of what value a we chose.

Now, consider any two groups a, b. We have,

$$r = \mathbb{P}\{Y = 1 \mid \ell(R) = r, A = a\}$$
$$= \mathbb{P}\{Y = 1 \mid R \in \ell^{-1}(r), A = a\}$$
$$= \mathbb{P}\{Y = 1 \mid R \in \ell^{-1}(r), A = b\}$$
$$= \mathbb{P}\{Y = 1 \mid \ell(R) = r, A = b\},$$

thus showing that $\ell(R)$ is calibrated by group. ∎

We conclude that sufficiency and calibration by group are essentially equivalent notions.

In practice, there are various heuristics to achieve calibration. For example, Platt scaling takes a possibly uncalibrated score, treats it as a single feature, and fits a one-variable regression model against the target variable based on this feature [110]. We also apply Platt scaling for each of the groups defined by the sensitive attribute.

Calibration by Group as a Consequence of Unconstrained Learning

Sufficiency is often satisfied by the outcome of unconstrained supervised learning without the need for any explicit intervention. This should not come as a surprise. After all, the goal of supervised learning is to approximate an optimal score function. The optimal score function we saw earlier, however, is calibrated for any group as the next fact states formally.

Fact 3 *The optimal score $r(x) = \mathbb{E}[Y \mid X = x]$ satisfies group calibration for any group. Specifically, for any set S we have*

$$\mathbb{P}\{Y = 1 \mid R = r, X \in S\} = r.$$

We generally expect a learned score to satisfy sufficiency in cases where the group membership is either explicitly encoded in the data or can be predicted from the other attributes. To illustrate this point we look at the calibration values of a standard machine learning model, a *random forest ensemble*, on an income classification task derived from the American Community Survey (ACS) of the US Census Bureau [111]. We restrict the dataset to the three most populous states, California, Texas, and Florida.

After splitting the data into training and testing data, we fit a random forest ensemble using the standard Python library *sklearn* on the training data. We then examine how well calibrated the model is out of the box on test data (figure 3.9).

We see that the calibration curves for the three largest racial groups in the dataset, which the Census Bureau codes as "White alone," "Black or African American alone," and "Asian alone," are very close to the main diagonal. This means that the scores derived from our random forest model satisfy calibration by group up to small error. The same is true when looking at the two groups "Male" and "Female" in the dataset.

These observations are not coincidental. Theory shows that under certain technical conditions, unconstrained supervised learning does, in fact, imply group calibration [112]. Note, however, that for this to be true the classifier must be able to detect group membership. If detecting group membership is impossible, then group calibration generally fails.

The lesson is that sufficiency often comes for free (at least approximately) as a consequence of standard machine learning practices. The flip side is that

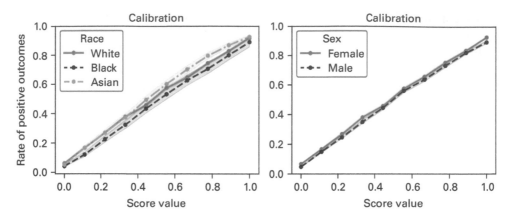

Figure 3.9
Group calibration curves on Census ACS data.

imposing sufficiency as a constraint on a classification system may not be much of an intervention. In particular, it would not effect a substantial change in current practices.

How to Satisfy a Nondiscrimination Criterion

Now that we have formally introduced three nondiscrimination criteria, it is worth asking how we can achieve them algorithmically. We distinguish between three different techniques. While they generally apply to all the criteria and their relaxations that we review in this chapter, our discussion here focuses on independence.

- **Preprocessing**: Adjust the feature space to be uncorrelated with the sensitive attribute.

- **In-training**: Work the constraint into the optimization process that constructs a classifier from training data.

- **Post-processing**: Adjust a learned classifier so as to be uncorrelated with the sensitive attribute.

The three approaches have different strengths and weaknesses.

Preprocessing is a family of techniques to transform a feature space into a representation that as a whole is independent of the sensitive attribute. This approach is generally agnostic to what we do with the new feature space in downstream applications. After the preprocessing transformation ensures independence, any deterministic training process on the new space will also satisfy independence. This is a formal consequence of the well-known data-processing inequality from information theory. [113]

Achieving independence at training time can lead to the highest utility since we get to optimize the classifier with this criterion in mind. The disadvantage is that we need access to the raw data and training pipeline. We also give up a fair bit of generality as this approach typically applies to specific model classes or optimization problems.

Postprocessing refers to the process of taking a trained classifier and adjusting it possibly depending on the sensitive attribute and additional randomness in such a way that independence is achieved. Formally, we say a *derived classifier* $\widehat{Y} = F(R, A)$ is a possibly randomized function of a given score R and the sensitive attribute. Given a cost for false negatives and false positives, we can find the derived classifier that minimizes the expected cost of false positive and false negatives subject to the fairness constraint at hand. Postprocessing has the advantage that it works for any *black-box* classifier regardless of its inner workings. There's no need for retraining, which is useful in cases where the training pipeline is complex. It's often also the only available option when we have access only to a trained model with no control over the training process.

Postprocessing sometimes even comes with an optimality guarantee: If we postprocess the Bayes optimal score to achieve separation, then the resulting classifier will be optimal among all classifiers satisfying separation [114]. Conventional wisdom has it that certain machine learning models, like gradient boosted decision trees, are often nearly Bayes optimal on tabular datasets with many more rows than columns. In such cases, postprocessing by adjusting thresholds is nearly optimal.

A common objection to postprocessing, however, is that the resulting classifier uses group membership quite explicitly by setting different acceptance thresholds for different groups.

Relationships between Criteria

The criteria we reviewed constrain the joint distribution in nontrivial ways. We should therefore suspect that imposing any two of them simultaneously overconstrains the space to the point where only degenerate solutions remain. We will now see that this intuition is largely correct. What this shows, in particular, is that if we observe that one criterion holds, we expect others to be violated.

Independence versus Sufficiency

We begin with a simple proposition that shows how in general independence and sufficiency are mutually exclusive. The only assumption needed here is that the sensitive attribute A and the target variable Y are *not* independent. This is a different way of saying that group membership has an effect on the statistics of the target variable. In the binary case, this means one group has a higher rate of positive outcomes than another. Think of this as the typical case.

Proposition 2 *Assume that A and Y are not independent. Then sufficiency and independence cannot both hold.*

Proof. By the contraction rule for conditional independence,

$$A \perp R \quad \text{and} \quad A \perp Y \mid R \quad \implies \quad A \perp (Y, R) \quad \implies \quad A \perp Y.$$

To be clear, $A \perp (Y, R)$ means that A is independent of the pair of random variables (Y, R). Dropping R cannot introduce a dependence between A and Y.

In the contrapositive,

$$A \not\perp Y \quad \implies \quad A \not\perp R \quad \text{or} \quad A \not\perp Y \mid A. \qquad \blacksquare$$

Independence versus Separation

An analogous result of mutual exclusion holds for independence and separation. The statement in this case is a bit more contrived and requires the additional assumption that the target variable Y is binary. We also additionally need that the score is not independent of the target. This is a rather mild assumption, since any useful score function should have correlation with the target variable.

Proposition 3 *Assume Y is binary, A is not independent of Y, and R is not independent of Y. Then, independence and separation cannot both hold.*

Proof. Assume $Y \in \{0, 1\}$. In its contrapositive form, the statement we need to show is

$$A \perp R \quad \text{and} \quad A \perp R \mid Y \quad \implies \quad A \perp Y \quad \text{or} \quad R \perp Y.$$

By the law of total probability,

$$\mathbb{P}\{R = r \mid A = a\} = \sum_y \mathbb{P}\{R = r \mid A = a, Y = y\} \mathbb{P}\{Y = y \mid A = a\}.$$

Applying the assumption $A \perp R$ and $A \perp R \mid Y$, this equation simplifies to

$$\mathbb{P}\{R = r\} = \sum_y \mathbb{P}\{R = r \mid Y = y\} \mathbb{P}\{Y = y \mid A = a\}.$$

Applied differently, the law of total probability also gives

$$\mathbb{P}\{R = r\} = \sum_y \mathbb{P}\{R = r \mid Y = y\} \mathbb{P}\{Y = y\}.$$

Combining this with the previous equation, we have

$$\sum_y \mathbb{P}\{R = r \mid Y = y\} \mathbb{P}\{Y = y\} = \sum_y \mathbb{P}\{R = r \mid Y = y\} \mathbb{P}\{Y = y \mid A = a\}.$$

Careful inspection reveals that when y ranges over only two values, this equation can be satisfied only if $A \perp Y$ or $R \perp Y$.

Indeed, we can rewrite the equation more compactly using the symbols $p = \mathbb{P}\{Y = 0\}$, $p_a = \mathbb{P}\{Y = 0 \mid A = a\}$, $r_y = \mathbb{P}\{R = r \mid Y = y\}$, as:

$$pr_0 + (1 - p)r_1 = p_a r_0 + (1 - p_a)r_1.$$

Equivalently, $p(r_0 - r_1) = p_a(r_0 - r_1)$.

This equation can be satisfied only if $r_0 = r_1$, in which case $R \perp Y$, or if $p = p_a$ for all a, in which case $Y \perp A$. ∎

The claim is not true when the target variable can assume more than two values, which is a natural case to consider.

Separation versus Sufficiency

Finally, we turn to the relationship between separation and sufficiency. Both ask for a nontrivial conditional independence relationship between the three variables A, R, Y. Imposing both simultaneously leads to a degenerate solution space, as our next proposition confirms.

Proposition 4 *Assume that all events in the joint distribution of (A, R, Y) have positive probability, and assume $A \not\perp Y$. Then, separation and sufficiency cannot both hold.*

Proof. A standard fact (Theorem 17.2 in Larry Wasserman's text [115]) about conditional independence shows

$$A \perp R \mid Y \quad \text{and} \quad A \perp Y \mid R \quad \Longrightarrow \quad A \perp (R, Y).$$

Moreover,

$$A \perp (R, Y) \quad \Longrightarrow \quad A \perp R \quad \text{and} \quad A \perp Y.$$

Taking the contrapositive completes the proof. ∎

For a binary target, the nondegeneracy assumption in the previous proposition states that in all groups, at all score values, we have both positive and negative instances. In other words, the score value never fully resolves uncertainty regarding the outcome. Recall that sufficiency holds for the Bayes optimal score function. The proposition therefore establishes an important fact: optimal scores generally violate separation.

The proposition also applies to binary classifiers. Here, the assumption says that within each group the classifier must have nonzero true positive, false positive, true negative, and false negative rates. We can weaken this assumption a bit and require only that the classifier is imperfect in the sense of making at least one false positive prediction. What's appealing about the resulting claim is that its proof essentially only uses a well-known relationship between true positive rate (recall) and positive predictive value (precision). This trade-off is often called *precision-recall trade-off*.

Proposition 5 *Assume Y is not independent of A and assume $\widehat{Y}$ is a binary classifier with nonzero false positive rate. Then, separation and sufficiency cannot both hold.*

Proof. Since Y is not independent of A there must be two groups, call them 0 and 1, such that

$$p_0 = \mathbb{P}\{Y = 1 \mid A = 0\} \neq \mathbb{P}\{Y = 1 \mid A = 1\} = p_1.$$

Now suppose that separation holds. Since the classifier is imperfect this means that all groups have the same nonzero false positive rate $\mathrm{FPR} > 0$, and the same true positive rate $\mathrm{TPR} \geq 0$. We will show that sufficiency does not hold.

Recall that in the binary case, sufficiency implies that all groups have the same positive predictive value. The positive predictive value in group a, denoted PPV_a satisfies

$$\mathrm{PPV}_a = \frac{\mathrm{TPR}p_a}{\mathrm{TPR}p_a + \mathrm{FPR}(1 - p_a)}.$$

From the expression we can see that $\mathrm{PPV}_a = \mathrm{PPV}_b$ only if $\mathrm{TPR} = 0$ or $\mathrm{FPR} = 0$. The latter is ruled out by assumption. So it must be that $\mathrm{TPR} = 0$. However, in this case, we can verify that the negative predictive value NPV_0 in group 0 must be different from the negative predictive value NPV_1 in group 1. This follows from the expression

$$\mathrm{NPV}_a = \frac{(1 - \mathrm{FPR})(1 - p_a)}{(1 - \mathrm{TPR})p_a + (1 - \mathrm{FPR})(1 - p_a)}.$$

Hence, sufficiency does not hold. ∎

In the proposition we just proved, separation and sufficiency both refer to the binary classifier $\widehat{Y}$. The proposition does *not* apply to the case where separation refers to a binary classifier $\widehat{Y} = \mathbb{1}\{R > t\}$ and sufficiency refers to the underlying score function R.

Case Study: Credit Scoring

We now apply some of the notions we saw to credit scoring. Credit scores support lending decisions by giving an estimate of the risk that a loan applicant will default on a loan. Credit scores are widely used in the United States and other countries when allocating credit, ranging from micro loans to jumbo mortgages. In the United

Table 3.4
Credit score distribution by ethnicity

Race or ethnicity	Samples with both score and outcome
White	133,165
Black	18,274
Hispanic	14,702
Asian	7,906
Total	174,047

States, there are three major credit-reporting agencies that collect data on various borrowers. These agencies are for-profit organizations that each offer risk scores based on the data they collected. FICO scores are a well-known family of proprietary scores developed by the FICO corporation and sold by the three credit reporting agencies.

Regulation of credit agencies in the United States started with the Fair Credit Reporting Act, first passed in 1970, that aims to promote the accuracy, fairness, and privacy of consumer of information collected by the reporting agencies. The Equal Credit Opportunity Act, a United States law enacted in 1974, makes it unlawful for any creditor to discriminate against any applicant the basis of race, color, religion, national origin, sex, marital status, or age.

Score Distribution

Our analysis relies on data published by the Federal Reserve [116]. The dataset provides aggregate statistics from 2003 about a credit score, demographic information (race or ethnicity, gender, marital status), and outcomes (to be defined shortly). We will focus on the joint statistics of score, race, and outcome, where the race attributes assume four values (table 3.4).

The score used in the study is based on the TransUnion TransRisk score. TransUnion is a US credit-reporting agency. The TransRisk score is in turn based on FICO scores. The Federal Reserve renormalized the scores for the study to vary from 0 to 100, with 0 being *least creditworthy.*

The information on race was provided by the Social Security Administration, thus relying on self-reported values. The cumulative distribution of these credit scores strongly depends on the racial group, as figure 3.10 reveals.

Performance Variables and ROC Curves

As is often the case, the outcome variable is a subtle aspect of this dataset. Its definition is worth emphasizing. Since the score model is proprietary, it is not clear what target variable was used during the training process. What is it then that the score

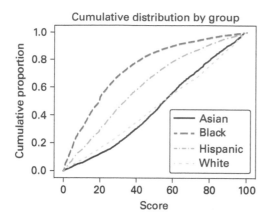

Figure 3.10
Cumulative density of scores by group.

is trying to predict? In a first reaction, we might say that the goal of a credit score is to predict a *default* outcome. However, that's not a clearly defined notion. Defaults vary in the amount of debt recovered and the amount of time given for recovery. Any single binary performance indicator is typically an oversimplification.

What is available in the Federal Reserve data is a so-called *performance* variable that measures a *serious delinquency in at least one credit line of a certain time period*. More specifically, the Federal Reserve states

> [the] measure is based on the performance of new or existing accounts and measures whether individuals have been late 90 days or more on one or more of their accounts or had a public record item or a new collection agency account during the performance period.

With this performance variable at hand, we can look at the ROC curve to get a sense of how predictive the score is in different demographics (figure 3.11).

The meaning of true positive rate is *the rate of predicted positive performance given positive performance*. Similarly, false positive rate is *the rate of predicted negative performance given a positive performance*.

We see that the shapes appear roughly visually similar in the groups, although the "White" group encloses a noticeably larger area under the curve than the "Black" group. Also note that even two ROC curves with the same shape can correspond to very different score functions. A particular trade-off between true positive rate and false positive rate achieved at a threshold t in one group could require a different threshold t' in the other group.

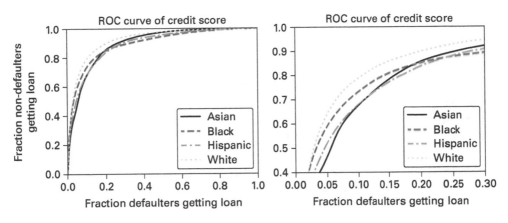

Figure 3.11
ROC curve of credit score by group.

Comparison of Different Criteria

With the score data at hand, we compare four different classification strategies:

- **Maximum profit**: Pick possibly group-dependent score thresholds in a way that maximizes profit.

- **Single threshold**: Pick a single uniform score threshold for all groups in a way that maximizes profit.

- **Independence**: Achieve an equal acceptance rate in all groups. Subject to this constraint, maximize profit.

- **Separation**: Achieve an equal true/false positive rate in all groups. Subject to this constraint, maximize profit.

To make sense of maximizing profit, we need to assume a reward for a true positive (correctly predicted positive performance), and a cost for false positives (negative performance predicted as positive). In lending, the cost of a false positive is typically many times greater than the reward for a true positive. In other words, the interest payments resulting from a loan are relatively small compared with the loan amount that could be lost. For illustrative purposes, we imagine that the cost of a false positive is six times greater than the return on a true positive. The absolute numbers don't matter. Only the ratio matters. This simple cost structure glosses over a number of details that are likely relevant for the lender, such as the terms of the loan.

There is another major caveat to the kind of analysis we're about to do. Since we're given only aggregate statistics, we cannot retrain the score with a particular classification strategy in mind. The only thing we can do is to define a setting of

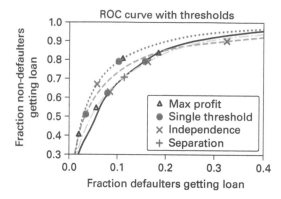

Figure 3.12
ROC curves with optimal thresholds for different criteria.

thresholds that achieves a particular criterion. This approach may be overly pessimistic with regard to the profit achieved subject to each constraint. For this reason and the fact that our choice of cost function was rather arbitrary, we do not state the profit numbers. The numbers can be found in the original analysis [114], which reports that "single threshold" achieves higher profit than "separation," which in turn achieves higher profit than "independence."

What we do instead is to look at the different trade-offs between true and false positive rate that each criterion achieves in each group (figure 3.12).

We can see that even though the ROC curves are somewhat similar, the resulting trade-offs can differ widely by group for some of the criteria. The true positive rate achieved by *max profit* for the Asian group is twice of what it is for the Black group. The separation criterion, of course, results in the same trade-off in all groups. Independence equalizes acceptance rate, but leads to widely different trade-offs. For instance, the Black group has a false positive rate more than three times higher than the false positive rate of the Asian group.

Calibration Values

Finally, we consider the nondefault rate by group. This corresponds to the calibration plot by group (figure 3.13).

We see that the performance curves by group are reasonably well aligned. This means that a monotonic transformation of the score values would result in a score that is roughly calibrated by group according to our earlier definition. Due to the differences in score distribution by group, it could nonetheless be the case that thresholding the score leads to a classifier with different positive predictive values in each group. Calibration is typically lost when taking a multivalued score and making it binary.

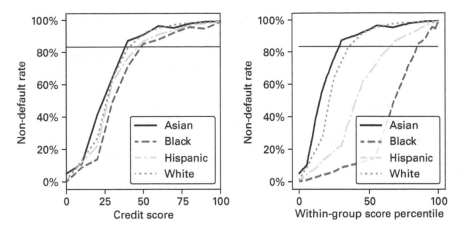

Figure 3.13
Calibration values of credit score by group.

Inherent Limitations of Observational Criteria

The criteria we've seen so far have one important aspect in common. They are properties of the joint distribution of the score, sensitive attribute, and the target variable. In other words, if we know the joint distribution of the random variables (R, A, Y), we can without ambiguity determine whether this joint distribution satisfies one of these criteria or not. For example, if all variables are binary, there are eight numbers specifying the joint distributions. We can verify each of the criteria we discussed in this chapter by looking at only these eight numbers and nothing else. We can broaden this notion a bit and also include all other features in X, not just the group attribute. So, let's call a criterion *observational* if it is a property of the joint distribution of the features X, the sensitive attribute A, a score function R and an outcome variable Y. Intuitively speaking, a criterion is observational if we can write it down unambiguously using probability statements involving the random variables at hand.

Observational definitions have many appealing aspects. They're often easy to state and require only a lightweight formalism. They make no reference to the inner workings of the classifier, the decision maker's intent, the impact of the decisions on the population, or any notion of whether and how a feature actually influences the outcome. We can reason about them fairly conveniently as we saw earlier. In principle, observational definitions can always be verified given samples from the joint distribution—subject to statistical sampling error.

This simplicity of observational definitions also leads to inherent limitations. What observational definitions hide are the mechanisms that created an observed disparity. In one case, a difference in acceptance rate could be due to spiteful consideration of group membership by a decision maker. In another case, the difference in acceptance rates could reflect an underlying inequality in society that gives one group an advantage in getting accepted. While both are cause for concern, in the first case discrimination is a direct action of the decision maker. In the other case, the locus of discrimination may be outside the agency of the decision maker.

Observational criteria cannot, in general, give satisfactory answers as to what the causes and mechanisms of discrimination are. Subsequent chapters, in particular our chapter on causality, develop tools to go beyond the scope of observational criteria.

Chapter Notes

For the early history of probability and the rise of statistical thinking, turn to books by Ian Hacking [117, 118], Theodore Porter [103], and Alain Desrosières [102].

The statistical decision theory we covered in this chapter is also called (signal) detection theory and is the subject of various textbooks. What we call classification is also called prediction in other contexts. Likewise, classifiers are often called predictors. For a graduate introduction to machine learning, see the text by Moritz Hardt and Benjamin Recht [119]. Larry Wasserman's textbook [115] provides additional statistical background, including an exposition of conditional independence that is helpful in understanding some of the material of the chapter.

Similar fairness criteria to the ones reviewed in this chapter were already known in the 1960s and 70s, primarily in the education testing and psychometrics literature [120]. The first and most influential fairness criterion in this context is identified by Anne Cleary [121, 122]. A score passes Cleary's criterion if knowledge of group membership does not help in predicting the outcome from the score with a linear model. This condition follows from sufficiency and can be expressed by replacing the conditional independence statement with an analogous statement about partial correlations [123].

Hillel Einhorn and Alan Bass [124] considered equality of precision values, which is a relaxation of sufficiency as we saw earlier. Robert Thorndike [125] considered a weak variant of calibration by which the frequency of positive predictions must equal the frequency of positive outcomes in each group, and proposed achieving it via a postprocessing step that sets different thresholds in different groups. Thorndike's criterion is incomparable to sufficiency in general.

Richard Darlington [123] stated four different criteria in terms of succinct expressions involving the correlation coefficients between various pairs of random variables. These criteria include independence, a relaxation of sufficiency, a relaxation of separation, and Thorndike's criterion. Darlington included an intuitive visual argument showing that the four criteria are incompatible except in degenerate cases. Mary Lewis [126] reviewed three fairness criteria including equal precision and equal true/false positive rates.

These important early works were rediscovered later in the machine learning and data mining communities [120]. Numerous works considered variants of independence as a fairness constraint. [127, 128]. Michael Feldman, Sorelle Friedler, John Moeller et al. [107] studied a relaxation of demographic parity in the context of disparate impact law. Richard Zemel, Yu Wu, Kevin Swersky et al. [129] adopted the mutual information viewpoint and proposed a heuristic preprocessing approach for minimizing mutual information. As early as 2012, Cynthia Dwork, Moritz Hardt, Toniann Pitassi et al. [130] argued that the independence criterion was inadequate as a fairness constraint. In particular, this work identified the problem with independence we discussed in this chapter.

The separation criterion appeared under the name *equalized odds* [114], alongside the relaxation to equal false negative rates, called *equality of opportunity.* These criteria also appeared in an independent work [131] under different names. Blake Woodworth, Suriya Gunasekar, Mesrob Ohannessian et al. [132] studied a relaxation of separation stated in terms of correlation coefficients. This relaxation corresponds to the third criterion studied by Darlington [123].

ProPublica [133] implicitly adopted equality of false positive rates as a fairness criterion in its article on COMPAS scores. Northpointe, the maker of the COMPAS software, emphasized the importance of calibration by group in their rebuttal [134] to ProPublica's article. Similar arguments were made quickly after the publication of ProPublica's article by bloggers including Abe Gong. There has been extensive scholarship on actuarial risk assessment in criminal justice that long predates the ProPublica debate; Richard Berk, Hoda Heidari, Shahin Jabbari et al. [135] provide a survey with commentary.

Variants of the trade-off between separation and sufficiency were shown by Alexandra Chouldechova [136] and Jon Kleinberg, Mullainathan, and Raghaven [137]. Each of them considered somewhat different criteria to trade-off. Chouldechova's argument is very similar to the proof we presented that invokes the relationship between positive predictive value and true positive rate. Subsequent work [138] considers trade-offs between relaxed and approximate criteria. The other trade-off results presented in this chapter are new to this book. The proof of the proposition relating separation and independence for binary classifiers, as well as the counterexample for ternary classifiers, is due to Shira Mitchell and Jackie Shadlen, pointed out to us in personal communication.

The credit score case study is from Moritz Hardt, Eric Price, and Nati Srebro [114]. However, we highlight the independence criterion in our plots, whereas the authors of the paper highlight the equality of opportunity criterion instead. The

Table 3.5
List of statistical nondiscrimination criteria

Name	Criterion	Note	Reference
Independence	Indep.	Equiv.	Calders et al. (2009) [127]
Group fairness	Indep.	Equiv.	
Demographic parity	Indep.	Equiv.	
Conditional statistical parity	Indep.	Relax.	Corbett-Davies et al. (2017)
Darlington criterion (4)	Indep.	Relax.	Darlington (1971) [123]
Equal opportunity	Separ.	Relax.	Hardt, Price, Srebro (2016) [114]
Equalized odds	Separ.	Equiv.	Hardt, Price, Srebro (2016) [114]
Conditional procedure accuracy	Separ.	Equiv.	Berk et al. (2017) [135]
Avoiding disparate mistreatment	Separ.	Equiv.	Zafar et al. (2017) [131]
Balance for the negative class	Separ.	Relax.	Kleinberg et al. (2016)
Balance for the positive class	Separ.	Relax.	Kleinberg et al. (2016)
Predictive equality	Separ.	Relax.	Corbett-Davies et al. (2017)
Equalized correlations	Separ.	Relax.	Woodworth et al. (2017) [132]
Darlington criterion (3)	Separ.	Relax.	Darlington (1971) [123]
Cleary model	Suff.	Relax.	Cleary (1966) [121]
Conditional use accuracy	Suff.	Equiv.	Berk et al. (2017) [135]
Predictive parity	Suff.	Relax.	Chouldechova (2016) [136]
Calibration within groups	Suff.	Equiv.	Chouldechova (2016) [136]
Darlington criterion (1), (2)	Suff.	Relax.	Darlington (1971) [123]

numbers about the racial composition of the dataset come from the "Estimation sample" column of table 9 on the webpage for the Federal Reserve report [116].

A Dictionary of Criteria

For convenience we collect some demographic fairness criteria below that have been proposed in the past (not necessarily including the original reference). We will match them to their closest relative among the three criteria independence, separation, and sufficiency. This table is meant as a reference only and is not exhaustive. There is no need to memorize these different names.

4

Relative Notions of Fairness

In chapter 3, we considered a range of statistical criteria that help to highlight group-level differences in both the treatments and the outcomes that might be brought about by the use of a machine learning model. But why should we be concerned with group-level differences? And how should we decide which groups we should be concerned with? In this chapter, we'll explore the many different normative reasons we might have to object to group-level differences. This is a subtle, but important, shift in focus from chapter 2, where we considered some of the normative reasons why *individuals* might object to decision-making schemes that distribute desirable resources or opportunities. In this chapter, we'll focus on why we might be concerned with the uneven allocation of resources and opportunities across *specific groups* and *society overall*. In particular, we'll review the normative foundations that ground claims of discrimination and calls for distributive justice. We'll then try to connect these arguments more directly to the statistical criteria developed in chapter 3, with the aim of giving those criteria greater normative substance.

A point of terminology: we use the terms *unfairness* and *discrimination* roughly synonymously. Linguistically, the term "discrimination" puts more emphasis on the agency of the decision maker. We also specifically avoid using the terminology "disparate treatment" and "disparate impact" in this chapter as these are legal terms of art with more precise meanings and legal significance; we'll address these in chapter 6.

Systematic Relative Disadvantage

Discussions of discrimination in the context of machine learning can seem odd if you consider that the very point of many machine learning applications is to figure out how to treat different people differently—that is, to discriminate between them. However, what we call discrimination in this chapter is not different treatment in and of itself, but rather treatment that systematically imposes a disadvantage on one social group relative to others. The systematicity in the differences in treatment and outcomes is what gives discrimination its normative force as a concept.

To better appreciate this point, consider three levels at which people might be subject to unfair treatment. First, a person might be subject to the prejudice of an individual decision maker—for example, a specific hiring manager whose decisions are influenced by racial animus. Second, a person might encounter systematic barriers to entering certain occupations, perhaps because members of the group to which they belong are not viewed as fit to be engineers, doctors, lawyers, and so on, regardless of their true capabilities or potential. For example, in some occupations women might have limited employment opportunities across the board due to their gender. Finally, certain personal characteristics might be an organizing principle for society overall such that members of certain groups are systematically excluded from opportunities across multiple spheres of life. For example, race and gender might limit people's access not only to employment, but to education, credit, housing, and so on.

In the examples from the previous paragraph, we have relied on race and gender precisely because both have served, historically, as organizing principles for many societies; they are not just the idiosyncratic bases upon which specific employers or professions have denied members of these groups important opportunities [139].

This helps to explain why these are characteristics of particular concern and why others might not be. For example, we might not care that a particular employer or profession has systematically rejected left-handed applicants beyond the fact that we might find the decision arbitrary and thus irrational, given that handedness might not have anything to do with job performance.

But if handedness became the basis for depriving people of opportunities across the board and not just by this one decision maker or in this one domain, we might begin to view it as problematic. To the extent that handedness dictates people's standing and position in society overall, it would rise to the level of a characteristic worthy of special concern [140].

Race and gender—among others enumerated in discrimination law and described in more detail in chapter 6—rise to such a level because they have served as the basis for perpetuating systematic relative disadvantage across the board. In extreme cases, certain characteristics can provide the foundations for a strict social hierarchy in which members of different groups are slotted into more or less desirable positions in society. Such conditions can create the equivalent of a caste system [141], in which certain groups are confined to a permanent position of relative disadvantage.

It is also important to note the unique threat posed by differential treatment on the basis of characteristics that persist intergenerationally. For example, children are often assumed to belong to the same racial group as their biological parents, making the relative disadvantage that people may experience due to their race especially systematic: children born into families that have been unfairly deprived of resources and opportunities will have less access to these resources and opportunities, thereby limiting from the start of their lives how effectively they might realize their potential—even before they might be subject to discrimination themselves.

Six Accounts of the Wrongfulness of Discrimination

Scholars have developed many normative theories to account for the wrongfulness of discrimination—specifically the wrongfulness of treating people differently according to characteristics like race, gender, or disability. While each of these theories is concerned with how such differential treatment gives rise to systematic relative disadvantage, they differ in how they understand what makes decision making on the basis of these characteristics morally objectionable.

Relevance: One reason—and perhaps the most common reason—to object to discrimination is because it relies on characteristics that bear little to no relevance to the outcome or quality that decision makers might be trying to predict or assess [141, 142]. For example, one reason why it might be wrong to base employment decisions on characteristics like race or gender is that these characteristics bear no relevance to determinations about job applicants' capacity to perform the job. Note that this is a variant of the objection that we covered in chapter 2, where individuals might contest decisions because they were rendered on the basis of irrelevant information. In this case, it is important not only that the reliance on irrelevant factors leads to mistakes but also that those mistakes lead to systematic relative disadvantage for particular social groups.

Generalizations: Or we might argue that the harm lies in the needlessly coarse groupings perpetuated by decisions made on the basis of race or gender, even if these can be shown to possess some statistical relevance to the decision at hand [82]. This harkens back to another idea in chapter 2, that people deserved to be treated as individuals and assessed on their unique merits. As you will recall, the intuitive idea of a perfectly individualized assessment is unattainable. Any form of judgment based on individual characteristics must draw on some generalizations and past experience. Yet we might still object to the coarseness of the generalizations, especially if there is obviously additional information that might provide a more granular—and thus more accurate—way to draw distinctions. For example, we might object if women were excluded as firefighters based on the assumption that women as a group are less likely to meet the strength requirements, as opposed to administering a fitness test to applicants of all genders.

Prejudice: Another common argument for why discrimination is wrongful is that it amounts to a form of prejudicial decision making, in which members of certain groups are presumed to be of inferior status. Rather than being merely a problem of relevance or granularity, as in the perspectives discussed above, it is a problem of beliefs, in which decision makers hold entire groups in lesser regard than others. For example, the problems with decisions guided by racial animus or misogyny is not merely that they may result in inaccurate predictions, but that decision makers hold these views in the first place [139, 143, 144].

Disrespect: A related, but distinct, idea is that making decisions on the basis of such characteristics is wrongful when it demeans those who possess such characteristics [139, 145]. On this account, the problem with discrimination is that it casts certain groups as categorically inferior to others and thus not worthy of equal respect. This objection differs from those based on prejudice because the harm is not located in decision makers' belief about the inferiority of members of particular groups, but in what decision makers' actions communicate about the social status of the groups. For example, the problem with sexist hiring practices is not merely that they confine women to particular roles in society or that they are based on prejudiced beliefs, but that they suggest that women are inferior to men. Understood in this way, discrimination is harmful not merely to the specific person subject to an adverse decision but to the entire group to which the person belongs because it harms the group's social standing in the community.

Immutability: An entirely different argument for why discrimination is wrongful is because it involves treating people differently according to characteristics over which they possess no control. On this account, the reason we should care about differences in the treatment of, for example, people with and without a disability is because people may not have any control over their disability [146, 147, 148]. As explored in chapter 2, decisions that rest on immutable characteristics deny people that possess these characteristics the agency to achieve different outcomes from the decision-making process, effectively condemning all such people to adverse outcomes. This amounts to wrongful discrimination specifically when the immutable characteristic in question is one whose use in decision making will give rise to systematic relative disadvantage.

Compounding injustice: Perhaps the problem with discrimination stems from the fact that it may compound existing injustice [149]. In many ways, this is an extension of the previous argument about control, but with a specific focus on the effects of past injustice. In particular, it is an argument that people cannot be morally culpable for certain facts about themselves that are not the result of their own actions, especially if these facts are the result of having been subject to some past injustice. Failing to take into account the fact that members of certain groups might have been, in the past, subject to many of the types of problems described above could lead decision makers to feel perfectly justified in treating them differently. Yet the reason people might appear differently at the time of the decision might be some past injustice, including past discrimination [150].

Ignoring this fact would mean that decision makers subject specific groups to worse decisions simply because they have already suffered some earlier harm. Note that this objection has nothing to do with concerns with accuracy; in fact, it suggests, as we first discussed in chapter 2, that we might have a moral obligation to sometimes discount the effects of factors over which people have no control, even if that means making less accurate predictions. In chapter 2, we considered this principle without any particular concern for distributional outcomes; here, we can adapt the principle to account for the wrongfulness of discrimination by pointing out that suffering some past mistreatment due to someone's membership in a particular group might be outside that person's control.

None of the above six accounts is a complete theory of the wrongfulness of discrimination. Some situations that we might view as obviously objectionable can be caught by certain theories, but missed by others. For example, objections to religious discrimination cannot be grounded on the idea that people lack control over their religious affiliations, but could be supported by reference to concerns about prejudice or disrespect [151]. Or to take another example, even if decision makers' actions are not prejudicial or demeaning, their decisions may still be based on irrelevant characteristics—a possibility that we'll consider in the next section. While there is unlikely to be a single answer to the question of why discrimination is wrong, these theories are still helpful because they highlight that we often need to consider many factors when deciding whether subjecting particular groups to systematic relative disadvantage is morally justified.

Intentionality and Indirect Discrimination

So far we have focused on why taking certain characteristics into account when making consequential decisions can be normatively objectionable. According to each of these accounts of the wrongfulness of discrimination, the harm originates from the choice to rely on these characteristics when making such decisions. But what about when decisions do not rely on these characteristics? Does removing these characteristics from the decision-making process ensure that it is fair?

An easy case is when the decision maker purposefully relies on proxies for these characteristics (e.g., relying on a person's name as a proxy for their gender) in order to indirectly discriminate against members of a specific group (e.g., women). The fact that such decisions do not consider the characteristics explicitly may not render them any less problematic, given that the decision maker does so only with the goal of treating members of these different groups differently. Thus, the wrongfulness of discrimination is not limited to the mere use of certain characteristics in decision making, but extends to any intentional efforts to subject members of specific groups to systematically disfavorable treatment, even if this is achieved via the use of proxies for such characteristics [152]. With that in mind, we might want to check whether any decision-making process leads to disparate outcomes for different groups as a way of potentially smoking out intentional discrimination pursued with the aid of proxies. If we discover disparities in outcomes, we might want to check whether the decision-making process was purposefully orchestrated to achieve this, even if the decision maker didn't seem to take these characteristics into account explicitly.

But what about decisions that are not designed purposefully to discriminate? Does the mere fact that a decision-making process can lead to quite disparate outcomes for different groups mean that it is unfair? What if the decision maker can offer some reason for making decisions in this particular manner (e.g., that the employer needs people with a specific accounting credential and that such credentials happen to be held more commonly among certain groups than others)?

We can extend some of the reasoning first introduced in the previous section to try to answer these questions. In this case, rather than asking whether the criteria under consideration are serving merely as proxies for some characteristic of concern, we could instead ask whether the choice of criteria can be justified by demonstrating that they actually serve the stated goals of the decision maker. Decision-making processes that do little to serve these goals, but nevertheless subject specific groups to systematically less favorable outcomes raise the same question about relevance that arise in cases of intentional and direct discrimination. If the chosen criteria lack relevance to the decision at hand, but result in systematic relative disadvantage for a specific group, then relying on them can easily become functionally equivalent to relying on group membership directly despite its lack of relevance to the decision at hand. In both cases, the reliance on irrelevant criteria is objectionable because it results in systematic relative disadvantage for particular groups.

Equality of Opportunity

What about a process that produces disparities in outcomes but does, in fact, serve to advance the goals of the decision maker? This is a harder case to reason about. But before doing so, let's take a step back.

Equality of opportunity is an idea that many scholars see as the goal of limiting discrimination. Equality of opportunity can be understood in both narrow and broad terms. The narrow view holds that we should treat similar people similarly given their current degree of similarity. The broad view holds that we should organize society in such a way that people of similar ability and ambition can achieve similar outcomes. A position somewhere in the middle holds that we should treat seemingly dissimilar people similarly, on the belief that their current degree of dissimilarity is the result of something that we should discount (e.g., past injustice or misfortune). Let's tackle each view in turn, and see what each would imply about the question above.

The Narrow View

Let's illustrate the narrow view of equality of opportunity with the notion of "individual fairness": that people who are similar with respect to a task should be treated similarly [130]. For now, we take *similar* to mean closeness in features deemed relevant to the task at hand. Individual fairness is a comparative notion of fairness in that it asks whether there are any differences in the way that similar people are being treated. It is not directly concerned with the way members of different *groups* might be treated. Instead, the comparison is between all people as individuals, not between the members of specific groups. Of course, if we agree in advance that people's race, gender, and so forth are irrelevant to a task at hand, then satisfying individual fairness will also limit the degree to which people who differ according to these characteristics will receive different treatment. But this is true only to the extent that such characteristics are deemed irrelevant; it is not an inherent part of the definition of individual fairness.

Individual fairness is related to consistency and some of the concerns with arbitrariness that we explored in chapter 2. We often expect consistent treatment in the absence of differences that would seem to justify differential treatment, especially when the treatment determines access to important opportunities. These expectations can be so strong that a failure to meet them will provoke a visceral reaction: why did I not get the desired treatment or outcome even though I am seemingly similar along the relevant dimensions to someone that did?

What does it mean for people to be similar? The answer is not a given. In the philosophical literature, the common answer to this question is that people should be treated similarly to those who are understood to be similarly meritorious—that is, that people should be judged according to their abilities and ambitions [153]. This understanding is so common that the concept of equality of opportunity— in this narrow formulation—is often taken to be synonymous with the concept of a meritocracy. Access to desirable resources and opportunities should be dictated not by the social group to which someone happens to belong, but rather by the characteristics that are legitimately relevant to the institution seeking to advance its goals in allocating these resources and opportunities.

Much depends on what we decide are the goals of the decision-making process. They could be defined as maximizing some outcome of interest to the decision maker, such as job performance. When applicants who are predicted to perform similarly well on the job are treated similarly in the hiring process, it is often interpreted as meritocracy. Or we might say that a firm has a legitimate interest in hiring workers with the necessary training to effectively perform a job, so it might hire only those who have completed the necessary training. If making decisions on this basis leads to uneven hiring rates across groups, under the narrow view of equality of opportunity, the decision maker is blameless and under no obligation to adjust the decision-making process.

But note that employers could just as easily decide on goals that bear no obvious relationship to what we perceive as merit, such as hiring applicants who would be likely to accept a particularly low salary. With this target in place, decision makers would only have an obligation to treat people similarly who possess similar sensitivities to pay, not those who are likely to perform similarly well on the job. Would it make sense to describe this as a case in which employers ensure equality of opportunity [139]? This reveals that the narrow view of equality of opportunity does not dictate what the abiding normative principle should be that determines how we view people as similar; it commands only that similar people be treated similarly. It thus possesses little normative substance beyond consistency. (And even then, there may be practical limits to this principle. For example, an employer may still need to choose only one applicant among the many that are predicted to perform similarly well on the job—and those who did not get the job might not object that they had been treated unfairly.)

The people subject to decisions might have their own ideas about how to define similarity, ideas that might be very different from those of the decision maker. This might be because they do not share the same goals as the decision maker, but it might also be because they believe that there are reasons completely independent of the goals of the decision-making process to view certain people as similar. Perhaps the reason we might view two different people as deserving of some opportunity is because they are equally likely to make the most of the opportunity or because they are equally needy—not only equally meritorious. In other words, we might judge job applicants as similar because they are likely to benefit similarly from the job, not only because they are likely to perform similarly well on the job. In many cases, the exact basis on which we might view people as relevantly similar in any given context can be quite challenging for us to articulate because our conception of similarity might rely on varied normative considerations.

The Broad View

A broad view of equality of opportunity sets aside questions about the fairness of any given decision-making process and instead focuses on the degree to which society overall is structured to allow people of similar ability and ambition to achieve similar success. This perspective has been most famously developed by philosopher John Rawls under the banner of "fair equality of opportunity." To simplify the argument considerably, the only defensible reason for why people might experience different outcomes over the course of their lives is if they possess different ability or ambition [154].

Anything about the design of particular institutions in society that prevents such people from realizing their potential violates this broader understanding of equality of opportunity because it deprives equally deserving people of the same chance at success. For example, a society that fails to provide a means for similarly capable individuals born into different circumstances—one into a wealthy family and another into a poor family—to achieve similar levels of success would violate this understanding of equality of opportunity.

According to this view, the basic institutions that help to cultivate people's potential over the course of their lives must be structured to ensure that people of similar ability and ambition have similar chances of obtaining desirable positions in society—along with the many benefits that come with such positions. Thus, if education is an important mechanism by which people's potential might be fostered, a broad view of equality of opportunity would command that schools be funded such that students of equal ability and ambition—whether from wealthy or poor families—face the same prospects of long-term success. Thus, any advantage that such children might receive from their wealthy families must be offset by a corresponding intervention to ensure that such children from poor families may flourish to the same degree. If wealthier children benefit from a local tax base that can fund a high-quality public school, then society must put in place policies that make available similar amounts of funding to the public schools that educate poorer children.

Note that this is an intervention that aims to equalize the quality of the education to which wealthy and poor students will have access; it is not an intervention into the admissions policy of any particular school. This helps to highlight the fact that the broad view of equality of opportunity is not really about fairness in decision making; it is about the design of society's basic institutions, with the goal of preventing unjust inequalities from arising in the first place. In theory, abiding by such a principle of equality of opportunity would result in a casteless society in which no one is permanently confined to a position of disadvantage despite having the potential to succeed under different circumstances [155]. Society would be structured to ensure social mobility for those who possess the relevant ability and ambition to achieve certain goals.

The Middle View
Somewhere between the two poles we have just explored is a middle view that is narrowly concerned with the fairness of decision making, yet sensitive to the dynamics

by which disadvantage might be perpetuated in society more broadly. This view holds that decision makers have an obligation to avoid perpetuating injustice [150]. Specifically, they must, to some degree, treat seemingly dissimilar people similarly when the causes of these dissimilarities are themselves problematic. For example, those who adopt this view might argue that universities should not simply rank order applicants by GPA or standardized test scores; instead, they must assess applicants with respect to the opportunities that applicants have been afforded over the course of their childhoods, recognizing that performance in school and standardized tests might differ according to past opportunity not only according to innate ability and ambition.

To give an example, the state of Texas has a law guaranteeing admission to state-funded universities to all students who graduate in the top 10 percent of their high school class. This can be seen as in keeping with the middle view. If access to opportunity varies geographically, the 10 percent rule identifies individuals with ability and ambition without systematically disadvantaging those who had the misfortune of growing up without access to well-funded high schools.

This middle view differs from the broad view insofar as it accepts that students of equal potential will not receive equally high-quality education leading up to the moment when they finally apply to college. Yet it also differs from the narrow view insofar as it refuses to allow colleges to ignore what might account for applicants' current dissimilarity at the time that they submit their applications. Instead, the middle view suggests that there is some burden on colleges to attempt to compensate for the disadvantages that some applicants may have faced over their lifetimes such that they might appear less competitive than other applicants from more privileged backgrounds.

As a result, the middle view calls for interventions not at the level of the design of institutions but at the level of the design of decision-making processes. It suggests that ensuring equality of opportunity requires assessing people as they *would have been* had they been afforded in the past opportunities comparable to other people of equal potential seeking the current opportunity. In certain respects, the middle view seems to be trying to realize the goals of the broad view via a much more limited intervention: while the broad view would seem to demand that children from wealthier and poorer families have access to equally high-quality education throughout their lives, the middle view seeks to compensate only for the disadvantages experienced by poorer students relative to their wealthier peers at specific decision-making junctures that are thought to be particularly high-stakes—in this case, in college admissions. The middle view tends to focus on these junctures because they seem to be where there is an opportunity to greatly alter a child's life course and to allow them to much more effectively realize their potential [156]. Indeed, this is often why they are perceived as high-stakes.

While the interventions imagined by the middle view might seem narrower than those in the broad view because they do not require a more radical restructuring of the basic institutions of society, it's worth noting that the more discrete interventions of the middle view are designed to bring about much greater change than any one of the more continuous interventions required of the broad view. The middle view targets specific decisions that can create a sudden step change in people's life prospects, whereas the broad view aims to obviate the need for such dramatic interventions in decision making by ensuring equality throughout people's lives. In other words, the middle view will require sudden and substantial change at specific moments of decision making, while the broad view will require a significant redistribution of resources on an ongoing basis.

While the middle view clearly prohibits ignoring the reasons for differences in merit between people, it does not offer a clear prescription for how to take them into account. Taking it to its logical conclusion would result in interventions that seem extreme: it could require imagining people without the effects of centuries of oppression that they and their ancestors might have endured, suggesting, for instance, that a bank should approve a large loan to someone who does not in reality have the ability to repay it. That said, there are other areas of decision making where this view might seem more reasonable. For example, in employment, we might expect hiring managers to adopt an approach similar to admissions officers at universities, assessing people according to the opportunities they have been afforded, discounting certain differences in qualifications that might be due to factors outside their control, especially if these are qualifications that the employer could help cultivate on the job. The middle view has particular purchase in the case of insurance, where we really might want insurers to ignore the additional costs that they are likely to face in setting the price of a policy for someone with an expensive preexisting condition outside the person's control. The extent to which we expect decision makers to bear such responsibility tends to be context-specific and contested. We will return to it shortly.

Tensions between the Different Views

There is an obvious conflict between the view that decision makers should treat people similarly according to how they appear at the time of decision making and the view that decision makers should treat people similarly according to how they would have appeared had they enjoyed privileges and advantages similar to others of equal ability and ambition. Thus, a person who at present seems to be more meritorious with respect to some opportunity might object if they are passed over in favor of someone who at present seems less meritorious—even if the decision maker believes the other person would be more meritorious than the person objecting if they had both enjoyed the same privileges and advantages.

A similar tension arises in the way we might try to deal with discrimination. The narrow view of equality of opportunity suggests that the way to deal with discrimination is to ensure that decisions are made only on the basis of factors that are genuinely relevant to the task at hand. In other words, treating similar people similarly with respect to the task should, in most cases, rule out treating people differently according to their gender, race, and so on, because these characteristics are not likely to be relevant to the task at hand. Thus, committing to the narrow view of equality of opportunity should help to keep these factors from entering the decision-making process. In contrast, the middle view suggests that we might want to deal with discrimination by explicitly considering these characteristics when making decisions because it is likely that these characteristics would help to explain a good deal of the deprivation and disadvantage that people might have faced over the course of their lives. In other words, in order to understand how people who possess these characteristics might have appeared under counterfactual circumstances, decisions must take these characteristics into account. This again seems to set up a conflict because realizing a commitment to the middle view of equality of opportunity necessitates violating the requirements imposed by the narrow view of equality of opportunity.

John Roemer says that these tensions boil down to different views on "when the competition starts" for desirable positions in society: at what point in the course of our lives are we ultimately responsible for how we might compare to others [157]? Given that we have no control over the wealth of the families into which we are born or the quality of the education we might receive, we might discount any differences between people that are due to such differences. In other words, we might say that we don't think it's reasonable to adopt a narrow view of equality of opportunity when assessing applicants to college because many of the relevant differences between applicants might not have emerged from a fair competition. In contrast, we might think that employers are justified in assessing job applicants, especially those for more senior roles that require many years of experience, according to the ability and ambition that they have demonstrated over the course of their careers. That is, we might agree that a narrow view of equality of opportunity is appropriate in this case because people who are well into their careers have had a fair chance to cultivate their ability and demonstrate their ambition. Tensions arise when there is disagreement over where this transition occurs in people's lives. Someone who has been passed over in favor of another person who seems less meritorious might consider it unfair because she thinks that whatever differences exist between the two of them have emerged during a period of fair competition between them. The decision maker might disagree, believing that the differences actually owe to advantages that the passed-over individual accrued during a period prior to the start of a fair competition (table 4.1).

Table 4.1
Some key differences between the three views of equality of opportunity

	Goal	Intervention point	Who bears the cost of uplifting historically disadvantaged groups
Narrow view	Ensure that people who are similarly qualified for an opportunity have similar chances of obtaining it	Decision making	No one
Middle view	Discount differences due to past injustice that accounts for current differences in qualifications	Decision making, especially critical life opportunities	Decision maker (who may pass on the cost to decision subjects)
Broad view	Ensure people of equal ability and ambition are able to realize their potential equally well	Government, on a continuous basis	Taxpayers

Merit and Desert

Even if we buy into the middle and broad view of equality of opportunity, we may want some normative principles that allow us to decide just how far decision makers and the government must go in seeking to counteract inequality. More concretely, what differences between people, if any, actually justify differences in outcomes? So far, we have tended to answer this question by describing those differences that cannot or should not serve as a justification for differences in outcomes. But it's also worth reflecting more deeply on the principles that seem to allow for—or perhaps even require—these differences in outcomes. In this section, we'll discuss two principles that help answer this question. The first, which already has played an important role in our discussion, is the principle of merit. The second is the principle of desert, meaning *that which is deserved.*

Merit plays an important role in all three views of equality of opportunity. In our discussion of the narrow view of equality of opportunity, merit is one way to establish how people are similar and thus who should be treated similarly. In the broad view, the idea of merit—in the form of abilities and ambitions—allows for the possibility that people of differing merit might not achieve comparable outcomes in life, but it also dictates the amount of support that must be provided to people who have the same potential as their more privileged peers but who would be unable to realize their potential as effectively in the absence of that support. Merit is crucial to the idea that there is a moral obligation to help people realize their potential—but no obligation to go further than that. Finally, merit plays a similar role in the middle view insofar as decision makers are expected to evaluate people according to how meritorious they would have been under counterfactual circumstances. Understood in this way, all three views are perhaps more similar than they might first appear: each is calling for people of similar merit to have the same chances of success.

But what, exactly, is merit? Merit is not an objective property possessed by any given individual. Instead, merit concerns the qualities possessed by a specific person that are expected to help advance the goal of the institution that is offering the sought-after opportunity [153]. Thus, what makes a particular job applicant meritorious is how well that applicant is likely to advance the goals of the employer. While it is tempting to think that there are some universal answers to what makes any given job applicant more or less meritorious than others (e.g., how smart they are, how hardworking they are, etc.), this is not the case. Instead, merit, on this account, is purely a function of what an employer views as relevant to advancing its goals, whatever they might be. And different employers might have very different goals and very different ideas about what would do the most to help advance them. The goals of the employer might not be the goals that the job applicant would like them to be or what outsiders would want them to be. Others' conceptions of merit might differ from employers' because they simply have different ideas about the goals that employers should have in the first place.

The subjectiveness of this view of merit seems to be in conflict with earlier discussions of abilities and ambitions, which are presented as universal properties that are not tied to the goals of any particular institution that is providing the sought-after opportunity. In suggesting that people of similar ability and ambition should have similar chances of obtaining desirable positions in society, there seems to be an implicit belief that people can be compared on their merits regardless of the opportunity in question. This reflects the fact that there are often widely held and well-entrenched beliefs about the relevance of certain criteria in determining whether someone is deserving of a particular opportunity. An employer that assesses people according to their intelligence and industriousness is commonly understood to be assessing people according to their merit because these are the properties that can be safely assumed to help the employer advance its own interest. But there is no reason why these need to be the properties according to which job applicants must be assessed in order to ensure that the employer's decisions are based on merit.

This observation anticipates a related notion: desert. Unlike merit, desert is not tied to how well a person might help to advance the interests of the decision maker, but how well a person has performed along the dimensions according to which a decision maker is *expected* to evaluate people. For example, we might say that a person who has worked diligently throughout school to obtain high grades is more meritorious with respect to a job opportunity than a person who blew off classes and received middling grades, even if both people are likely to advance the goals of an employer equally well. In this account, people deserve certain opportunities given that they might have good reason to believe that certain investments would help them gain access to the sought-after opportunity. In other words, decision makers have an obligation to provide opportunities to people who have taken actions for which they deserve to be rewarded.

This principle can help to explain why we believe that people who plan to start a family should have the same chance of securing a job as others who do not when they demonstrate equal ability and ambition, even if starting a family requires employees to go on leave for extended periods and even if it increases the likelihood that employees will quit, thereby imposing costs on employers that might otherwise be avoided. While the employer's goal might be to recruit people who are likely to work diligently without interruption and who are likely to remain at the company indefinitely, selecting among applicants on this basis might cause an employer to disfavor applicants who deserve to be hired in light of their ability and ambition. Notably, would-be mothers, who are more likely than their peers to take time off or quit their jobs to start a family (due to entrenched gender norms around the division of parental responsibilities), should not be passed over in favor of others if they have all made the same investments in preparing themselves to apply for these positions. The principle of desert says that those who have cultivated the necessary skills to succeed on the job should all be assessed similarly, regardless of differences in the likelihood that applicants will need to take time off or give up their jobs.

Discussions of merit and desert also help to highlight that there can be quite different justifications for the constraints or demands that we might place on decision makers. In some cases, we might argue that people are simply morally entitled to certain treatment. For example, we might say that it is wrong to hold people responsible for characteristics about themselves over which they have no control, even if doing so would be in the rational interest of a decision maker. Likewise, we might say that people are owed certain opportunities in light of their ability and ambition, even if the decision maker would prefer to judge people on a different basis. These are what philosophers call deontological arguments: moral reasons why some actions are preferable to others regardless of the consequences of these actions. We must discount the effects of bad luck and take merit into account because that is what fairness demands.

In contrast, we might argue that the way decision makers treat people should be dictated by the consequences of such actions. For example, we might say that merit-based decision making is justified on the grounds that allocating opportunities according to merit helps to advance the interest of society, not just the individual seeking a particular opportunity or the decision maker providing the opportunity. Hiring on the basis of ability and ambition may have the consequences of enhancing

overall welfare if it means people who are particularly well prepared to undertake some activity are more likely to obtain the opportunity to do so. Merit-based
decision making is thus justified because it puts individuals' talents to good use for
society's collective benefit—not because any given individual is morally entitled to
a particular opportunity in light of their merits. We might further argue that differential treatment on the basis of merit incentivizes and rewards productive investment
that benefits all of society.

Of course, there are also consequentialist arguments in favor of interventions
designed to uplift those who have experienced disadvantage or discrimination in
the past. For example, society suffers overall when members of specific groups are
denied the opportunity to realize their true potential because society forgoes the
collective benefits that might be brought about by the contributions of such groups.

The Cost of Fairness

Different views of equality of opportunity—as well as the notions of merit and
desert on which such views depend—allocate the responsibility and associated costs
of dealing with unfairness quite differently. Notably, the middle view places the burden on individual decision makers and specific institutions regardless of the speed
with which or the extent to which a person is able to realize their potential. For
example, we might expect universities to incur some up-front costs in admitting
students from less privileged backgrounds because universities may have to invest
additional resources in helping bring those students up to speed with their more
privileged peers. This could take the form of providing classes over the summer
leading up to the start of formal undergraduate programs. Or it could take the form
of designing introductory courses without taking much background knowledge as
a given, which might spend some time covering material that is familiar to students
from better-funded high schools, but perhaps new to those who come from less
well-funded school districts. Universities might even invest in programs that seek
to limit the degree to which the inequalities that exist between students prior to
enrolling in college *carry through* their college experiences. For example, universities might offer financial scholarships to poorer students with the goal of allowing
them to avoid having to work in order to support themselves, thereby allowing these
students to devote a similar amount of time to their studies as their more privileged peers. Such scholarships could also help to avoid saddling poorer students
with significant debt, which might suppress future earnings and negatively influence career choices—burdens that richer students without significant debt need not
navigate. Interventions along these lines blur the distinction between the middle and
broad view of equality of opportunity because they seem targeted not at remedying

some past unjust inequality but at preventing an unjust inequality from reemerging. Chapter 8 covers such interventions in greater depth.

Despite these efforts, universities may find that their investments in these kinds of interventions may take many years to pay off: students from less privileged backgrounds might trail their peers from more privileged backgrounds in the grades that they obtain over the course of their undergraduate careers, but ultimately achieve comparable success once they enter the labor market.

Likewise, employers that hire candidates that they recognize as having great potential, but also the need for additional support, might not be the employers that enjoy the payoff of such investments. Employees might take another job before the original employer feels that it has recouped its investment. This is an important aspect of the middle view of equality of opportunity because it highlights that it might not always be in the rational interest of decision makers to behave in these ways. (This might cut the other way as well, though: an unconstrained decision maker might discount someone who seems meritorious because the decision maker recognizes that the person has benefited from good luck, and is thus lacking in the ability or ambition that they are actually searching for.)

The middle view is thus not simply an argument that decision makers must attend to their long-term self interests, it is an argument that certain institutions are the right actors to incur some cost in the service of remedying inequality and injustice, even if there is no guarantee of obtaining a reward of at least equivalent value.

This contrasts with the broad view of equality of opportunity, where the government is understood to be the appropriate actor to facilitate the necessary redistribution to compensate for unjust disparities, likely through direct taxation and transfers. According to the broad view, the government—which is to say, everyone who pays taxes to the government—bears the burden of counteracting the advantages that would otherwise be enjoyed by, for example, students from more privileged backgrounds. To the extent that interventions by employers or other institutions are necessary, the government should subsidize their efforts with tax money. In contrast, the middle view places the burden on specific decision makers to compensate for the disadvantages that people have already experienced.

This all suggests a number of difficult questions: To what extent should the burden for past discrimination fall on individual decision makers? On what time scale should we attempt to correct the effects of historical injustice? And is it even possible to offset the cumulative result of the thousands of moments in which people treat each other unequally over the course of a lifetime? We'll return to these questions in chapter 6 when we consider, from a legal and practical perspective, who we might view as best positioned to incur these costs.

Connecting Statistical and Moral Notions of Fairness

We now attempt to map some of the moral notions we've discussed so far in this chapter to the statistical criteria from chapter 3. Of course, many of the concepts in this chapter, such as whether a decision subject has control over an attribute used for decision making, cannot be expressed in the statistical language of conditional probabilities and expectations. Further, even for notions that do seem to translate to statistical conditions, we reiterate our usual note of caution that statistical criteria alone cannot certify that a system is fair. As just one reason for this, the criteria in chapter 3 are invariant to the application rates of different groups. For example, if 50 percent of loan applicants from a particular group decided not to apply for some reason, a classifier that satisfied independence/demographic parity before the change in application rates would still satisfy independence/demographic parity after the change. The same is true of sufficiency/calibration and separation/error rate parity. Yet, we wouldn't consider a bank, employer, or another institution to be fair if it discouraged applications from certain people or groups. This is related to Andrew Selbst, Danah Boyd, Sorelle Friedler et al.'s *framing trap*: a "failure to model the entire system over which [fairness] will be enforced" [158].

But we must also resist the opposite extreme, which is the view that statistical criteria have no normative content. We take the position that statistical criteria are one facet of what it means for a sociotechnical system to be fair and, combined with procedural protections, they can help us achieve different moral goals.

Demographic Parity

With those caveats out of the way, let's start with a relatively simple statistical criterion: demographic parity. It has a tenuous but discernible relationship to the broad view of equality of opportunity insofar as it aims to equalize outcomes. The high-level similarity between the two is the idea of proportional distribution of resources. But moral notions never map exactly on to technical criteria. Let us look at the differences between them as a way of understanding the relationship.

The broad view of equality of opportunity is concerned with people's *life* outcomes (such as wealth) rather than discrete moments of decision making. Still, we may hope that imposing some notion of equality in decisions that affect the outcome of interest (such as jobs in the case of wealth) will lead to equality in the corresponding outcome. Empirically, however, it is far from clear that imposing equality in the short term will lead to equality in the long term. In fact, theoretical work has shown that this is not always the case [159].

Further, equality of outcome—that is, enforcing equal life outcomes—ignores differences in ability and ambition between people that might reasonably justify differences in outcomes. This also happens to be the most common criticism of equality of outcome: it rules out the particular understanding of merit-based decision making that underpins the application of machine learning in many settings.

Even though this is often considered a fatal objection to equality of outcome, the criticism loses much of its force when applied to demographic parity. To try to justify demographic parity despite individual differences in ability and ambition, we acknowledge those differences but argue that these cancel out at the level of groups. Thus, while decisions made about individuals can be attuned to the differences between them, we require the benefits and burdens of those decisions to be equally distributed among groups, on average.

But which groups should we pay attention to? As before, we pay special attention to group differences when we consider them especially likely to arise due to unjust historical conditions or to compound over time. These correspond to the dimensions along which society was historically and is currently stratified. In this view, we may care about equality of outcome not just for its own sake but also because inequality of outcome is a good indicator that there might be inequality of opportunity in the broad sense of the term [160]. In other words, certain inequalities in outcomes might not have arisen had there not been some past inequalities in opportunity.

There are many other gaps between demographic parity and equality of outcome. We'll mention just one more: not all decision subjects (and groups) may value the resource equally. Targeted ads may be helpful to wealthier individuals by informing them about things their money can buy, but prey on the economic insecurities of poorer individuals (e.g., payday loans [161]). Policing may be helpful to some communities but put a burden on others, depending on the prejudices of police officers. In these cases, actual outcomes—benefits and harms—can be vastly different despite statistical parity in allocation.

Calibration

Recall from chapter 3 that if group membership is redundantly encoded in the features, which is roughly true in sufficiently rich datasets, then calibration is a consequence of unconstrained supervised learning. Thus, it can be achieved without paying explicit attention to group membership. In other words, imposing calibration as a requirement is not much of an intervention.

Still, the notion has intuitive appeal: if a score is calibrated by group, then we know that a score value (say, 10 percent risk of default) indicates the same rate of positive outcomes (e.g., default rate) in all groups. By the same token, it has some diagnostic usefulness from a fairness perspective, such as flagging "irrational" discrimination. If the classifier explicitly encodes a preference for one group or a prejudice against another (or a human decision maker exercises such a preference or prejudice), the resulting distribution will not be calibrated by group.

Calibration can also be viewed as a sanity check for optimization. Precisely because calibration is implied by unconstrained optimization, we can detect optimization failures from violations of calibration. But that's all it is: a sanity check. A model can be egregiously inaccurate and still satisfy calibration. Indeed, a model with no discriminative power that always simply outputs the mean outcome of the population is perfectly calibrated. A model that is highly accurate for one group (optimal as defined in chapter 3) while always predicting the mean for another group is also perfectly calibrated.

Calibration by group fits with a narrow view of equality of opportunity. Suppose that a decision maker uses only features deemed relevant, while group membership is deemed irrelevant. Then calibration by group says that the decision maker does not consider group membership beyond the extent to which it is encoded in task-relevant features. The decision may justify group differences in outcomes by appeal to differences in relevant features.

Some nontrivial normative justification is required for violating calibration in models used for decision making. We have discussed many such justifications, such as a belief that the risk arises partly from factors for which the decision subject should not be held accountable.

The Similarity Criterion

Let's return to the similarity criterion: treating similar people similarly. As we discussed, the normative substance of this notion largely comes down to what we mean by similar. One common view is to think of it as closeness with respect to features that relate to qualifications for the task at hand, interpreting features at face value.

To translate this to a technical notion, we can imagine defining a task-specific similarity function or metric between two feature vectors representing individuals. We can then insist that for any two individuals who are sufficiently similar, the decisions they receive be correspondingly similar. We call this the *similarity criterion*. This notion was made precise and analyzed by Cynthia Dwork, Moritz Hardt, Toniann Pitassi et al. [130]. Once we have a metric, we can solve a constrained optimization problem. The optimization objective is as usual (e.g., minimizing the difference between predicted and observed job performance) and the similarity criterion is the constraint.

We can illustrate this approach in the context of online behavioral advertising. Our discussion assumes that we view advertisements as allocating access to opportunity (e.g., through targeting job openings or credit offers). Ad networks collect demographic information about individuals, such as their browsing history, geographical location, and shopping behavior, and utilize this to assign a person to one of a few dozen segments. Segments have names such as "White Picket Fences," a market category with median household income of just over $50,000, aged twenty-five to forty-four with kids, with some college education, and so on. Individuals in a segment are considered similar for marketing purposes, and advertisers are allowed to target ads only at the level of segments and not individuals.

This reflects the narrow view of equality of opportunity. If two individuals differ only on dimensions that are deemed irrelevant to the advertiser's commercial interests, say religion, they will be in the same segment and thus are expected to see the same ads. On the other hand, if some people or social groups have had advantages throughout their lives that have enabled a certain income level, then the similarity criterion allows the benefits of those advantages to be reflected in the ads that they see.

Targeted advertising is a particularly good domain to apply these ideas. There is an intermediary—the ad network—that collapses feature vectors into categories (i.e., ad segments), and exposes only these categories to the advertisers, rather than directly allowing advertisers to target individuals. The ad network should construct the segments in such a way that members who are similar for advertising purposes must be in the same segment. For example, it would not be acceptable to include a segment corresponding to disability, because disability is not a relevant targeting criterion for the vast majority of types of ads. In domains other than targeted advertising, say college admissions, applying these ideas is more challenging. Absent an intermediary like the ad network, it is up to each decision maker to provide transparency into their similarity metric.

This narrow interpretation of the similarity criterion relates to other formal definitions of individual fairness, such as, the notion of *meritocratic fairness* in the context of bandit learning [162]. The normative content of the similarity criterion, however, extends beyond the narrow view of equality of opportunity if we broaden the principles from which we construct a similarity metric. For example, the notion of similarity might explicitly adjust features based on consideration of past injustice and disadvantage. We might agree at the outset that an SAT test score of 1200 under certain circumstances corresponds to a score of 1400 under more favorable background conditions.

Comparisons such as these are closely related to Roemer's formal definition of equality of opportunity [157]. Roemer envisions a partition of the population into *types* based on "easily observable and nonmaniupulable" features that relate to "differential circumstances of individuals for which we believe they should not be held accountable." The formal definition then compares individuals who expend the same quantile of *effort* relative to their type.

Randomization, Thresholding, and Fairness
If we think about applying the similarity criterion to a task like hiring, we run into another problem: pairs of candidates who are extremely similar may fall on opposite

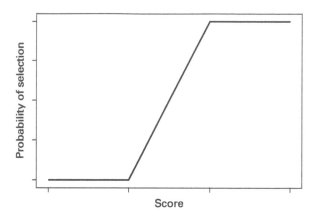

Figure 4.1
A randomized classifier. Only randomized classifiers can satisfy the similarity criterion. Two similar individuals would have similar scores and thus similar probabilities of selection.

sides of the score threshold, because we have to draw the line somewhere. This would violate the similarity criterion. One way to overcome this is to insist that the classifier be randomized (figure 4.1).

Randomization sometimes offends deeply held moral intuitions, especially in domains such as criminal justice, by conjuring the specter of decisions made based on a coin flip. But there are several reasons why randomization may not just be acceptable but necessary for fairness in some cases (in addition to the fact that it allows us to treat similar people similarly, at least in a probabilistic sense). In fact, Ronen Perry and Tal Zarsky present numerous examples of cases where the law requires that consequential decisions be based on lotteries [163].

To understand the justification for randomized decision making, we must recognize that precisely controlled and purposeful randomness is not the same as arbitrariness or capriciousness. Suppose these three conditions hold: a resource to be allocated is indivisible, there are fewer units of it than claimants, and there is nothing that entitles one claimant to the resource any more or any less than other claimants. Then randomization may be the *only* egalitarian way to break the tie. This is the principle behind lotteries for allocation of low-rent public housing and immigration visas. The same principle applies to burdens rather than valued resources, as seen in the random selection of people for jury duty or tax audits [163].

But the whole point of employing machine learning is that there does exist a way to rank the claims of the applicants, so the scenarios of interest to us are more complicated than the above examples. The complication is that there is a conflict between the goals of treating similar people similarly (which requires randomization) and minimizing unpredictability in the decision (which requires avoiding randomization).

One critical distinction that affects the legitimacy of randomization is whether there are equivalent opportunities for which an applicant might be eligible. This is generally true in the case of hiring or lending, and not true in the case of pretrial detention. Randomization is more justifiable in the former case because it avoids the problem of an applicant perpetually falling just short of the selection threshold. If randomization is employed, a reasonably qualified but not stellar job applicant might have to apply to several jobs, but will eventually land one [164].

Another way to avoid the problem of similar candidates falling on opposite sides of a cutoff is to redesign the system so that decisions are not binary. This is again easier for some institutions than others. A lender can account for different levels of risk by tailoring the interest rate for a loan rather than rejecting the loan altogether. In contrast, a binary notion of determination of guilt is built into the criminal justice system. This is not easy to change. Note that determinations of guilt are not predictions; they are meant to reflect some binary ground truth and the goal of the criminal justice system is to uncover it.

The Normative Underpinnings of Error Rate Parity

Of the three main families of statistical criteria in chapter 3, we have discussed independence / demographic parity and sufficiency / calibration, leaving separation / error rate parity. Error rate parity is the hardest criterion to rigorously connect to any moral notion. At the same time, it is undeniable that it taps into some widely held fairness intuition. ProPublica's study of the COMPAS criminal risk prediction system was so powerful because of the finding that Black defendants had "twice the false positive rate" of White defendants [133].

But there is no straightforward justification for this intuition, which has led to error rate parity becoming a topic of fierce debate [165, 166]. Building on this scholarship, we provide our view of why we should care about error rate parity.

We'll assume a prediction-based resource allocation problem such as lending that has a substantial degree of inherent uncertainty with respect to the predictability of outcomes. In contrast, error rate disparity often comes up in perception problems like facial recognition or language detection where there is little or no inherent uncertainty [167, 168]. The crucial normative difference is that in face recognition, language detection, and similar applications there is no notion of a difference in qualification between individuals that could potentially justify dissimilar treatment. Thus, assuming that misclassification imposes a cost on the subject, it is much more straightforward to justify why unequal error rates are problematic.

Another observation to set the stage: the moral significance of error rate is asymmetric. One type of error, roughly speaking, corresponds to unjust denial (of freedom or opportunity) and the other corresponds to overly lenient treatment. In most domains, the first type is much more significant as a normative matter than the second. For example, in the context of bail decisions, it is primarily the disparity in the rates of pretrial detention of nonrecidivists that's worrisome, rather than disparities in the rates of pretrial release of recidivists. While it is true that the release of would-be recidivists has a cost in the form of a threat to public safety, that cost depends on the *total* error and not the distribution of that error between groups. Thus, it is not necessarily meaningful to simply compare error rates between groups.

Error Rate Parity and the Middle View of Equality of Opportunity

Recall that the middle view of equality of opportunity takes into account historical and present social conditions that may affect why people's qualifications may differ. To understand a decision-making system with respect to the middle view, it is critical to know if the effects of the decisions might themselves perpetuate these conditions in society.

Unfortunately, this is hard to do with the data available at the moment of decision making, especially if the features (that encode decision subjects' qualifications) are not available. One thing we can do even without the features is to look at differences in base rates (i.e., rates at which different groups achieve desired outcomes, such as loan repayment or job success). If the base rates are significantly different—and if we assume that individual differences in ability and ambition cancel out at the level of groups—it suggests that people's qualifications may differ due to circumstances beyond the individual.

But base rates alone don't shed light on whether the classifier might perpetuate existing inequalities. For this analysis, what's important is whether the classifier imposes an unequal *burden* on different groups. There are many reasonable ways to measure the burden, but since we consider one type of error—mistakenly classifying someone as undeserving or high-risk—to be especially egregious, we can consider the rate of such misclassification among members of a group as a proxy for the burden placed on that group. This is especially true when we consider the possibility of spillover effects: for example, denying pretrial release has effects on defendants' families and communities.

When a group is burdened by disproportionately high error rates, it suggests that the system may perpetuate cycles of inequality. Indeed, Aziz Huq argues that for this reason, the criminal justice system entrenches racial stratification, and this is the primary racial inequity in algorithmic criminal justice [169]. To be clear, the effect of institutions on communities is an empirical and causal question that cannot be boiled down to error rates, but given the limitations of observational data available in typical decision making scenarios, error rates are a starting point for investigating this question. This yields a distinct reason why error rates carry some moral significance. But note a finding of error rate disparity, by itself, doesn't suggest any particular intervention.

What to Do about Error Rate Disparity

Collecting more data and investing in improving classification technology are ways to potentially mitigate error rate disparity. Normally, we give significant deference to the decision maker on the tradeoff between data collection cost and model accuracy. This deference, especially in private sector applications, is based on the idea that the interests of the decision maker and decision subjects are generally aligned. For example, we defer to lenders on how accurate their predictions should be. If, instead, lenders were required to be highly accurate in their predictions, they might lend only in the safest of cases, depriving many people of the ability to obtain loans, or they might go to great lengths to collect data about borrowers, raising the cost of operating the system and pushing some of that cost to the borrowers.

The argument above considers only total welfare and not how the benefits and costs are distributed among people and groups. When we introduce distributional considerations, there are many scenarios where it is justifiable to lower the deference to decision makers, and the presence of error rate disparity is one such scenario. In this case, requiring the decision maker to mitigate error rate disparity can be seen as asking them to bear some of the cost that's being pushed onto some individuals and groups (figures 4.2, 4.3).

While improving the overall accuracy of the classifier may close the disparity in some cases, in other cases it may leave the disparity unchanged or even worsen it. Accuracy is bounded by that of the optimal classifier, and recall that the optimal

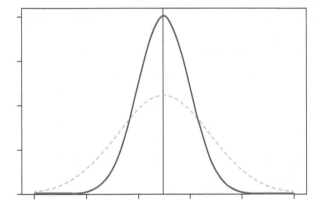

Figure 4.2
Probability density of risk scores for two groups, and a classification threshold. Throughout the illustrations in this section, we assume that the score is perfectly calibrated. The group shown with a solid line has a higher error rate. Intuitively this is because the probability mass is more concentrated (i.e., the score function is worse at distinguishing among members of this group). Collecting more data would potentially bring the solid curve closer to the dashed curve, mitigating the error rate disparity.

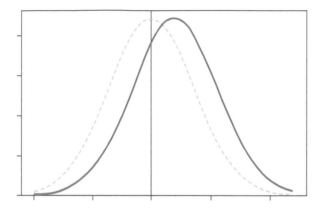

Figure 4.3
Probability density of risk scores for two groups, and a classification threshold. Again the solid group has a higher error rate—specifically, a higher false positive rate, where false positives are people incorrectly classified as high risk. But this time it is because the solid group has a higher base rate (the curve is shifted to the right compared to the dashed group). Collecting more data is unlikely to mitigate the error rate disparity.

classifier doesn't necessarily satisfy error rate parity. As a concrete example, assume that loan defaults primarily arise due to unexpected job loss, one group of loan applicants holds more precarious jobs that are at a greater risk of layoff, and layoffs are not predictable at decision time. In this scenario, improvements in data collection and classification will not yield error rate parity.

Faced with this intrinsic limitation, it may be tempting to perform an adjustment step that achieves error rate parity, such as different risk thresholds for different groups. One way would be to do this without making anyone worse off compared to an unconstrained classifier. For example, a lender could use a more lenient risk threshold for one group to lower its error rate. This would violate the narrow view of equality of opportunity, as people from different groups with the same risk score may be treated differently. Whether the intervention is still justified is a difficult normative question that lacks a uniform answer (figure 4.4).

In other situations, even this might not be possible. For example, the intervention may increase the lender's risk so much that it goes out of business.

In fact, if base rates are so different that we expect large disparities in error rates that cannot be mitigated by interventions like data collection, then it suggests that the use of predictive decision making is itself problematic, and perhaps we should scrap the system or apply more fundamental interventions.

In summary, error rate parity lacks a direct relationship to any single normative principle. But it captures something about both the narrow and the middle views of equality of opportunity. It is also a way to incentivize decision makers to invest in fairness and to question the appropriateness of predictive decision making.

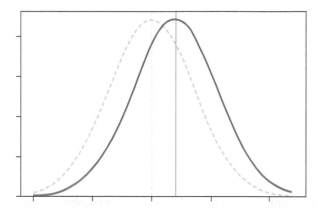

Figure 4.4
Probability density of risk scores for two groups, and two different classification thresholds resulting in equal error rates.

Alternatives for Realizing the Middle View of Equality of Opportunity

We've discussed how error rate parity bears some relationship to the middle view of equality of opportunity. But there are many other possible interventions that decision makers might adopt to try to realize the middle view of equality of opportunity that do not map onto any of the criteria discussed in chapter 3. The middle view is an inherently fuzzy notion, leaving a lot of room to decide the extent to which we want to discount people's apparent differences and the manner in which to do so. Here are a few other ways in which we can try to operationalize it. Unsurprisingly, all of these violate the narrow view of equality of opportunity.

Decision makers could reconsider the goals that they are pursuing such that the decision-making process that seeks to meet these goals generates less disparate outcomes. For example, employers might choose a different target variable that is perceived to be an equally good proxy for their goal, but whose accurate prediction leads to a less significant disparity in outcomes for different groups [170].

They might explore whether it is possible to train alternative models with a similar degree of accuracy as their original model, but which produces smaller disparities in the rates at which members of different groups achieve the desired outcome or are subject to erroneous assessment [171]. Empirically, this appears to be possible in many cases, including for particularly high-stakes decision making [172].

They might sacrifice a good deal of apparent accuracy on the belief that there is serious measurement error and that people from some groups are actually far more qualified than they might appear (we assume that it is not possible

Table 4.2
Views of equality of opportunity and their formal relatives

	Goal	Related formal criteria
Narrow view	Ensure that people who are similarly qualified for an opportunity have similar chances of obtaining it	Similarity criterion, meritocratic fairness, calibration by group
Middle view	Discount differences due to past injustice that accounts for current differences in qualifications	Similarity criterion, Roemer's formal equality of opportunity, error rate parity
Broad view	Ensure people of equal ability and ambition are able to realize their potential equally well	Demographic parity

to explicitly correct the measurement error and that group membership is not sufficiently redundantly encoded in the features, preventing the optimal classifier from automatically accounting for measurement error) [173].

Finally, they could forgo some of the benefits they might have achieved under the original decision-making process so as to provide important benefits to the groups that have been subject to past mistreatment. To do so, they might treat members of certain groups counterfactually, as if they hadn't experienced the injustice that makes them less qualified by the time of decision making.

Summary

Fairness is most often conceptualized as equality of opportunity. But in this chapter, we have seen that there are a variety of ways to understand equality of opportunity. The differences among them are at the heart of why fairness is such a contested topic. All three views can be seen in contemporary political debates. The narrow view aligns with what is often meant by the term "meritocracy." The middle view drives diversity, equity, and inclusion (DEI) efforts at many workplaces. The broad view is too sweeping to find much support for a full-throated implementation, but the ideas behind it come up in debates around topics such as reparations [174].

The views differ along many axes, including what they seek to achieve; how they understand the causes of current differences between groups (and whether they seek to understand them at all); and how to distribute the cost of uplifting historically disadvantaged groups (table 4.2).

In the latter part of the chapter, we attempted to connect these moral notions to the statistical criteria from chapter 3. Loose connections emerged through this exercise, but, ultimately, none of the statistical criteria are strongly anchored in normative foundations.

But even these rough similarities illustrate one important point about the impossibility results from chapter 3. The impossibility results aren't some kind of artifact of statistical decision making; they simply reveal moral dilemmas. Once we recognize the underlying moral difficulties, these mathematical tensions seem much less surprising.

For example, an approach that makes accurate predictions based on people's currently observable attributes, and then makes decisions based on those predictions (calibration) won't result in equality of outcomes (independence) as long as different groups have different qualifications on average. Similarly, its results also differ from an approach that is willing to treat seemingly similar people differently in order to attempt to equalize the burden on different groups (error rate parity). The approaches also differ in the extent to which measurement errors are seen as the responsibility of the decision maker and in who should bear the costs of fairness interventions.

One reason the normative foundations of statistical fairness criteria are shaky is that conditional independence doesn't give us a vocabulary to reason about the causes of disparities between groups or the effects of interventions. We attempt to address these limitations in the next chapter.

5

Causality

Our starting point is the difference between an observation and an action. What we see in passive observation is how individuals follow their routine behavior, habits, and natural inclination. Passive observation reflects the state of the world projected to a set of features we choose to highlight. Data that we collect from passive observation show a snapshot of our world as it is.

There are many questions we can answer from passive observation alone. For example, do sixteen-year-old drivers have a higher rate of traffic accidents than eighteen-year-old drivers? Formally, the answer corresponds to a difference of conditional probabilities assuming we model the population as a distribution as we did in the last chapter. We can calculate the conditional probability of a traffic accident given that the driver's age is sixteen years and subtract from it the conditional probability of a traffic accident given the age is eighteen years. Both conditional probabilities can be estimated from a large enough sample drawn from the distribution, assuming that there are both sixteen-year-old and eighteen-year-old drivers. The answer to the question we asked is solidly in the realm of observational statistics.

But important questions often are not observational in nature. Would traffic fatalities decrease if we raised the legal driving age by two years? Although the question seems similar on the surface, we quickly realize that it asks for a fundamentally different insight. Rather than asking for the frequency of an event in our manifested world, this question asks for the effect of a hypothetical action.

As a result, the answer is not so simple. Even if older drivers have a lower incidence rate of traffic accidents, this might simply be a consequence of additional driving experience. There is no obvious reason why an eighteen-year-old with two months on the road would be any less likely to be involved in an accident than, say, a sixteen-year-old with the same experience. We can try to address this problem by holding the number of months of driving experience fixed, while comparing individuals of different ages. But we quickly run into subtleties. What if eighteen-year-olds with two months of driving experience correspond to individuals who are exceptionally cautious and hence—by their natural inclination—not only drive less but also drive more cautiously? What if such individuals predominantly live in regions where traffic conditions differ significantly from those in areas where people feel a greater need to drive at a younger age?

We can think of numerous other strategies to answer the original question of whether raising the legal driving age reduces traffic accidents. We could compare countries with different legal driving ages, say the United States and Germany. But again, these countries differ in many other possibly relevant ways, such as the legal drinking age.

At the outset, causal reasoning is a conceptual and technical framework for addressing questions about the effect of hypothetical actions or *interventions*. Once we understand what the effect of an action is, we can turn the question around and ask what action plausibly *caused* an event. This gives us a formal language to talk about cause and effect.

Not every question about cause is equally easy to address. Some questions are overly broad, such as, "What is the cause of success?" Other questions are too specific: "What caused your interest in nineteenth-century German philosophy?" Neither question might have a clear answer. Causal inference gives us a formal language to ask these questions, in principle, but it does not make it easy to choose the right questions. Nor does it trivialize the task of finding and interpreting the answer to a question. Especially in the context of fairness, the difficulty is often in deciding what the question is that causal inference is the answer to.

In this chapter, we develop sufficient technical understanding of causality to support at least three different purposes. The first is to conceptualize and address some limitations of the observational techniques we saw in chapter 3. The second is to provide tools that help in the design of interventions that reliably achieve a desired effect. The third is to engage with the important normative debate about when and to which extent reasoning about discrimination and fairness requires causal understanding.

The Limitations of Observation

Before we develop any new formalism, it is important to understand why we need it in the first place. To see why we turn to the venerable example of graduate admissions at the University of California, Berkeley, in 1973 [175]. Historical data show that 12,763 applicants were considered for admission to one of 101 departments and interdepartmental majors. Of the 4,321 women who applied, roughly 35 percent were admitted, while 44 percent of the 8,442 men who applied were admitted. Standard statistical significance tests suggest that the observed difference would be

Table 5.1
UC Berkeley admissions data from 1973

Department	Men Applied	Admitted (%)	Women Applied	Admitted (%)
A	825	62	108	**82**
B	520	60	25	**68**
C	325	**37**	593	34
D	417	33	375	**35**
E	191	**28**	393	24
F	373	6	341	**7**

highly unlikely to be the outcome of sample fluctuation if there were no difference in underlying acceptance rates.

A similar pattern exists if we look at the aggregate admission decisions of the six largest departments. The acceptance rate across all six departments for men is about 44 percent, while it is roughly only 30 percent for women, again a significant difference. Recognizing that departments have autonomy over who to admit, we can look at the gender bias of each department.

We can see from table 5.1 that four of the six largest departments show a higher acceptance ratio for women, while two show a higher acceptance rate for men. However, these two departments cannot account for the large difference in acceptance rates that we observed in aggregate. So, it appears that the higher acceptance rate for men that we observed in aggregate seems to have reversed at the department level.

Such reversals demonstrate what is sometimes called *Simpson's paradox*, even though mathematically they are no surprise. It's a fact of conditional probability that there can be an event Y (here, acceptance), an attribute A (here, female gender taken to be a binary variable), and a random variable Z (here, department choice) such that:

1. $\mathbb{P}\{Y \mid A\} < \mathbb{P}\{Y \mid \neg A\}$
2. $\mathbb{P}\{Y \mid A, Z = z\} > \mathbb{P}\{Y \mid \neg A, Z = z\}$ for all values z that the random variable Z assumes.

Simpson's paradox nonetheless causes discomfort to some, because intuition suggests that a trend that holds for all subpopulations should also hold at the population level.

The reason why Simpson's paradox is relevant to our discussion is that it's a consequence of how we tend to misinterpret what information conditional probabilities encode. Recall that a statement of conditional probability corresponds to passive observation. What we see here is a snapshot of the normal behavior of women and men applying to graduate school at UC Berkeley in 1973.

What is evident from the data is that gender influences department choice. Women and men appear to have different preferences for different fields of study. Moreover, different departments have different admission criteria. Some have lower acceptance rates, some higher. Therefore, one explanation for the data we see is that women *chose* to apply to more competitive departments, and hence were rejected at a higher rate than men.

Indeed, this is the conclusion the original study drew:

> The bias in the aggregated data stems not from any pattern of discrimination on the part of admissions committees, which seems quite fair on the whole, but apparently from prior screening at earlier levels of the educational system. Women are shunted by their socialization and education toward fields of graduate study that are generally more crowded, less productive of completed degrees, and less well funded, and that frequently offer poorer professional employment prospects. [175]

In other words, the article concluded that the source of gender bias in admissions was a *pipeline problem*: Without wrongdoing by the admissions committee, women were "shunted by their socialization," which happened at an earlier stage in their lives.

It is difficult to debate this conclusion on the basis of the available data alone. The question of discrimination, however, is far from resolved. We can ask why women applied to more competitive departments in the first place. There are several possible reasons. Perhaps less competitive departments, such as engineering schools, were unwelcoming of women at the time. This may have been a general pattern at the time or specific to the university. Perhaps some departments had a track record of poor treatment of women that was known to the applicants. Perhaps the department advertised the program in a manner that discouraged women from applying.

The data we have also shows no measurement of *qualification* of an applicant. It's possible that due to self-selection women applying to engineering schools in 1973 were overqualified relative to their peers. In this case, an equal acceptance rate between men and women might actually be a sign of discrimination.

There is no way of knowing what was the case from the data we have. There are multiple possible scenarios with different interpretations and consequences that we cannot distinguish from the data at hand. At this point, we have two choices. One is to design a new study and collect more data in a manner that might lead to a more conclusive outcome. The other is to argue over which scenario is more likely, based on our beliefs and plausible assumptions about the world. Causal inference is helpful in both cases. On the one hand, it can be used as a guide in the design of new studies. It can help us choose which variables to include, which to exclude, and which to hold constant. On the other hand, causal models can serve as a mechanism to incorporate scientific domain knowledge and exchange plausible assumptions for plausible conclusions.

Causal Models

We will develop just enough formal concepts to engage with the technical and normative debate around causality and discrimination. The topic is much deeper than we can explore in this chapter.

We choose *structural causal models* as the basis of our formal discussion as they have the advantage of giving a sound foundation for various causal notions we will encounter. The easiest way to conceptualize a structural causal model is as a program for generating a distribution from independent noise variables through a sequence of formal instructions. Let's unpack this statement. Imagine instead of samples from a distribution, somebody gave you a step-by-step computer program to generate samples on your own starting from a random seed. The process is not unlike how you would write code. You start from a simple random seed and build up increasingly more complex constructs. That is basically what a structural causal model is, except that each assignment uses the language of mathematics rather than any concrete programming syntax.

A First Example

Let's start with a toy example not intended to capture the real world. Imagine a hypothetical population in which an individual exercises regularly with probability $\frac{1}{2}$. With probability $\frac{1}{3}$, the individual has a latent disposition to become overweight that manifests in the absence of regular exercise. Similarly, in the absence of exercise, heart disease occurs with probability $\frac{1}{3}$. Denote by X the indicator variable of regular exercise, by W that of excessive weight, and by H the indicator of heart disease. Below is a structural causal model to generate samples from this hypothetical population. To ease the description, we let $B(p)$ denote a Bernoulli random variable with bias p, that is, a biased coin toss that assumes value 1 with probability p and value 0 with probability $1 - p$.

1. Sample independent Bernoulli random variables $U_1 \sim B(\frac{1}{2}), U_2 \sim B(\frac{1}{3}), U_3 \sim B(\frac{1}{3})$
2. $X := U_1$
3. $W :=$ if $X = 1$ then 0 else U_2
4. $H :=$ if $X = 1$ then 0 else U_3.

Contrast this generative description of the population with a random sample drawn from the population. From the program description, we can immediately see that in our hypothetical population *exercise* averts both *overweight* and *heart disease*, but in the absence of exercise the two are independent. At the outset, our program generates a joint distribution over the random variables (X, W, H). We can calculate probabilities under this distribution. For example, the probability of heart disease under the distribution specified by our model is $\frac{1}{2} \cdot \frac{1}{3} = \frac{1}{6}$. We can also calculate the conditional probability of heart diseases given overweight. From the event $W = 1$ we can infer that the individual does not exercise so that the probability of heart disease given overweight increases to $\frac{1}{3}$ compared with the baseline of $\frac{1}{6}$.

Does this mean that overweight causes heart disease in our model? The answer is *no* as is intuitive given the program to generate the distribution. But let's see how we would go about arguing this point formally. Having a program to generate a distribution is substantially more powerful than just having sampling access. One reason is that we can manipulate the program in whichever way we want, assuming we still end up with a valid program. We could, for example, set $W := 1$, resulting in a new distribution. The resulting program looks like this:

2. $X := U_1$
3. $W := 1$
4. $H := $ if $X = 1$ then 0 else U_3.

This new program specifies a new distribution. We can again calculate the probability of heart disease under this new distribution. We still get $\frac{1}{6}$. This simple calculation reveals a significant insight. The substitution $W := 1$ does not correspond to a conditioning on $W = 1$. One is an action, albeit inconsequential in this case. The other is an observation from which we can draw inferences. If we observe that an individual is overweight, we can infer that they have a higher risk of heart disease (in our toy example). However, this does not mean that lowering body weight would avoid heart disease. It wouldn't in our example. The active substitution $W := 1$ in contrast creates a new hypothetical population in which all individuals are overweight, with all that it entails in our model.

Let us labor this point a bit more by considering another hypothetical population, specified by the equations:

2. $W := U_2$
3. $X := $ if $W = 0$ then 0 else U_1
4. $H := $ if $X = 1$ then 0 else U_3.

In this population exercise habits are driven by body weight. Overweight individuals choose to exercise with some probability, but that's the only reason anyone would exercise. Heart disease develops in the absence of exercise. The substitution $W := 1$ in this model leads to an increased probability of exercise, hence lowering the probability of heart disease. In this case, the conditioning on $W = 1$ has the same affect. Both lead to a probability of $\frac{1}{6}$.

What we see is that fixing a variable by substitution may or may not correspond to a conditional probability. This is a formal rendering of our earlier point that observation isn't action. A substitution corresponds to an action we perform. By substituting a value we break the natural course of action our model captures. This is the reason why the substitution operation is sometimes called the *do-operator*, written as $\mathrm{do}(W := 1)$.

Structural causal models give us a formal calculus to reason about the effect of hypothetical actions. We will see how this creates a formal basis for all the different causal notions that we will encounter in this chapter.

Structural Causal Models, More Formally

Formally, a structural causal model is a sequence of assignments for generating a joint distribution starting from independent noise variables. By executing the sequence of assignments we incrementally build a set of jointly distributed random variables. A structural causal model therefore not only provides a joint distribution, but also a description of how the joint distribution can be generated from elementary noise variables. The formal definition is a bit cumbersome compared with the intuitive notion.

Definition 4 *A structural causal model M is given by a set of variables $X_1, \ldots, X_d$ and corresponding assignments of the form*

$$X_i := f_i(P_i, U_i), \qquad i = 1, \ldots, d.$$

Here, $P_i \subseteq \{X_1, \ldots, X_d\}$ is a subset of the variables that we call the parents *of X_i. The random variables $U_1, \ldots, U_d$ are called* noise variables, *which we require to be jointly independent. The* causal graph *corresponding to the structural causal model is the directed graph that has one node for each variable X_i with incoming edges from all the parents P_i.*

Let's walk through the formal concepts introduced in this definition in a bit more detail. The noise variables that appear in the definition model *exogenous factors* that influence the system. Consider, for example, how the weather influences the delay on a traffic route you choose. Due to the difficulty of modeling the influence of weather more precisely, we could take the weather-induced delay to be an exogenous factor that enters the model as a noise variable. The choice of exogenous variables and their distribution can have important consequences for what conclusions we draw from a model.

The parent nodes P_i of node i in a structural causal models are often called the *direct causes* of X_i. Similarly, we call X_i the direct effect of its direct causes P_i. Recall our hypothetical population in which weight gain was determined by lack of exercise via the assignment $W := \min\{U_1, 1 - X\}$. Here we would say that exercise (or lack thereof) is a direct cause of weight gain.

Structural causal model are a collection of formal *assumptions* about how certain variables interact. Each assignment specifies a *response function*. We can think of nodes as receiving messages from their parents and acting according to these messages as well as to the influence of an exogenous noise variable.

To what extent a structural causal model conforms to reality is a separate and difficult question that we will return to in more detail later. For now, think of a structural causal model as formalizing and exposing a set of assumptions about a data-generating process. As such, different models can expose different hypothetical scenarios and serve as a basis for discussion. When we make statements about cause and effect in reference to a model, we don't mean to suggest that these relationships necessarily hold in the real world. Whether they do depends on the scope, purpose, and validity of our model, which may be difficult to substantiate.

It's not hard to show that a structural causal model defines a unique joint distribution over the variables $(X_1, \ldots, X_d)$ such that $X_i = f_i(P_i, U_i)$. It's convenient to introduce a notion for probabilities under this distribution. When M denotes a structural causal model, we write the probability of an event E under the entailed joint distribution as $\mathbb{P}_M\{E\}$. To gain familiarity with the notation, let M denote the structural causal model for the hypothetical population in which both weight gain and heart disease are directly caused by an absence of exercise. We calculated earlier that the probability of heart disease in this model is $\mathbb{P}_M\{H\} = \frac{1}{6}$.

In what follows we will derive from this single definition of a structural causal model all the different notions and terminology that we need in this chapter. Throughout, we restrict our attention to acyclic assignments. Many real-world systems are naturally described as stateful dynamical systems with closed feedback loops. There are some ways of dealing with such closed loop systems. For example, often cycles can be broken up by introducing time dependent variables, such as, investments at time 0 grow the economy at time 1 which in turn grows investments at time 2, continuing so forth until some chosen time horizon t. This processing is called *unrolling* a dynamical system.

Causal Graphs

We saw how structural causal models naturally give rise to *causal graphs* that represent the assignment structure of the model graphically. We can go the other way as well by simply looking at directed graphs as placeholders for an unspecified structural causal model which has the assignment structure given by the graph. Causal graphs are often called *causal diagrams*. We'll use these terms interchangeably.

The causal graphs for the two hypothetical populations from our heart disease example each have two edges and the same three nodes (figure 5.1). They agree on the link between exercise and heart disease but they differ in the direction of the link between exercise and weight gain.

Causal graphs are convenient when the exact assignments in a structural causal model are of secondary importance, but what matters are the paths present and absent in the graph. Graphs also let us import the established language of graph theory to discuss causal notions. We can say, for example, that an *indirect cause* of a node is any ancestor of the node in a given causal graph. In particular, causal graphs allow us to distinguish cause and effect based on whether a node is an ancestor or descendant of another node.

Let's take a first glimpse at a few important graph structures.

Forks

A *fork* is a node Z in a graph that has outgoing edges to two other variables X and Y. Put differently, the node Z is a common cause of X and Y. We already saw an example of a fork in our weight and exercise example: $W \leftarrow X \rightarrow H$. Here, exercise X influences both weight and heart disease. We also learned from the example that Z has a *confounding* effect: ignoring exercise X, we saw that W and H appear to be positively correlated. However, the correlation is a mere result of confounding. Once we hold exercise levels constant (via the do-operation), weight has no effect on heart disease in our example (figure 5.2).

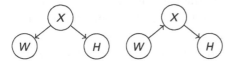

Figure 5.1
Causal diagrams for the heart disease examples.

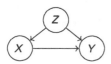

Figure 5.2
Example of a fork (confounder).

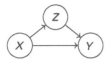

Figure 5.3
Example of a chain (mediator).

Confounding leads to a disagreement between the calculus of conditional probabilities (observation) and do-interventions (actions). Real-world examples of confounding are a common threat to the validity of conclusions drawn from data. For example, in a well-known medical study a suspected beneficial effect of *hormone replacement therapy* in reducing cardiovascular disease disappeared after identifying *socioeconomic status* as a confounding variable [176].

Mediators
The case of a fork is quite different from the situation where Z lies on a directed path from X to Y. In this case, the path $X \to Z \to Y$ contributes to the total effect of X on Y. It's a causal path and thus one of the ways in which X causally influences Y. That's why Z is not a confounder. We call Z a *mediator* instead (figure 5.3).

We saw a plausible example of a mediator in our UC Berkeley admissions example. In one plausible causal graph, department choice mediates the influences of gender on the admissions decision. The notion of a mediator is particularly relevant to the topic of discrimination analysis, since mediators can be interpreted as the mechanism behind a causal link.

Colliders
Finally, let's consider another common situation: the case of a *collider*. Colliders aren't confounders. In fact, in figure 5.4, X and Y are unconfounded, meaning that we can replace do-statements by conditional probabilities. However, something interesting happens when we condition on a collider. The conditioning step can create correlation between X and Y, a phenomenon called *explaining away*. A good example of the explaining away effect, or *collider bias*, is due to Berkson. Two independent diseases can become negatively correlated when we analyze hospitalized patients. The reason is that when either disease (X or Y) is sufficient for admission to the hospital (indicated by variable Z), observing that a patient has one disease makes the other statistically less likely [177].

Berkson's law is a cautionary tale for statistical analysis when we're studying a cohort that has been subjected to a selection rule. For example, there's an ongoing debate about the effectiveness of GRE scores in higher education. Some studies [178, 179] argue that GRE scores are not predictive of various success outcomes in a

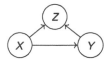

Figure 5.4
Example of a collider.

graduate student population. However, care must be taken when studying the effectiveness of educational tests, such as the GRE, by examining a sample of admitted students. After all, students were in part admitted on the basis of the test score. It's the selection rule that introduces the potential for collider bias.

Interventions and Causal Effects

Structural causal models give us a way to formalize the effect of hypothetical actions or interventions on the population within the assumptions of our model. As we saw earlier, all we needed was the ability to do substitutions.

Substitutions and the Do-Operator

Given a structural causal model M we can take any assignment of the form

$$X := f(P, U)$$

and replace it by another assignment. The most common substitution is to assign X a constant value x:

$$X := x.$$

We denote the resulting model by $M' = M[X := x]$ to indicate the surgery we performed on the original model M. Under this assignment we hold X constant by removing the influence of its parent nodes and thereby any other variables in the model.

Graphically, the operation corresponds to eliminating all incoming edges to the node X. The children of X in figure 5.5 now receive a fixed message x from X when they query the node's value. The assignment operator is also called the *do-operator* to emphasize that it corresponds to performing an action or intervention. We already have notation to compute probabilities after applying the do-operator, namely, $\mathbb{P}_{M[X:=x]}(E)$. Another notation is popular and common:

$$\mathbb{P}\{E \mid \mathrm{do}(X := x)\} = \mathbb{P}_{M[X:=x]}(E).$$

This notation analogizes the do-operation with the usual notation for conditional probabilities, and is often convenient when doing calculations involving the

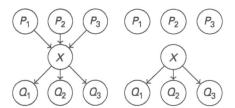

Figure 5.5
Graph before and after substitution.

do-operator. Keep in mind, however, that the do-operator (action) is fundamentally different from the conditioning operator (observation).

Causal Effects

The *causal effect* of an action $X := x$ on a variable Y refers to the distribution of the variable Y in the model $M[X := x]$. When we speak of the causal effect of a variable X on another variable Y we refer to all the ways in which setting X to any possible value x affects the distribution of Y.

Often we think of X as a binary treatment variable and are interested in a quantity such as

$$\mathbb{E}_{M[X:=1]}[Y] - \mathbb{E}_{M[X:=0]}[Y].$$

This quantity is called the *average treatment effect*. It tells us how much treatment (action $X := 1$) increases the expectation of Y relative to no treatment (action $X := 0$). Causal effects are population quantities. They refer to effects averaged over the whole population. Often the effect of treatment varies greatly from one individual or group of individuals to another. Such treatment effects are called *heterogeneous*.

Confounding

Important questions in causality relate to when we can rewrite a do-operation in terms of conditional probabilities. When this is possible, we can estimate the effect of the do-operation from conventional conditional probabilities that we can estimate from data.

The simplest question of this kind asks when a causal effect $\mathbb{P}\{Y = y \mid \text{do} (X := x)\}$ coincides with the condition probability $\mathbb{P}\{Y = y \mid X = x\}$. In general, this is not true. After all, the difference between observation (conditional probability) and action (interventional calculus) is what motivated the development of causality.

The disagreement between interventional statements and conditional statements is so important that it has a well-known name: *confounding*. We say that X and Y are confounded when the causal effect of action $X := x$ on Y does not coincide with the corresponding conditional probability.

When X and Y are confounded, we can ask if there is some combination of conditional probability statements that gives us the desired effect of a do-intervention. This is generally possible given a causal graph by conditioning on the parent nodes PA of the node X:

$$\mathbb{P}\{Y = y \mid \mathrm{do}(X := x)\} = \sum_z \mathbb{P}\{Y = y \mid X = x, PA = z\}\mathbb{P}\{PA = z\}.$$

This formula is called the *adjustment formula*. It gives us one way of estimating the effect of a do-intervention in terms of conditional probabilities.

The adjustment formula is one example of what is often called *controlling for* a set of variables: We estimate the effect of X on Y separately in every slice of the population defined by a condition $Z = z$ for every possible value of z. We then average these estimated subpopulation effects weighted by the probability of $Z = z$ in the population. To give an example, when we control for age, we mean that we estimate an effect separately in each possible age group and then average out the results so that each age group is weighted by the fraction of the population that falls into the age group.

Controlling for more variables in a study isn't always the right choice. It depends on the graph structure. Let's consider what happens when we control for the variable Z in the three causal graphs we discussed above.

- Controlling for a confounding variable Z in a fork $X \leftarrow Z \rightarrow Y$ will deconfound the effect of X on Y.

- Controlling for a mediator Z on a chain $X \rightarrow Z \rightarrow Y$ will eliminate some of the causal influence of X on Y.

- Controlling for a collider will create correlation between X and Y. That is the opposite of what controlling for Z accomplishes in the case of a fork. The same is true if we control for a descendant of a collider.

The Backdoor Criterion

At this point, we might worry that things get increasingly complicated. As we introduce more nodes in our graph, we might fear a combinatorial explosion of possible scenarios to discuss. Fortunately, there are simple sufficient criteria for choosing a set of deconfounding variables that is safe to control for.

A well known graph-theoretic notion is the *backdoor* criterion [180]. Two variables are confounded if there is a so-called *backdoor* path between them.

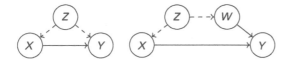

Figure 5.6
Two cases of unobserved confounding.

A *backdoor path* from X to Y is any path starting at X with a backward edge "←" into X such as:

$$X \leftarrow A \rightarrow B \leftarrow C \rightarrow Y.$$

Intuitively, backdoor paths allow information flow from X to Y in a way that is not causal. To deconfound a pair of variables we need to select a *backdoor set* of variables that "blocks" all backdoor paths between the two nodes. A backdoor path involving a chain $A \rightarrow B \rightarrow C$ can be blocked by controlling for B. Information by default cannot flow through a collider $A \rightarrow B \leftarrow C$. So we only have to be careful not to open information flow through a collider by conditioning on the collider, or the descendant of a collider.

Unobserved Confounding
The adjustment formula might suggest that we can always eliminate confounding bias by conditioning on the parent nodes. However, this is true only in the absence of *unobserved confounding*. In practice often there are variables that are hard to measure or that were simply left unrecorded. We can still include such unobserved nodes in a graph, typically denoting their influence with dashed lines, instead of solid lines.

Figure 5.6 shows two cases of unobserved confounding. In the first example, the causal effect of X on Y is unidentifiable. In the second case, we can block the confounding backdoor path $X \leftarrow Z \rightarrow W \rightarrow Y$ by controlling for W even though Z is not observed. The backdoor criterion lets us work round unobserved confounders in some cases where the adjustment formula alone wouldn't suffice.

Unobserved confounding nonetheless remains a major obstacle in practice. The issue is not just lack of measurement, but often lack of anticipation or awareness of a counfounding variable. We can try to combat unobserved confounding by increasing the number of variables under consideration. But as we introduce more variables into our study, we also increase the burden of coming up with a valid causal model for all variables under consideration. In practice, it is not uncommon to control for as many variables as possible in the hope of disabling confounding bias. However, as we saw, controlling for mediators or colliders can be harmful.

Randomization

The backdoor criterion gives a nonexperimental way of eliminating confounding bias given a causal model and a sufficient amount of observational data from the joint distribution of the variables. An alternative experimental method of eliminating confounding bias is the well-known *randomized controlled trial*.

In a *randomized controlled trial*, a group of subjects is randomly partitioned into a *control group* and a *treatment group*. Participants do not know which group they were assigned to and neither does the staff administering the trial. The treatment group receives an actual treatment, such as a drug that is being tested for efficacy, while the control group receives a placebo identical in appearance. An outcome variable is measured for all subjects.

The goal of a randomized controlled trial is to break natural inclination. Rather than observing who chose to be treated on their own, we assign treatment randomly. Thinking in terms of causal models, what this means is that we eliminate all incoming edges into the treatment variable. In particular, this closes all backdoor paths and hence avoids confounding bias.

There are many reasons why often randomized controlled trials are difficult or impossible to administer. Treatment might be physically or legally impossible, too costly, or too dangerous. As we saw, randomized controlled trials are not always necessary for avoiding confounding bias and for reasoning about cause and effect. Nor are they free of issues and pitfalls [181].

Graphical Discrimination Analysis

We now explore how we can bring causal graphs to bear on discussions of discrimination. We return to the example of graduate admissions at Berkeley and develop a causal perspective on the earlier analysis.

The first step is to come up with a plausible causal graph consistent with the data that we saw earlier. The data contained only three variables, sex A, department choice Z, and admission decision Y. It makes sense to draw two arrows $A \to Y$ and $Z \to Y$, because both features A and Z are available to the institution when making the admissions decision. We'll draw one more arrow, for now, simply because we have to. If we only included the two arrows $A \to Y$ and $Z \to Y$, our graph would claim that A and Z are statistically independent. However, this claim is inconsistent with the data. We can see from the table that several departments have a statistically significant gender bias among applicants. This means we need to include either the arrow $A \to Z$ or $Z \to A$. Deciding between the two isn't as straightforward as it might first appear.

If we interpreted A in the narrowest possible sense as the applicant's *reported sex*, that is, literally which box they checked on the application form, we could imagine

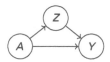

Figure 5.7
Possible causal graph for the UC Berkeley graduate admissions scenario.

a scenario where some applicants choose to (mis)report their sex in a certain way
that depends in part on their department choice. Even if we assume no misreporting
occurs, it's hard to substantiate *reported sex* as a plausible cause of department
choice. The fact that an applicant checked a box labeled *male* certainly isn't the
cause for their interest in engineering.

The proposed causal story in the study is a different one. It alludes to a socialization and preference formation process that took place in the applicant's life before
they applied. It is this process that, at least in part, depended on the applicant's
sex. To align this story with our causal graph, we need the variable A to reference
whatever ontological entity it is that through this "socialization process" influences
intellectual and professional preferences and hence, department choice. It is difficult to maintain that this ontological entity coincides with sex as a biological
trait. There is no scientific basis to support the idea that the biological trait *sex* is
what determines our intellectual preferences. Few scholars (if any) would currently
attempt to maintain a claim such as *two X chromosomes cause an interest in English
literature.*

The truth is that we don't know the exact mechanism by which the thing referenced by A influences department choice. In drawing the arrow A to Z we
assert—perhaps with some naivety or ignorance—that there exists such a mechanism. We will discuss the important difficulty we encountered here in depth later
on. For now, we commit to this modeling choice and thus arrive at figure 5.7.

In this figure, department choice mediates the influence of gender on admissions.
There's a direct path from A to Y and an indirect path that goes through Z. We
will use this model to put pressure on the claim that *there is no evidence of sex
discrimination*. In causal language, the argument had two components:

1. There appears to be no direct effect of sex A on the admissions decision Y that
   favors men.

2. The indirect effect of A on Y that is mediated by department choice should not
   be counted as evidence of discrimination.

We will discuss both arguments in turn.

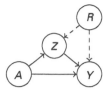

Figure 5.8
Alternative causal graph for the UC Berkeley graduate admissions scenario showing influence of residence.

Direct Effects

To obtain the direct effect of A on Y we need to disable all paths between A and Y except for the direct link. In our model, we can accomplish this by holding department choice Z constant and evaluating the conditional distribution of Y given A (figure 5.8). Recall that holding a variable constant is generally not the same as conditioning on the variable. Specifically, a problem would arise if department choice and admissions outcome were confounded by another variable, such as, state of residence R.

Department choice is now a collider between A and R. Conditioning on a collider opens the backdoor path $A \to Z \leftarrow R \to Y$. In this figure, conditioning on department choice does *not* give us the desired direct effect. The real possibility that state of residence confounds department choice and decision was the subject of an exchange between Judea Pearl and Dana Mackenzie [182].

If we assume, however, that department choice and admissions decisions are unconfounded, then the approach Peter Bickel, Eugene Hammel, and William O'Connell took indeed supports the first claim. Unfortunately, the direct effect of a protected variable on a decision is a poor measure of discrimination on its own. At a technical level, it is rather brittle as it cannot detect any form of *proxy discrimination*. The department could, for example, use the applicant's personal statement to make inferences about their gender, which are then used to discriminate.

We can think of the direct effect as corresponding to the explicit *use* of the attribute in the decision rule. The absence of a direct effect loosely corresponds to the somewhat troubled notion of a *blind* decision rule that doesn't have explicit access to the sensitive attribute. As we argued in preceding chapters, blind decision rules can still be the basis of discriminatory practices.

Indirect Paths

Let's turn to the indirect effect of sex on admission that goes through department choice. It's tempting to think of the the node Z as referencing the applicant's inherent department preferences. In this view, the department is not responsible

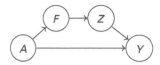

Figure 5.9
Alternative causal graph for the UC Berkeley graduate admissions scenario where department preferences are shaped by fear of discrimination.

for the applicant's preferences. Therefore the mediating influence of department preferences is not interpreted as a sign of discrimination. This, however, is a substantive judgment that may not be a fact. There are other plausible scenarios, consistent with both the data and our causal model, in which the indirect path encodes a pattern of discrimination.

For example, the admissions committee may have advertised the program in a manner that strongly discouraged women from applying. In this case, department preference in part measures exposure to this hostile advertising campaign. Alternatively, the department could have a track record of hostile behavior against women and it is awareness of such that shapes preferences in an applicant (figure 5.9). Finally, even blatant discriminatory practices, such as compensating women at a lower rate than equally qualified male graduate students, correspond to an indirect effect mediated by department choice.

Accepting the indirect path as *nondiscriminatory* is to assert that all these scenarios we described are deemed implausible. Fundamentally, we are confronted with a substantive question. The path $A \to Z \to Y$ could be either where discrimination occurs or what explains the absence thereof. Which case we're in isn't a purely technical matter and cannot be resolved without subject matter knowledge. Causal modeling gives us a framework for exposing these questions, but not necessarily one for resolving them.

Path Inspection

To summarize, discrimination may not occur only on the direct pathway from the sensitive category to the outcome. Seemingly innocuous mediating paths can hide discriminatory practices. We have to carefully discuss what pathways we consider evidence for or against discrimination.

To appreciate this point, contrast our Berkeley scenario with the important legal case *Griggs v. Duke Power Co.*, which was argued before the US Supreme Court in 1970. Duke Power Company had introduced the requirement of a high school diploma for certain higher-paying jobs. We could draw a causal graph for this scenario not unlike the one for the Berkeley case. There is a mediating variable (here, level of education), a sensitive category (here, race), and an employment outcome (here, employment in a higher-paying job). The company didn't directly make employment decisions based on race, but rather used the mediating variable. The court ruled that the requirement of a high school diploma was not justified by business necessity, but rather had adverse impact on ethnic minority groups where the prevalence of high school diplomas is lower. Put differently, the court decided that the use of this mediating variable was not an argument against, but rather for discrimination.

Maria Glymour [183] makes another related and important point about the moral character of mediation analysis:

> Implicitly, the question of what mediates observed social effects informs our view of which types of inequalities are socially acceptable and which types require remediation by social policies. For example, a conclusion that women are "biologically programmed" to be depressed more than men may ameliorate the social obligation to try to reduce gender inequalities in depression. Yet if people get depressed whenever they are, say, sexually harassed—and women are more frequently sexually harassed than men—this suggests a very strong social obligation to reduce the depression disparity by reducing the sexual harassment disparity.

Ending on a technical note, we currently do not have a method to estimate indirect effects. Estimating an indirect effect somehow requires us to *disable* the direct influence. There is no way of doing this with the do-operation that we've seen so far. However, we will shortly introduce *counterfactuals*, which among other applications will give us a way of estimating path-specific effects.

Structural Discrimination

There's an additional problem we have neglected so far. Imagine a spiteful university administration that systematically defunds graduate programs that attract more female applicants. This structural pattern of discrimination is invisible from the causal model we drew. There is a kind of type mismatch here. Our model talks about individual applicants, their department preferences, and their outcomes. Put differently, individuals are the *units* of our investigation. University policy is not one of the mechanisms that our model exposes. We cannot *featurize* university policy to make it an attribute of the individual. As a result we cannot talk about university policy as a cause of discrimination in our model.

The model we chose commits us to an individualistic perspective that frames discrimination as the consequence of how decision makers respond to information about individuals. An analogy is helpful. In epidemiology, scientists can seek the cause of health outcomes in biomedical aspects and lifestyle choices of individuals, such as whether or not an individual smokes, exercises, maintains a balanced diet, and so on. The growing field of social epidemiology criticizes the view of individual choices as causes of health outcomes, and instead draws attention to social and structural causes [184], such as poverty and inequality.

Similarly, we can contrast the individualistic perspective on discrimination with structural discrimination. Causal modeling can in principle be used to study the causes of structural discrimination, as well, but it requires a different perspective than the one we chose for our Berkeley scenario.

Counterfactuals

Fully specified structural causal models allow us to ask causal questions that are more delicate than the mere effect of an action. Specifically, we can ask *counterfactual* questions such as: Would I have avoided the traffic jam had I taken a different route this morning? Counterfactual questions are common and relevant for questions of discrimination. We can compute the answer to counterfactual questions given a structural causal model. The procedure for extracting the answer from the model looks a bit subtle at first. We'll walk through the formal details starting from a simple example before returning to our discussion of discrimination.

A Simple Counterfactual

To understand counterfactuals, we first need to convince ourselves that they aren't quite as straightforward as a single substitution in our model.

Assume every morning we need to decide between two routes $X = 0$ and $X = 1$. On bad traffic days, indicated by $U = 1$, both routes are bad. On good days, indicated by $U = 0$, the traffic on either route is good unless there was an accident on the route. Let's say that $U \sim B(\frac{1}{2})$ follows the distribution of an unbiased coin toss. Accidents occur independently on either route with probability $\frac{1}{2}$. So, choose two Bernoulli random variables $U_0, U_1 \sim B(\frac{1}{2})$ that tell us if there is an accident on route 0 and route 1, respectively. We reject all external route guidance and instead decide on which route to take uniformly at random. That is, $X := U_X \sim B(\frac{1}{2})$ is also an unbiased coin toss.

Introduce a variable $Y \in \{0, 1\}$ that tells us whether the traffic on the chosen route is good ($Y = 0$) or bad ($Y = 1$). Reflecting our discussion above, we can express Y as

$$Y := X \cdot \max\{U, U_1\} + (1 - X) \max\{U, U_0\}.$$

In other words, when $X = 0$ the first term disappears and so traffic is determined by the larger of the two values U and U_0. Similarly, when $X = 1$ traffic is determined by the larger of U and U_1 (figure 5.10).

Now, suppose one morning we have $X = 1$ and we observe bad traffic $Y = 1$. Would we have been better off taking the alternative route this morning?

Figure 5.10
Causal diagram for our traffic scenario.

Table 5.2
Possible noise settings
after observing evidence

U	U_1
0	1
1	1
1	0

A natural attempt to answer this question is to compute the likelihood of $Y = 0$ after the do-operation $X := 0$, that is, $\mathbb{P}_{M[X:=0]}(Y = 0)$. A quick calculation reveals that this probability is $\frac{1}{2} \cdot \frac{1}{2} = \frac{1}{4}$. Indeed, given the substitution $X := 0$ in our model, for the traffic to be good we need that $\max\{U, U_0\} = 0$. This can happen only when both $U = 0$ (probability $\frac{1}{2}$) and $U_0 = 0$ (probability $\frac{1}{2}$).

But this isn't the correct answer to our question. The reason is that we took route $X = 1$ and observed that $Y = 1$. From this observation, we can deduce that certain background conditions did not manifest for they are inconsistent with the observed outcome. Formally, this means that certain settings of the noise variables (U, U_0, U_1) are no longer feasible given the observed event $\{Y = 1, X = 1\}$. Specifically, if U and U_1 had both been zero, we would have seen no bad traffic on route $X = 1$, but this is contrary to our observation. In fact, the available evidence $\{Y = 1, X = 1\}$ leaves only the settings for U and U_1 shown in table 5.2.

We leave out U_0 from the table, since its distribution is unaffected by our observation. Each of the remaining three cases is equally likely, which in particular means that the event $U = 1$ now has probability $\frac{2}{3}$. In the absence of any additional evidence, recall, $U = 1$ had probability $\frac{1}{2}$. What this means is that the observed evidence $\{Y = 1, X = 1\}$ has biased the distribution of the noise variable U toward 1. Let's use the letter U' to refer to this biased version of U. Formally, U' is distributed according to the distribution of U conditional on the event $\{Y = 1, X = 1\}$.

Working with this biased noise variable, we can again entertain the effect of the action $X := 0$ on the outcome Y. For $Y = 0$, we need that $\max\{U', U_0\} = 0$. This means that $U' = 0$, an event that now has probability $\frac{1}{3}$, and $U_0 = 0$ (probability $\frac{1}{2}$ as before). Hence, we get the probability $\frac{1}{6} = \frac{1}{2} \cdot \frac{1}{3}$ for the event that $Y = 0$ under our do-operation $X := 0$, and after updating the noise variables to account for the observation $\{Y = 1, X = 1\}$.

To summarize, incorporating available evidence into our calculation decreased the probability of no traffic ($Y = 0$) when choosing route 0 from $\frac{1}{4}$ to $\frac{1}{6}$. The intuitive reason is that the evidence made it more likely that it was generally a bad traffic day, and even the alternative route would have been clogged. More formally, the event that we observed biases the distribution of exogenous noise variables.

We think of the result we just calculated as the *counterfactual* of choosing the alternative route given that the route we chose had bad traffic.

The General Recipe

We can generalize our discussion of computing counterfactuals from the previous example to a general procedure. There were three essential steps. First, we incorporated available observational evidence by biasing the exogenous noise variables through a conditioning operation. Second, we performed a do-operation in the structural causal model after we substituted the biased noise variables. Third, we computed the distribution of a target variable. These three steps are typically called *abduction*, *action*, and *prediction*, and can be described as follows.

Definition 5   *Given a structural causal model M, an observed event E, an action $X := x$, and target variable Y, we define the* counterfactual $Y_{X:=x}(E)$ *by the following three-step procedure:*

1. ***Abduction:*** *Adjust noise variables to be consistent with the observed event. Formally, condition the joint distribution of $U = (U_1, \ldots, U_d)$ on the event E. This results in a biased distribution U'.*

2. ***Action:*** *Perform do-intervention $X := x$ in the structural causal model M resulting in the model $M' = M[X := x]$.*

3. ***Prediction:*** *Compute target counterfactual $Y_{X:=x}(E)$ by using U' as the random seed in M'.*

It's important to realize that this procedure *defines* what a counterfactual is in a structural causal model. The notation $Y_{X:=x}(E)$ denotes the outcome of the procedure and is part of the definition. We haven't encountered this notation before. Put in words, we interpret the formal counterfactual $Y_{X:=x}(E)$ as the value Y would have taken had the variable X been set to value x in the circumstances described by the event E.

In general, the counterfactual $Y_{X:=x}(E)$ is a random variable that varies with U'. But counterfactuals can also be deterministic. When the event E narrows down the distribution of U to a single point mass, called a *unit*, the variable U' is constant and hence the counterfactual $Y_{X:=x}(E)$ reduces to a single number. In this case, it's common to use the shorthand notation $Y_x(u) = Y_{X:=x}(\{U = u\})$, where we make the variable X implicit and let u refer to a single unit.

The motivation for the name *unit* derives from the common situation where the structural causal model describes a population of entities that form the atomic units of our study. It's common for a unit to be an individual (or the description of a single individual). However, depending on the application, the choice of units can vary. In our traffic example, the noise variables dictate which route we take and what the road conditions are.

Answers to counterfactual questions strongly depend on the specifics of the structural causal model, including the precise model of how the exogenous noise variables come into play. It's possible to construct two models that have identical graph structures and behave identically under interventions, yet give different answers to counterfactual queries [185].

Potential Outcomes

The *potential outcomes* framework is a popular formal basis for causal inference, which goes about counterfactuals differently. Rather than deriving them from a structural causal model, we assume their existence as ordinary random variables, albeit some unobserved.

Specifically, we assume that for every unit u there exist random variables $Y_x(u)$ for every possible value of the assignment x. In the potential outcomes model, it's customary to think of a binary *treatment* variable X so that x assumes only two values, 0 for *untreated* and 1 for *treated*. This gives us two potential outcome variables, $Y_0(u)$ and $Y_1(u)$, for each unit u. There is some potential for notational confusion here. Readers familiar with the potential outcomes model may be used to the notation "$Y_i(0), Y_i(1)$" for the two potential outcomes corresponding to unit i. In our notation, the unit (or, more generally, set of units) appears in the parentheses and the subscript denotes the substituted value for the variable we intervene on.

The key point about the potential outcomes model is that we observe the potential outcome $Y_1(u)$ only for units that were treated. For untreated units we observe $Y_0(u)$. In other words, we can never simultaneously observe both, although they're both assumed to exist in a formal sense. Formally, the outcome $Y(u)$ for unit u that we observe depends on the binary treatment $T(u)$ and is given by the expression

$$Y(u) = Y_0(u) \cdot (1 - T(u)) + Y_1(u) \cdot T(u).$$

It's often convenient to omit the parentheses from our notation for counterfactuals, so that this expression would read $Y = Y_0 \cdot (1 - T) + Y_1 \cdot T$.

We can revisit our traffic example in this framework. Table 5.3 summarizes what information is observable in the potential outcomes model. We think of the route we choose as the treatment variable and the observed traffic as reflecting one of the two potential outcomes.

Often this information comes in the form of samples. For example, we might observe the traffic on different days. With sufficiently many samples, we can estimate the above frequencies with arbitrary accuracy (table 5.4).

Table 5.3
Traffic example in the potential outcomes model

Route X	Outcome Y_0	Outcome Y_1	Probability
0	0	?	1/8
0	1	?	3/8
1	?	0	1/8
1	?	1	3/8

Table 5.4
Traffic data in the potential outcomes model

Day	Route X	Outcome Y_0	Outcome Y_1
1	0	1	?
2	0	0	?
3	1	?	1
4	0	1	?
5	1	?	0
...	...	...	...

A typical query in the potential outcomes model is the *average treatment effect* $\mathbb{E}[Y_1 - Y_0]$. Here, the expectation is taken over the properly weighted units in our study. If units correspond to equally weighted individuals, the expectation is an average over these individuals.

In our original traffic example, there were *sixteen* units corresponding to the background conditions given by the four binary variables U, U_0, U_1, U_X. When the units in the potential outcome model agree with those of a structural causal model, then causal effects computed in the potential outcomes model agree with those computed in the structural equation model. The two formal frameworks are perfectly consistent with each other.

As is intuitive from table 5.4, causal inference in the potential outcomes framework can be thought of as filling in the missing entries ("?"). This is sometimes called *missing data imputation* and there are numerous statistical methods for this task. If we could *reveal* what's behind the question marks, estimating the average treatment effect would be as easy as counting rows.

There is a set of established conditions under which causal inference becomes possible:

1. **Stable Unit Treatment Value Assumption** (SUTVA): The treatment that one unit receives does not change the effect of treatment for any other unit.

2. **Consistency**: Formally, $Y = Y_0(1 - T) + Y_1 T$. That is, $Y = Y_0$ if $T = 0$ and $Y = Y_1$ if $T = 1$. In other words, the outcome Y agrees with the potential outcome corresponding to the treatment indicator.

3. **Ignorability**: The potential outcomes are independent of treatment given some deconfounding variables Z, that is, $T \perp (Y_0, Y_1) \mid Z$. In other words, the potential outcomes are conditionally independent of treatment given some set of deconfounding variables.

The first two assumptions automatically hold for counterfactual variables derived from structural causal models according to the procedure described above. This assumes that the units in the potential outcomes framework correspond to the atomic values of the background variables in the structural causal model.

The third assumption is a major one. It's easiest to think of it as aiming to formalize the guarantees of a perfectly executed randomized controlled trial. The assumption on its own cannot be verified or falsified, since we never have access to samples with both potential outcomes manifested. However, we can verify if the assumption is consistent with a given structural causal model by checking if the set Z blocks all backdoor paths from treatment T to outcome Y.

There's no tension between structural causal models and potential outcomes and there's no harm in having familiarity with both. It nonetheless makes sense to say a few words about the differences of the two approaches.

We can derive potential outcomes from a structural causal model as we did above, but we cannot derive a structural causal model from potential outcomes alone. A structural causal model in general encodes more assumptions about the relationships of the variables. This has several consequences. On one hand, a structural causal model gives us a broader set of formal concepts (causal graphs, mediating paths, counterfactuals for every variable, and so on). On the other hand, coming up with a plausibly valid structural causal model is often a daunting task that might require knowledge that is simply not available. We will dive deeper into questions of validity below. Difficulty in coming up with a plausible causal model often exposes unsettled substantive questions that require resolution first.

The potential outcomes model, in contrast, is generally easier to apply. There's a broad set of statistical estimators of causal effects that can be readily applied to observational data. But the ease of application can also lead to abuse. The assumptions underpinning the validity of such estimators are experimentally unverifiable. Frivolous application of causal effect estimators in situations where crucial assumptions do not hold can lead to false results, and consequently to ineffective or harmful interventions.

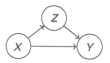

Figure 5.11
Causal graph with mediator Z.

Counterfactual Discrimination Analysis

Counterfactuals serve at least two purposes for us. On the technical side, counter-factuals give us a way to compute path-specific causal effects. This allows us to make path analysis a quantitative matter. On the conceptual side, counterfactuals let us engage with the important normative debate about whether discrimination can be captured by counterfactual criteria. We will discuss each of these in turn.

Quantitative Path Analysis

Mediation analysis is a venerable subject dating back decades [186]. Generally speaking, the goal of mediation analysis is to identify a mechanism through which a cause has an effect. We will review some recent developments and how they relate to questions of discrimination.

In the language of our formal framework, mediation analysis aims to decompose a total causal effect into path-specific components. We will illustrate the concepts in the basic three-variable case of a mediator, although the ideas extend to more complicated structures.

There are two different paths from X to Y. A direct path and a path through the mediator Z (figure 5.11). The conditional expectation $\mathbb{E}[Y \mid X = x]$ lumps together influence from both paths. If there were another confounding variable in our graph influencing both X and Y, then the conditional expectation would also include whatever correlation is the result of confounding. We can eliminate the confounding path by virtue of the do-operator $\mathbb{E}[Y \mid \mathrm{do}(X := x)]$. This gives us the total effect of the action $X := x$ on Y. But the total effect still conflates the two causal pathways, the direct effect and the indirect effect. We will now see how we can identify the direct and indirect effects separately.

We dealt earlier with the direct effect as it did not require any counterfactuals. Recall, we can hold the mediator fixed at level $Z := z$ and consider the effect of treatment $X := 1$ compared with no treatment $X := 0$ as follows:

$$\mathbb{E}\left[Y \mid \mathrm{do}(X := 1, Z := z)\right] - \mathbb{E}\left[Y \mid \mathrm{do}(X := 0, Z := z)\right].$$

We can rewrite this expression in terms of counterfactuals equivalently as:

$$\mathbb{E}\left[Y_{X:=1,Z:=z} - Y_{X:=0,Z:=z}\right].$$

To be clear, the expectation is taken over the background variables in our structural causal models. In other words, the counterfactuals inside the expectation are invoked with an elementary setting u of the background variables, that is, $Y_{X:=1,Z:=z}(u) - Y_{X:=0,Z:=x}(u)$, and the expectation averages over all possible settings.

The formula for the direct effect above is usually called *controlled direct effect*, since it requires setting the mediating variable to a specified level. Sometimes it is desirable to allow the mediating variable to vary as it would had no treatment occurred. This too is possible with counterfactuals and it leads to a notion called *natural direct effect*, defined as

$$\mathbb{E}\left[Y_{X:=1,Z:=Z_{X:=0}} - Y_{X:=0,Z:=Z_{X:=0}}\right].$$

The counterfactual $Y_{X:=1,Z:=Z_{X:=0}}$ is the value that Y would obtain had X been set to 1 and had Z been set to the value Z would have assumed had X been set to 0.

The advantage of this slightly mind-bending construction is that it gives us an analogous notion of *natural indirect effect*:

$$\mathbb{E}\left[Y_{X:=0,Z:=Z_{X:=1}} - Y_{X:=0,Z:=Z_{X:=0}}\right].$$

Here we hold the treatment variable constant at level $X := 0$, but let the mediator variable change to the value it would have attained had treatment $X := 1$ occurred.

In our three-node example, the effect of X on Y is unconfounded. In the absence of confounding, the natural indirect effect corresponds to the following statement of conditional probability (involving neither counterfactuals nor do-interventions):

$$\sum_z \mathbb{E}\left[Y \mid X = 0, Z = z\right]\left(\mathbb{P}(Z = z \mid X = 1) - \mathbb{P}(Z = z \mid X = 0)\right).$$

In this case, we can estimate the natural direct and indirect effect from observational data.

The technical possibilities go beyond the case discussed here. In principle, counterfactuals allow us to compute all sorts of path-specific effects even in the presence of (observed) confounders. We can also design decision rules that eliminate path-specific effects we deem undesirable.

Counterfactual Discrimination Criteria

Beyond their application to path analysis, counterfactuals can also be used as a tool to put forward normative fairness criteria. Consider the typical setup of chapter 3. We have features X, a sensitive attribute A, an outcome variable Y, and a predictor $\widehat{Y}$.

One criterion that is technically natural would say the following: For every possible demographic described by the event $E := \{X := x, A := a\}$ and every possible setting a' of A, we ask that the counterfactual $\widehat{Y}_{A:=a}(E)$ and the counterfactual $\widehat{Y}_{A:=a'}(E)$ follow the same distribution.

Introduced as *counterfactual fairness* [187], we refer to this condition as *counterfactual demographic parity*, since it's closely related to the observational criterion *conditional demographic parity*. Recall, conditional demographic parity requires that in each demographic defined by a feature setting $X = x$, the sensitive attribute is independent of the predictor. Formally, we have the conditional independence relation $\widehat{Y} \perp A \mid X$. In the case of a binary predictor, this condition is equivalent to requiring for all feature settings x and groups a, a':

$$\mathbb{E}[\widehat{Y} \mid X = x, A = a] = \mathbb{E}[\widehat{Y} \mid X = x, A = a'].$$

The easiest way to satisfy counterfactual demographic parity is for the predictor $\widehat{Y}$ to use only nondescendants of A in the causal graph. This is analogous to the statistical condition of using only features that are independent of A.

In the same way that we defined a counterfactual analogue of demographic parity, we can explore causal analogues of other statistical criteria in chapter 3. In doing so, we need to be careful to separate technical questions about the difference between observational and causal criteria from the normative content of the criterion. Just because a causal variant of a criterion might get around some statistical issues of noncausal correlations does not mean that the causal criterion resolves normative concerns or questions with its observational cousin.

Counterfactuals in the Law

We'll now scratch the surface of a deep subject in legal scholarship that we return to in chapter 6 after developing greater familiarity with the legal background. The subject is the relationship of causal counterfactual claims and legal cases of discrimination. Many technical scholars see support for a counterfactual interpretation of United States discrimination law in various rulings by judges that seem to have invoked counterfactual language. Here's a quote from a popular textbook on causal inference [188]:

> U.S. courts have issued clear directives as to what constitutes employment discrimination. According to law makers, "The central question in any employment-discrimination case is whether the employer would have taken the same action had the employee been of a different race (age, sex, religion,

national origin etc.) and everything else had been the same." (In Carson vs Bethlehem Steel Corp., 70 FEP Cases 921, 7th Cir. (1996).)

Unfortunately, the situation is not so simple. This quotation invoked here—and in several other technical papers on the topic—expresses the opinion of judges in the 7th Circuit Court at the time. This court is one of thirteen United States courts of appeals. The case has little precedential value; the quotation cannot be considered a definitive statement on what employment discrimination means under either Title VII or Equal Protection law.

More significant in US jurisprudence is the standard of "but-for causation" that has gained support through a 2020 US Supreme Court decision relating to sex discrimination in the case *Bostock v. Clayton County*. In reference to the Title VII statute about employment discrimination in the Civil Rights Act of 1964, the court argued:

> While the statute's text does not expressly discuss causation, it is suggestive. The guarantee that each person is entitled to the "same right ... as is enjoyed by White citizens" directs our attention to the counterfactual—what would have happened if the plaintiff had been White? This focus fits naturally with the ordinary rule that a plaintiff must prove but-for causation. (*Comcast Corp. v. NAT. ASSN. AFRICAN AMERICAN-OWNED*, 140 S. Ct. 1009, 589 U.S., 206 L. Ed. 2d 356 [2020])

Although the language of counterfactuals appears here, the notion of but-for causation may not effectively correspond to a correct causal counterfactual. Expanding on how to interpret but-for causation, the court noted:

> a but-for test directs us to change one thing at a time and see if the outcome changes. If it does, we have found a but-for cause.

Changing one attribute while holding all others fixed is not in general a correct way of computing counterfactuals in a causal graph. This important issue was central to an major discrimination lawsuit.

Harvard College Admissions
In a trial dating back to 2015, the plaintiff Students for Fair Admissions (SFFA) alleged discrimination in Harvard undergraduate admissions against Asian Americans. Plaintiff SFFA is an offshoot of a legal defense fund which aims to end the use of race in voting, education, contracting, and employment.

The trial entailed unprecedented discovery regarding higher education admissions processes and decision making, including statistical analyses of individual-level applicant data from the past five admissions cycles.

The plaintiff's expert report by Peter S. Arcidiacono, professor of economics at Duke University, claims:

> Race plays a significant role in admissions decisions. Consider the example of an Asian-American applicant who is male, is not disadvantaged, and has other characteristics that result in a 25% chance of admission. Simply changing the race of the applicant to white—and leaving all his other characteristics the same—would increase his chance of admission to 36%. Changing his race to Hispanic (and leaving all other characteristics the same) would increase his chance of admission to 77%. Changing his race to African-American (again, leaving all other characteristics the same) would increase his chance of admission to 95%. (EXPERT REPORT OF PETER S. ARCIDIACONO Students for Fair Admissions, Inc. v. Harvard No. 14-cv-14176-ADB [D. Mass])

The plaintiff's charge, summarized above, is based technically on the argument that conditional statistical parity is not satisfied by a model of Harvard's admissions decisions. Harvard's decision process isn't codified as a formal decision rule. Hence, to talk about Harvard's decision rule formally, we first need to model Harvard's decision rule. The plaintiff's expert did so by fitting a logistic regression model against Harvard's past admissions decisions in terms of variables deemed relevant for the admission decision.

Formally, denoting by $\widehat{Y}$ the model of Harvard's admissions decisions, by X a set of applicant features deemed relevant for admission, and by A the applicant's reported race, we have that

$$\mathbb{E}[\widehat{Y} \mid X = x, A = a] < \mathbb{E}[\widehat{Y} \mid X = x, A = a'] - \delta,$$

for some groups a, a' and some significant value of $\delta > 0$.

The violation of this condition certainly depends on which features we deem relevant for admissions, formally, which features X we should condition on. Indeed, this point is to a large extent the basis of the response of the defendant's expert, David Card, professor of economics at the University of California, Berkeley. Card argues that under a different reasonable choice of X, one that includes among other features the applicant's interview performance and the year they applied, the observed disparity disappears.

The selection and discussion of what constitute relevant features is certainly important for the interpretation of conditional statistical parity. But arguably a bigger question is whether a violation of conditional statistical parity constitutes evidence of discrimination in the first place. This isn't merely a question of having selected the right features to condition on.

What does the plaintiff's expert report mean by "changing his race"? The literal interpretation is to "flip" the race attribute in the input to the model without changing any of the other features of the input. But a formal interpretation in terms of attribute swapping is not necessarily what triggers our moral intuition. As we know now, attribute flipping generally does not produce valid counterfactuals. Indeed, if we assume a causal graph in which some of the relevant features are influenced by race, then computing counterfactuals with respect to race would require adjusting downstream features. Changing the race attribute without a change in any other attribute only corresponds to a counterfactual in the case where race does not have any descendant nodes—an implausible assumption.

Attribute flipping is often mistakenly given a counterfactual causal interpretation. Obtaining valid counterfactuals is in general substantially more involved than flipping a single attribute independently of the others. In particular, we cannot meaningfully talk about counterfactuals without bringing clarity to what exactly we refer to in our causal model and how we can produce *valid* causal models. We turn to this important topic next.

Validity of Causal Modeling

Consider a claim of employment discrimination of the kind: *The company's hiring practices discriminated against applicants of a certain religion.* Suppose we want to interrogate this claim using the formal machinery developed in this chapter. At the outset, this requires that we formally introduce an attribute corresponding to the "religious affiliation" of an individual.

Our first attempt is to model *religious affiliation* as a personal trait or characteristic that someone either does or does not possess. This trait, call it A, may influence choices relating to one's appearance, social practices, and variables relevant to the job, such as the person's level of education Z. So, we might like to start with a model such as figure 5.12.

Religious affiliation A is a source node in this figure, which influences the person's level of education Z. Members of certain religions may be steered away from or encouraged toward obtaining a higher level of education by their social peer group. This story is similar to how in our Berkeley admissions graph *sex* influences *department choice.*

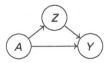

Figure 5.12
Religion as a root node.

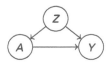

Figure 5.13
Religion as ancestor.

This view of religion places the burden on understanding the possible indirect pathways, such as $A \to Z \to Y$, through which religion can influence the outcome. There may be insufficient understanding of how a religious affiliation affects numerous other relevant variables throughout life. If we think of religion as a source node in a causal graph, changing it will potentially affect all downstream nodes. For each such downstream node we would need a clear understanding of the mechanisms by which religion influences the node. Where would such *scientific knowledge* of such relationships come from?

But the causal story around religion might also be different. It could be that obtaining a higher level of education causes an individual to lose their religious beliefs. In fact, this modeling choice has been put forward in technical work on this topic [32]. Empirically, data from the United States General Social Survey show that the fraction of respondents changing their reported religion at least once during a four-year period ranged from about 20 percent to about 40 percent [189]. Identities associated with sexuality and social class were found to be even more unstable. Changing one's identity to better align with one's politics appeared to explain some of this shift. From this perspective, religious affiliation is influenced by level of education and so the graph might look like figure 5.13.

This view of religion forces us to correctly identify the variables that influence religious affiliation and are also relevant to the decision. After all, these are the confounders between religion and outcome. Perhaps it is not just level of education, but also socioeconomic status and other factors that have a similar confounding influence.

What is troubling is that in our first graph education is a mediator, while in our second graph it is a confounder. The difference is important; to quote Pearl:

> As you surely know by now, mistaking a mediator for a confounder is one of the deadliest sins in causal inference and may lead to the most outrageous error. The latter invites adjustment; the former forbids it. [182]

The point is not that these are the only two possible modeling choices for how religious affiliation might interact with decision-making processes. Rather, the point is that there exist multiple plausible choices. Either of our modeling choices follows a natural causal story. Identifying which one is justified is no easy task. It's also not a task that we can circumvent by appeal to some kind of pragmatism. Different modeling choices can lead to completely different claims and consequences.

In order to create a valid causal model, we need to provide clarity about what the thing is that each node references and what relationships exist between these things. This is a problem of ontology and metaphysics. But we also need to know facts about the things we reference in causal models. This is a problem in epistemology, the theory of knowledge.

These problems might seem mundane for some objects of study. We might have strong scientifically justified beliefs on how certain mechanical parts in an airplane interact. We can use this knowledge to reliably diagnose the cause of an airplane crash. In other domains, especially ones relevant to disputes about discrimination, our subject matter knowledge is less stable and subject to debate.

Social Construction of Categories

The difficulties we encountered in our motivating example arise routinely when making causal statements involving human kinds and categories, such as race, religion, or gender, and how these interact with consequential decisions.

Consider the case of *race*. The metaphysics of race is a complex subject, highly debated, featuring a range of scholarly accounts today. A book by Joshua Glasgow, Sally Haslanger, Chike Jeffers, and Quayshawn Spencer represents four contemporary philosophical views of what race is [190]. The construction of racial categories and racial classification of individuals is inextricably tied to a long history of oppression, segregation, and discriminatory practices [191, 192, 193].

In the technical literature around discrimination and causality, it's common for researchers to model *race* as a source node in a causal graph, which is to say that race has no incoming arrows. As a source node it can directly and indirectly influence an outcome variable, say *getting a job offer*. Implicit in this modeling choice is a kind of naturalistic perspective that views race as a biologically grounded trait, similar to *sex*. The trait exists at the beginning of one's life. Other variables that come later in life, education and income for example, thus become ancestors in the causal graph.

This view of race challenges us to identify all the possible indirect pathways through which race can influence the outcome. But it's not just this modeling challenge that we need to confront. The view of race as a biologically grounded trait stands in contrast with the *social constructivist* account of race [190, 194, 195, 196]. In this view, roughly speaking, race has no strong biological grounding but rather is a social construct. Race stems from a particular classification of individuals by society, and the shared experiences that stem from the classification. As such, the surrounding social system of an individual influences what race is and how it is perceived. In the constructivist view, *race* is a socially constructed category into which individuals are assigned.

The challenge with adopting this view is that it is difficult to tease out a set of nodes that faithfully represent the influence that society has on race and perceptions of race. The social constructivist perspective does not come with a simple operational guide for identifying causal structures. In particular, socially constructed categories often lack the kind of modularity that a causal diagram requires. Suppose that group membership is constructed from a set of social facts about the group and practices of individuals within the group. We might have some understanding of how these facts and practices constitutively identify group membership, but we may not have an understanding of how each factor individually interacts with each other factor, or whether such a decomposition is even possible [197].

Ontological Instability

The previous arguments notwithstanding, pragmatist might accuse our discussion of adding unnecessary complexity to what might seem to some like a matter of common sense. Surely, we could also find subtlety in other characteristics, such as smoking habits or physical exercise. How is race different from other things we reference in causal models?

An important difference is a matter of ontological stability. When we say *rain caused the grass to be wet* we also refer to an implicit understanding of what rain is, what grass is, and what wet means. However, we find that acceptable in this instance, because all three things we refer to in our causal statement have *stable enough* ontologies. We know what we reference when we invoke them. To be sure, there could be subtleties in what we call grass. Perhaps the colloquial term *grass* does not correspond to a precise botanical category or that it corresponds to one that has changed over time and will again change in the future. However, by making the causal claim, we implicitly assert that these subtleties are irrelevant for the claim we make. We know that grass is a plant and that other plants would also get wet from rain. In short, we believe the ontologies we reference are *stable enough* for the claim we make.

This is not always an easy judgment to make. There are, broadly speaking, at least two sources of ontological instability. One stems from the fact that the world changes over time. Social progress, political events, and our own epistemic activities may make theories obsolete, create new categories, or disrupt existing ones [196]. Hacking's work describes another important source of instability. Categories lead people who putatively fall into such categories to change their behavior in possibly unexpected ways. Individuals might conform or disconform to the categories they are confronted with. As a result, the responses of people, individually or collectively, invalidate the theory underlying the categorization. Hacking calls this a "looping effect" [198]. As such, social categories are moving targets that need constant revision.

Certificates of Ontological Stability

The debate around human categories in causal models is by no means new. But it often surfaces in a seemingly unrelated yet long-standing discussion around causation and manipulation. One school of thought in causal inference aligns with the mantra *no causation without manipulation*, a view expressed by Paul Holland in an influential article from 1986:

> Put as bluntly and as contentiously as possible, in this article I take the position that causes are only those things that could, in principle, be treatments in experiments. [199]

Holland goes further by arguing that statements involving "attributes" are necessarily statements of association:

> The only way for an attribute to change its value is for the unit to change in some way and no longer be the same unit. Statements of "causation" that involve attributes as "causes" are always statements of association between the values of an attribute and a response variable across the units in a population. [199]

To give an example, Holland maintains that the sentence "She did well on the exam because she is a woman" means nothing but "the performance of women on the exam exceeds, in some sense, that of men" [199].

If we believed that there is no causation without manipulation, we would have to refrain from including immutable characteristics in causal models altogether. After all, there is by definition no experimental mechanism that turns immutable attributes into treatments.

Holland's view remains popular among practitioners of the potential outcomes model. The assumptions common in the potential outcomes model are easiest to conceptualize by analogy with a well-designed randomized trial. Practitioners in this framework are therefore used to conceptualizing causes as things that could, in principle, be a treatment in randomized controlled trials.

The desire or need to make causal statements involving race in one way or the other not only arises in the context of discrimination. Epidemiologists encounter the same difficulties when confronting health disparities [36, 200], as do social scientists when reasoning about inequality in poverty, crime, and education.

Practitioners facing the need of making causal statements about race often turn to a particular conceptual trick. The idea is to change object of study from the *effect of race* to the effect of *perceptions of race* [201]. What this boils down to is that we change the units of the study from individuals with a race attribute to *decision makers*. The treatment becomes *exposure to race* through some observable trait, like the name on a résumé in a job application setting. The target of the study is then how decision makers respond to such *racial stimuli* in the decision-making process. The hope behind this maneuver is that exposure to race, unlike race itself, may be something that we can control, manipulate, and experiment with.

While this approach superficially avoids the difficulty of conceptualizing manipulation of immutable characteristics, it shifts the burden elsewhere. We now have to sort out all the different ways in which we think that race could possibly be perceived: through names, speech, style, and all sorts of other characteristics and combinations thereof. But not only that. To make a counterfactual statements vis-à-vis *exposure to race*, we would have to be able to create the authentic background conditions under which all these perceptible characteristics would have come out in a manner that's consistent with a different racial category. There is no way to construct such counterfactuals accurately without a clear understanding of what we mean by the category of race [202]. Just as we cannot talk about witchcraft in a valid causal model for lack of any scientific basis, we also cannot talk about perceptions of witchcraft in a valid causal model for the very same reason. Similarly, if we lack the ontological and epistemic basis for talking about race in a valid causal model, there is no easy remedy to be found in moving to perceptions of race.

In opposition to Holland's view, other scholars, including Pearl, argue that causation does not require manipulability but rather an understanding of *interactions*. We can reason about hypothetical volcanic eruptions without being able to manipulate volcanoes. We can explain the mechanism that causes tides without being able to manipulate the moon by any feasible intervention. What is required is an understanding of the ways in which a variable interacts with other variables in the model. Structural equations in a causal model are *response functions*. We can think of a node in a causal graph as receiving messages from its parent nodes and responding to those messages. Causality is thus about who *listens* to whom. We can form a causal model once we know how the nodes in it interact.

But as we saw, the conceptual shift to *interaction*—who *listens* to whom—by no means makes it straightforward to come up with valid causal models. If causal models organize available scientific or empirical information, there are inevitably limitations to what constructs we can include in a causal model without running the danger of divorcing the model from reality. Especially in sociotechnical systems, scientific knowledge may not be available in terms of precise modular response functions.

We take the position that causes need not be experimentally manipulable. However, our discussion motivates that constructs referenced in causal models need a certificate of ontological and epistemic stability. Manipulation can be interpreted as a somewhat heavy-handed approach to clarifying the ontological nature of a node by specifying an explicit experimental mechanism for manipulating the node. This is one way, but not the only way, to clarify what it is that the node references.

Chapter Notes

There are several introductory textbooks on the topic of causality. For a short introduction to causality turn to the primer by Judea Pearl, Madelyn Glymour, and Nicholas Jewell [188], or the more comprehensive textbook by Pearl [180]. At the technical level, Pearl's text emphasizes causal graphs and structural causal models. Our exposition of Simpson's paradox and UC Berkeley was influenced by Pearl's discussion, updated for a new popular audience [182]. All of these texts touch on the topic of discrimination. In these books, Pearl takes the position that discrimination corresponds to the direct effect of the sensitive category on a decision.

The technically minded reader will enjoy complementing Pearl's book with the an open access text by Jonas Peters, Dominik Janzing, and Bernhard Schölkopf [185], which is also available online. The text emphasizes two variable causal models and applications to machine learning. See Peter Spirtes, Clark Glymour, and Richard Scheines et al. [203] for a general introduction based on causal graphs with an emphasis on *graph discovery*, that is inferring causal graphs from observational data.

Stephen Morgan and Christopher Winship [204] focus on applications in the social sciences. Guido Imbens and Donald Rubin [205] give a comprehensive overview of the technical repertoire of causal inference in the potential outcomes model. Joshua Angrist and Pischke Jörn-Steffen [206] focus on causal inference and potential outcomes in econometrics.

Miguel Hernán and James Robins [207] give another detailed introduction to causal inference that draws on the authors' experience in epidemiology.

Judea Pearl [180] already considered the example of gender discrimination in UC Berkeley's graduate admissions, which we discussed at length. In his discussion, he implicitly advocates for a view of discussing discrimination based on causal graphs by inspecting which paths in the graph go from the sensitive variable to the decision point. The UC Berkeley example has been discussed in various other writings, such as Pearl's discussion in the *Book of Why* [182]. However, the development in this chapter differs significantly in its arguments and conclusions.

For clarifications regarding the popular interpretation of Edward Simpson's original article [208], see Miguel Hernán's article [209] and Pearl's text [180].

The topic of causal reasoning and discrimination gained significant momentum in the computer science and statistics community around 2017. Lu Zhang, Yongkai Wu, and Xintao Wu [210] previously considered discrimination analysis via path-specific causal effects. Matt Kusner, Joshua Loftus, Chris Russell, and Ricardo Silva [187] introduced a notion of *counterfactual fairness*. The authors extend this line of thought in another work [211]. Silvia Chiappa introduces a path-specific notion of counterfactual fairness [212]. Niki Kilbertus, Mateo Rojas-Carulla, Giambattista

Parascandolo et al. [213] distinguish between two graphical causal criteria, called *unresolved discrimination* and *proxy discrimination*. Both notions correspond to either allowing or disallowing paths in causal models. Razieh Nabi and Ilya Shpitser [214] conceptualize discrimination as the influence of the sensitive attribute on the outcome along certain *disallowed* causal paths. Silvia Chiappa and William Isaac [215] give a tutorial on causality and fairness with an emphasis on the COMPAS debate. Atoosa Kasirzadeh and Andrew Smart expand on the discussion about the difficulties with constructing causal counterfactual claims about social categories in the context of machine learning problems [216].

There is also extensive relevant scholarship in other disciplines that we cannot fully survey here. Of relevance is the vast literature in epidemiology on health disparities. In particular, epidemiologists have grappled with race and gender in causal models. See, for example, the article by Tyler VanderWeele and Whitney Robinson [200], as well as Nancy Krieger's comment on the article [217], and Krieger's article on discrimination and health inequalities [218] for a starting point.

We retrieved the data about UC Berkeley admissions from http://www.random services.org/random/data/Berkeley.html on December 27, 2018. There is some discrepancy with the data displayed on the Wikipedia page for Simpson's paradox, which does not affect our discussion.

6

Understanding United States Antidiscrimination Law

In this chapter, we hope to give you an appreciation of what United States antidiscrimination law is and isn't. We use the US legal experience as a case study of how to regulate discrimination. Other countries take different approaches. We don't aim to describe US law comprehensively but rather to give a stylized description of the key concepts.

We start with a history of how the major civil rights statutes came to be, and draw lessons from this history that continue to be relevant today. Law represents one attempt to operationalize moral notions. It is an important and illustrative one. We will learn from the way in which the law navigates many tricky trade-offs. But we will also study its limitations and explain why we think algorithmic fairness shouldn't stop at legal compliance.

The final section addresses the specifics of regulating machine learning. Although US antidiscrimination law predates the widespread use of machine learning, it is just as applicable if a decision maker uses machine learning or other statistical techniques. That said, machine learning introduces many complications to the application of these laws, and existing law may be inadequate to address some types of discrimination that arise when machine learning is involved. At the same time, we believe that there is also an opportunity to exercise new regulatory tools to rein in algorithmic discrimination.

This chapter can be skipped on a first reading of the book, but a few connections are worth pointing out. The first section elaborates on a central viewpoint of the book, especially chapter 4, which is that attributes like race and gender are salient because they have historically served as organizing principles of many societies. That section also sets up chapter 8 that conceives of discrimination more broadly than in discrete moments of decision making. The section on the limitations of the law motivates another core theme of this book, which is that one can use the debates on machine learning and discrimination as an opportunity to revisit the moral foundations of fairness.

History and Overview of US Antidiscrimination Law

Every inch of civil rights protections built into law was fought for and hard won through decades of activism. In this section, we briefly describe these histories of oppression and discrimination, the movements that arose in response to them, and the legal changes that they accomplished.

Black Civil Rights

The Black civil rights movement, often simply called the civil rights movement, has its roots in slavery in the United States and the rampant racial discrimination that persisted after its abolition. The period immediately following the American Civil War and the abolition of slavery (roughly 1865–1877) is called the Reconstruction era. It resulted in substantial progress in civil rights. Notably, the Constitution was amended to abolish slavery (Thirteenth Amendment), require equal protection under the laws (Fourteenth Amendment), and guarantee voting rights regardless of race (Fifteenth Amendment).

However, these gains were rapidly undone as White supremacists gained political control in the Southern states, ushering in the so-called Jim Crow era, a roughly seventy-five-year period in which the Southern states orchestrated stark racial segregation, discrimination, and near-total disenfranchisement of Black people. Nearly every facet of life was racially segregated, including residential neighborhoods, schools, workplaces, and places of public accommodation such as restaurants and hotels. This segregation was blessed by the Supreme Court in 1896, when it ruled that laws mandating segregation did not violate the Equal Protection Clause under the "separate but equal" doctrine [219]. But in practice, things were far from equal. The jobs available to Black people usually paid far less, schools were underfunded and subject to closure, and accommodations were fewer and of inferior quality. As late as the 1950s, a cross-country drive by a Black person would have involved great peril, for instance, if they showed up at a small town at night they might be refused a place to stay [220]. Black people could not democratically challenge these laws as the states erected numerous practical barriers to voting—ostensibly race neutral, but with vastly different effects by race—and Black people at the polls were often met with violence. As a result, disenfranchisement was highly effective. For example, in Louisiana until the mid-1940s, less than 1 percent of African Americans were registered to vote [221]. (Data limitations preclude a nationwide assessment of the effectiveness of disenfranchisement.)

Meanwhile, in the Northern states, racial discrimination operated in more indirect ways. Residential zoning laws that prohibited higher-density, lower-cost housing were used to keep poorer Black residents out of White neighborhoods. The practice of "redlining" by banks, orchestrated to some extent by federal regulators, limited the availability of credit, especially mortgages, in specific neighborhoods [222]. The justification proffered was the level of risk, but it had the effect of discrimination against Black communities. Another prevalent technique to achieve segregation was the use of racially restrictive covenants, in which property owners in a neighborhood entered into a contract not to sell or rent to non-White people.

The civil rights movement emerged in the late 1800s and the early 1900s to confront these widespread practices of racism. Broadly, the movement adopted two complementary strategies: one was to challenge unjust laws and the other was to advance Black society within the constraints of segregation and discrimination. A key moment in the first prong was the formation of the National Association for the Advancement of Colored People in 1909. In addition to lobbying and litigation against Jim Crow laws, it sought to fight against lynching. Prominent efforts under the second prong included the Black entrepreneurship movement—1900–1930 has been called the golden age of Black businesses [223]—and notable achievements in education. Many of the Historically Black Colleges and Universities (HBCU) were founded during the Jim Crow era.

After decades of activism, an epochal moment was a Supreme Court ruling in 1954 that declared the segregation of public schools unconstitutional. This began the gradual dismantling of the Jim Crow system, a process that would take decades and whose effects we still feel today. The court victories further galvanized the movement, leading to more intense activism and mass protests. This led to major federal legislation in the following decade: the Civil Rights Act of 1960 and the Voting Rights Act of 1965, both of which targeted voter suppression efforts, and the Civil Rights Act of 1964 and the Fair Housing Act (FHA) of 1968, which targeted private discrimination. We will discuss the latter two in detail throughout this chapter.

Antidiscrimination laws were clearly a product of history and decades- or centuries-long trends—slavery, Jim Crow, and the civil rights movement. At the same time, their proximate causes were often specific, unpredictable events. For example, the assassination of Martin Luther King Jr. provided the impetus for the passage of the FHA. They also reflect political compromises that were necessary to secure their passage. For example, Title VII of the Civil Rights Act of 1964 created the Equal Employment Opportunity Commission (EEOC); was stripped of the enforcement powers that had been present in the original wording of the title [224].

Gender Discrimination

The struggle for gender equality also has a long and storied history of activism. In the 1800s, the law did not recognize basic rights of women, including voting and owning property. Changing this was the primary goal of *first-wave feminists* whose strategies included advocacy, civil disobedience, lobbying, and legal action. The culminating moment was the ratification of the Nineteenth Amendment in 1920, guaranteeing women the right to vote (yet, as discussed above, Black women's right to vote was still limited in the South). *Second-wave feminism* began in the 1960s. It targeted stereotypes about the role of women in society, private discrimination in education and employment, and bodily rights, including reproductive rights and domestic violence. In the early postwar years, gender norms regressed in some ways (e.g., women lost access to jobs that had been available to them because of the war), which was arguably an impetus for the movement [225]. Two early legislative victories were the Equal Pay Act of 1963 and Title VII of the Civil Rights Act of 1964, prohibiting employment discrimination.

However, these did not initially have much impact due to the aforementioned lack of enforcement, and the movement only intensified. An important milestone was the founding of the National Organization of Women in 1965. Borrowing strategies from the Black civil rights movement, the second-wave feminists adopted a plan to litigate in the courts to secure protections for women. A notable court victory in the following decade was the expansion of abortion rights by the Supreme Court [226]—a decision that would be overturned in 2022. On the legislative front, two major achievements for gender equality were Title IX of the Education Amendments Act of 1972, which prohibited sex discrimination in federally funded educational programs, and the Equal Credit Opportunity Act (ECOA), which prohibited sex discrimination in matters concerning credit.

Education, especially higher education, and credit were both important sectors for women's rights. Historically, many elite colleges simply did not accept women. Even in the 1970s, women faced many barriers in academia: sexual harassment, higher bars for admission, outright exclusion from some high-status fields such as law and medicine, and limited athletic opportunities. Similarly, credit discrimination in the 1970s was also stark, such as requiring women to reapply for credit upon marriage, usually in the husband's name [227]. After this period, the focus of the feminist movement expanded beyond major legislative victories to include the questioning of gender as a social construct.

LGBTQ Civil Rights
Discriminatory laws against LGBTQ people were historically numerous: prohibition of some sexual behavior (i.e., antisodomy laws [228]), lack of marriage rights, bans on military service and some other government positions, a failure to prohibit private discrimination and to treat hate crimes as such, and even a prohibition of literature advocating for gay rights under obscenity laws.

Tentative activism began in the 1950s, with the first legal changes coming in the early 1960s. A pivotal movement was the 1969 Stonewall riots, a series of protests in response to a police raid at a New York City gay bar. The aftermath of this event kickstarted the push for US LGBTQ rights, including the gay pride movement for visibility and acceptance. In 1973, the American Psychiatric Association's *Diagnostic and Statistical Manual of Mental Disorders* dropped homosexuality as a disorder, signaling (and furthering) a major shift in attitudes. The list of legal changes is long and ongoing. They include state-by-state changes to laws involving sodomy, marriage equality, private discrimination, and hate crimes; a 2003 Supreme Court decision ruling antisodomy laws unconstitutional [229]; and a 2015 Supreme Court decision guaranteeing the right to marry for same-sex couples nationwide [230]. In parallel, the push for LGBTQ rights in the private sector has progressed in part by interpreting existing statutory prohibitions on sex discrimination, such as Title VII of the Civil Rights Act of 1964, to encompass sexual orientation and gender identity discrimination [231].

Disability Laws

Another dimension of identity covered by antidiscrimination statutes is disability. Over a quarter of adults in the United States today have some type of disability, including mobility disabilities, blindness or other visual disability, deafness or other hearing disability, and cognitive disabilities [232]. These and other disabilities are distinct identities corresponding to different lived experiences and, sometimes, cultures. Still, the emergence of a cross-disability coalition and identity enabled more effective advocacy for disability rights. This movement gained steam in the decades following World War II. Activists aimed to make disability visible, rather than stigmatized, pitied, and hidden, and sought to achieve independent living. As with other rights movements, disabled people faced multiple, mutually reinforcing barriers: society's attitudes toward disability and disabled people, the lack of physical accommodations and assistive technologies, and discriminatory policies [233]. Attitudes that held back disabled people weren't just prejudices, but also mistaken views of disability as residing in the person (the *medical model*) instead of, or in addition to, being created by barriers in society (the *social model*). The first federal law protecting disability rights was the Rehabilitation Act of 1973 which prohibited disability discrimination in federally funded programs. Activism toward a broad civil rights statute continued, with the 1964 Civil Rights Act as a model. These efforts culminated in the Americans with Disabilities Act (ADA) of 1990. While the ADA has many similarities with the other civil rights statutes, it also has major differences due to its emphasis on accommodation in addition to formal nondiscrimination.

Lessons

The histories of the various civil rights movements hold several lessons that continue to be relevant today. First of all, the law is a political instrument: it can be used to discriminate, to create the conditions under which discrimination can flourish, or to challenge discrimnation. It can be a tool for subjugation or liberation. Laws may be facially neutral but they are created, interpreted, and enforced by actors that respond to the changing times and to activism. Court decisions are also influenced by contemporary activism and even scholarship.

Our brief historical discussion also helps explain why certain sectors are regulated and not others. Education, employment, housing, credit, and public accommodation are domains that are both highly salient to people's life courses and have had histories of discrimination that were deliberately used to subordinate some groups.

One consequence of this sector-specific approach is that the law can be tailored to the particularities of the sector in an attempt to avoid loopholes. For example, the FHA encompasses the full range of practices related to housing, including sales, rentals, advertising, and financing. It lists (and prohibits) various ways in which housing agents may subtly mislead or discourage clients belonging to protected classes. Recognizing the importance of financing for securing housing, it prohibits discrimination in financing with respect to "purchase, construction, improvement, repair, or maintenance." It even prohibits ads indicating a discriminatory preference. And that includes not just categorical statements such as "no children," but also targeting of ads to certain geographic regions in a way that correlates with race, and the selection of actors used in advertising.

In many cases these attempts to avoid loopholes have held up well in the face of recent technological developments. The prohibition on discriminatory advertising has forced online ad platforms to avoid discriminatory targeting of housing ads [234]. But this is not always so. Ride-hailing platforms are able to evade Title VII (employment discrimination) liability even though they terminate drivers based on the (potentially discriminatory) ratings given by passengers [235].

Even though laws are sector-specific, it is hard to understand discrimination by looking at any one set of institutions (such as employment or education, much less a single organization) in isolation. History shows us that there tend to be multiple interlocking systems of oppression operating in tandem, such as federal housing policy and private-sector discrimination. Similarly, the line between state and private discrimination is not always clear.

History also shows that when disrupted, hierarchies tend to reassert themselves by other means. For example, the end of de jure segregation accelerated the phenomenon of "White flight" from cities to the suburbs, exacerbating de facto segregation. Not only is progress fitful, regression is possible. For example, Woodrow Wilson and his administration segregated large parts of the federal workforce in the 1910s, eroding some of the gains Black people had made in previous decades. And as we were writing this chapter, the Supreme Court overturned *Roe v. Wade*, ending federal protection of abortion rights and enabling severe restrictions on abortion in many states.

Another important point that is not apparent from the laws themselves is that the various protected dimensions of identity have complex and distinct histories of discrimination and activism, even if statutes attempt to treat them all in a uniform and formal way. Even within a single dimension like ethnicity, the oppression and struggles of different groups take drastically different forms. Native Americans endured a century of attempts at forced assimilation in which children were sent to boarding schools and asked to abandon their culture. The Chinese Exclusion Act of 1882 all but eliminated the immigration of Chinese people for over half a century and made conditions inhospitable for the Chinese immigrant community that already existed. During World War II, over 100,000 people of Japanese ancestry, the majority of whom were US citizens, were interned in concentration camps under the pretense that they were disloyal to the country. These are just a few of the more gruesome episodes of discrimination on the basis of race, ethnicity, and national origin in US history, focusing on the actions of the government. National origin discrimination was often a thinly veiled form of racial discrimination. Thus, although the list of protected attributes in the law may grow over time, it is not arbitrary and is deeply informed by history.

Equality under the law remains a contested and evolving notion. This is especially the case when antidiscrimination runs up against some countervailing value or principle, such as religious freedom or limiting state authority. And because the law is intertwined with our lives and livelihoods in so many ways, equality under the law, in a broad sense, requires far more than formal nondiscrimination. Consider gender equality. The range of legal interventions necessary to achieve it is long and growing. Beyond voting rights and prohibition of sex and gender discrimination, it includes prohibition of pregnancy and marital status discrimination, curbing sexual harassment and sexual violence, abortion rights, maternity leave laws, and childcare subsidies. Each one of these battles has many fronts. For example, the #MeToo movement brought to light the role of nondisparagement clauses as used by employers in settlements to silence victims of workplace sexual harassment, and there is an ongoing effort to prohibit such clauses.

Finally, legal change is not the end of the road but in some ways is just the beginning. The effects of past discrimination tend to leave a lasting imprint. The law itself, given political realities, can do only so much to erase the effects of that history (table 6.1).

A Few Basics of the American Legal System

The US Constitution is the ultimate law of the land. The Constitution created the three branches of government: the legislature (Congress), the executive (the president, executive agencies, and others reporting to the president), and judiciary (the

Table 6.1
A summary of the major antidiscrimination statutes: Titles VI and VII of the Civil Rights Act of 1964, the Fair Housing Act (FHA), Title IX of the Education Amendments Act of 1972, the Equal Credit Opportunity Act (ECOA), and the Americans with Disabilities Act (ADA)

Law	Year	Covered entities and regulated activities	Protected categories (* = added later)
Title VI	1964	Any organization receiving federal funding (due to breadth, doesn't list regulated activities)	Race, color, national origin
Title VII	1964	Employers, employment agencies, labor unions	Race, color, religion, sex, national origin, pregnancy*
FHA	1968	Sales, rentals, advertising, and financing of housing	Race, color, religion, national origin, sex (*?), handicap, familial status
Title IX	1972	Educational programs receiving federal funding: hiring, pay, rank, sexual harassment, retaliation, segregation and same-sex education	Sex
ECOA	1974	Creditors (banks, small loan and finance companies, retail and department stores, credit card companies, and credit unions)	Race,* sex, age,* national origin,* marital status, receipt of public assistance*
ADA	1990	Employers, public services, public accommodation	Disability, record of disability, perception of disability

Supreme Court and other courts). All three branches have important roles when it comes to antidiscrimination law. State and local governments and laws also play important roles in antidiscrimination, but we will say less about them due to our pedagogical focus on federal law.

Before we get to the three branches, it is worth noting that the Constitution itself contains two elements relevant to discrimination law: the right to the due process of law (Fifth and Fourteenth Amendments) and the right to equal protection under the laws (Fourteenth Amendment). Both of these curtail the ability of the government to discriminate. Equal protection law has also sometimes been used to curtail private discrimination. Due process has been raised as a defense by *defendants* of discrimination lawsuits contending that laws that curtail their ability to discriminate violate their due process rights.

The Role of Congress (Legislative Branch)
Laws passed by Congress are called statutory laws, as opposed to constitutional law, case law, and other types of law. We encountered some of the major antidiscrimination statutes earlier. But there are many practical and political barriers to congressional action, and statutes or amendments are relatively rare. Thus, to stay relevant in a changing world, laws are generally broadly worded policies and do

not attempt to anticipate the nuances of every situation in which they might be applied. To interpret and enforce these policies, Congress delegates authority to federal agencies (such as the EEOC). The courts also perform a vital interpretive function, as well as keeping a check on the power of Congress itself.

There are three main law-making powers that Congress has used to enact antidiscrimination statutes within the limits of its constitutional authority. The first is the Commerce Clause, which allows Congress to regulate interstate commerce. The meaning of this clause has been interpreted expansively by the Supreme Court. Legislation pertaining to employment and credit antidiscrimination finds basis in the Commerce Clause. The second power comes from the Fourteenth Amendment, which guarantees to all citizens the equal protection of the laws, and further empowers Congress to enforce it through appropriate legislation. While the extent of state involvement needed for the Fourteenth Amendment to apply is not settled, the Fair Housing Act and the Americans with Disabilities Act owe their constitutional basis in part to this power. Finally, Congress has the "power of the purse": the ability to enact policy goals by controlling spending and by threatening to withhold federal funding for entities that fail to meet certain obligations. Title VI of the Civil Rights Act of 1964 and Title IX of the Educational Amendments Act of 1972 fall under this category, which is why they covered only organizations receiving federal funding.

Congress has used its power to enact antidiscrimination statutes covering a broad swath of activities. Still, there are many gaps and limitations in federal antidiscrimination law, in part because of constitutional limitations and in part because Congress has failed to act. As a result, state laws sometimes fill these gaps.

The Role of the Courts (Judicial Branch)

The United States adopts a common-law system, which means that courts have the power to make law that guides decisions in future cases. This is the concept of "precedent." In disputes where the facts or principles are similar to previous cases decided by relevant courts, judges are bound to follow the reasoning used in the past decision (the precedent). Similarly, courts are tasked with interpreting the statutory laws and the Constitution. This body of precedent is referred to as case law and can be as binding as any other law. For example, the important concept of disparate impact, under which decision-making practices may be unlawful if they have disproportionate effects even if facially neutral and without discriminatory intent, is the result of a Supreme Court decision interpreting the scope of a statutory law. Most of Europe, in contrast, adopts a civil-law system, which means that legislation is the primary source of law and judicial decisions have less value as precedent.

The hierarchical organization of courts determines which precedents are binding on a particular dispute. The federal courts are organized into three levels: the district courts at the bottom, thirteen appellate courts (also known as circuit courts) above them, and the Supreme Court at the top. Supreme Court decisions are binding on all lower courts and appellate court decisions are binding on the corresponding district courts. The appellate courts hear cases only on appeal—that is, when one of the parties alleges a material error in a district court's decision. The Supreme Court, in turn, usually hears only appeals of circuit court decisions. The Supreme Court is not required to accept petitions for review; in fact, it grants review in only a small fraction of requests.

When interpreting statutes, courts adopt both textual and contextual methods. The former confines itself to the plain meaning of the statute itself, while the latter looks to sources outside the text of the law, such as lawmakers' originally expressed intent and the law's purpose. The importance of contextual factors is a contested topic and judges differ in their approaches.

So far we have talked about the role of the courts in making law. Of course, the primary function of the courts is to adjudicate individual cases. So a few notes about court procedure are in order. Litigation can be *civil* or *criminal*. Civil cases involve wrongs against private individuals; most discrimination-related disputes fall into this category, with a few exceptions such as hate crimes. Criminal cases involve violations of criminal law and can be brought only by the government.

A central feature of US court procedure is the use of *adversary proceedings*. The two parties to a dispute are the *plaintiff* (who files a *complaint* alleging that they have been wronged) and the *defendant* (who is alleged to have committed the wrong). Both are typically represented by attorneys, who have a lot of power in determining how the case unfolds, with the judge having a relatively passive role as an arbiter and not an inquisitor.

An example will bring together the aspects of the court system that we have discussed so far. Consider the question of whether the websites of restaurants, retailers, and so on must be accessible to visually impaired people. The ADA prohibits excluding people with disabilities from availing themselves of the services of a place of public accommodation. But does this include websites? Congress could not have anticipated this question in 1991, so the statute (despite being unusually detailed, running to over 20,000 words) does not address this question directly. One set of circuit courts has looked at Congress's intent and purpose and found that websites themselves can be considered places of public accommodation in keeping with the ADA's "broad mandate," "sweeping purpose," and "comprehensive character." Another set of circuit courts took a more textual approach, and considered it crucial that the statute applies to the services *of* a place of public accommodation, not services *in* a place of public accommodation. Thus, as long as there is a sufficient "nexus" between the physical place and the website, the accessibility requirement extends to the website. In April 2021, one circuit court ruled differently, reading the text of the statute to mean that only physical places can be places of public accommodation and also rejecting the "nexus" standard adopted by the second set of courts. When circuit courts are split in this way, it usually takes the Supreme Court stepping in to resolve the inconsistency, but this may take many years.

One reason for this state of affairs is that the Department of Justice (DOJ), which is tasked with issuing regulations to implement the ADA, hasn't issued a final regulation on whether websites are places of public accommodation and, if so, what standard of accessibility they would need to satisfy. Therefore the courts had to exercise a greater degree of interpretive latitude than they otherwise would. It also gave rise to a concern in some courts that imposing accessibility requirements without setting a clear standard would deprive defendants of their constitutional right to due process. This highlights the importance of executive departments and federal agencies, to which we turn next.

The Role of the Federal Agencies (Executive Branch)

The main antidiscrimination functions of the federal agencies are rulemaking, guidance, and law enforcement. For example, the Equal Credit Opportunity Act broadly makes credit discrimination unlawful, but leaves it to the Federal Reserve to draft and interpret regulations that implement this mandate (Congress later transferred this authority to the Consumer Financial Protection Bureau). This process is called *rulemaking*. The resulting regulations constitute *administrative law* and have the force of law alongside statutory law and case law.

Rules differ slightly from guidelines. A group of agencies led by the Equal Employment Opportunity Commission issued the *Uniform Guidelines for Employee Selection Procedures* in 1978 that spells out a framework for ensuring that tests and other employee selection procedures are compliant with Title VII of the Civil Rights Act of 1964. The *Uniform Guidelines* are widely relied upon by employers. But the *Uniform Guidelines* do not constitute law. They are often referenced in court opinions, and courts generally give significant deference to agency guidelines, but courts are not bound by them.

It's hard to overstate the practical importance of the agencies. Whether or not a statute has real teeth depends in large part on the implementing agency. The EEOC initially refused to take up gender discrimination despite being empowered to do so. In fact, Title VII had no teeth for even racial discrimination until amended in 1972 to empower the EEOC to take action (the Equal Employment Opportunity Act of 1972) [236].

Agencies differ in their level of political independence; some are housed within the executive (such as the Department of Housing and Urban Development [HUD] and the Department of Labor) while others are more independent (such as the EEOC and the Federal Trade Commission [FTC]). The latter has enforcement powers in addition to rulemaking powers. They can conduct investigations and file suit in court; some even have their own judicial systems and are sometimes called a de facto fourth branch of government. In short, interpreting as well as enforcing the statutes are tasks shared by the federal agencies and the courts. They generally work harmoniously together, but the proliferation of sources of law and methods of enforcement can lead to inefficiency and confusion.

It is worth mentioning two other important sources of policy: executive orders and rules internal to institutions. Executive orders are directives issued by the president of the United States. They were originally intended as a way to manage the affairs of the government, but the vast reach of the federal government has meant that they are a powerful de facto instrument for enacting policy. For example, precursors to Title VII in the form of executive orders go as far back as 1941 [237]. Although much weaker in scope than the eventual legislation, they illustrate the ability of presidents to act quickly while Congress might be stalled.

Institutions, whether public or private, may set nondiscrimination rules or guidelines for their employees that may go beyond the requirements of the law. For example, asking a job candidate about their marital status is not per se unlawful. However, it would be construed as evidence of intent to discriminate in a legal dispute [238]. Considering this (and the fact that there is almost never a job-related reason for such an inquiry), many organizations prohibit their interviewers from asking such questions. On a day-to-day basis, these institutional guidelines are the most direct nondiscrimination rules that individuals are bound by.

Case Study: The Evolution of Title IX

Title IX of the Educational Amendments Act of 1972 prohibits sex discrimination in educational institutions that receive federal funds. In 1975, the Department of Health, Education and Welfare (HEW) published the final regulations detailing how Title IX would be enforced. Since 1975, the federal government has issued guidance clarifying how it interprets and enforces those regulations.

Two of the big questions surrounding Title IX were what constitutes receiving federal financial assistance and what constitutes sex discrimination. Each of these had the potential to vastly impact the reach of the law. In 1984, the Supreme Court ruled that Title IX was program specific: that is, only those programs and activities receiving direct federal funds needed to comply. This gutted the application of Title IX: for example, most athletic programs were no longer covered since they didn't directly receive federal funds. In response, Congress drafted a bill specifically intended to overturn this decision, restoring the broad scope of Title IX, which it retains to this day [239, 240].

Two other major disputes regarding the scope of Title IX are whether schools are responsible for sexual harassment that happens on campuses and whether discrimination on the basis of sexual orientation and gender identity is prohibited. Unlike the coverage question, these continue to be the topic of vociferous legal and political debate. On the sexual harassment front, the Supreme Court held that in the late 1990s that schools are responsible for creating a safe environment, including preventing harassment by other students, but "the student must show that an official of the school with authority to respond actually knew of and was deliberately indifferent to the harassment." The Obama, Trump, and Biden administrations have all introduced guidelines or regulations on this question, in turn expanding and contracting the scope of Title IX [241]. A similar political seesaw has played out with respect to LGBTQ protections. In 2021 the Department of Education interpreted Title IX expansively, buttressed by a 2020 Supreme Court ruling that also involved the relationship between sex discrimination and sexual orientation, though in the employment discrimination context [242].

How the Law Conceives of Discrimination

There are many possible ways to define discrimination and attempt to achieve nondiscrimination. In this section, we discuss how the law conceives of discrimination and how it tries to balance nondiscrimination with other ideals.

Disparate Treatment and Disparate Impact

Imagine an employer turning down a job candidate and explicitly informing them that this decision was on account of a protected characteristic. Such a case would be relatively straightforward to adjudicate based on the text of the statute itself ("It shall be an unlawful employment practice for an employer to fail or refuse to hire an individual ... because of such individual's race, color, religion, sex, or national origin"). However, in most cases of discrimination, the decision maker's behavior is less explicit and the evidence is more circumstantial. To deal with these, courts have created two main doctrines called *disparate treatment* and *disparate impact*.

Disparate treatment refers to intentional discrimination and roughly matches the average person's conception of discriminatory behavior. It subsumes the straightforward case described above. For more circumstantial cases, the Supreme Court has established a so-called burden-shifting framework under Title VII (employment law). First, the plaintiff must establish a "prima facie" case of discrimination by showing that they are a member of a protected class, were qualified for a position, were denied it, and the position then remained open or was given to someone not in the protected class. If the plaintiff is successful at this, the employer must produce a legitimate, nondiscriminatory reason for the adverse decision. The plaintiff then has the burden of proving that the proffered reason is merely a pretext for discrimination.

Disparate treatment usually involves reasoning about what action the defendant would have taken if the plaintiff's protected characteristic had been different, with all other facts of the case unchanged. Elsewhere in this book we argue why, from a technical perspective, these "attribute-flipping" counterfactuals are at odds with a nuanced understanding of causality and result in brittle tests for discrimination. In any event, the importance of causality in disparate treatment, especially so-called but-for causation, has increased following a 2020 Supreme Court decision [231] that held that it is impossible to engage in sexual orientation discrimination without engaging in sex discrimination by imagining a counterfactual in which the victim's sex is changed without affecting anything else, including gender preference. While celebrated from a civil rights perspective because of its implications for LGBTQ rights, we should keep in mind that this represents a narrow understanding of causality and its application in other scenarios may yield conclusions not so favorable to civil rights.

In contrast to disparate treatment, disparate impact is about practices that have a disproportionate effect on a protected class, even if unintentional. At a high level, disparate impact must be both unjustified and avoidable. This is again operationalized through a burden-shifting framework. First, the plaintiff must establish that there is a disproportionate difference in selection rates between different groups. If that can be shown, then the employer has the opportunity to explain if the reason for the different selection rates has a business justification. The burden then reverts to the plaintiff to show that there is an "alternative employment practice" that would have achieved the employer's aims while being less discriminatory.

One way to think about disparate impact is as a way to "sniff out" well-concealed intentional discrimination by putting the focus on its impacts, which are more readily observable. Indeed, the case that led to the doctrine involved an employer who introduced aptitude tests for promotion on the very day that the Civil Rights Act of 1964, which prohibited employment discrimination based on race, took effect [243].

But disparate impact is also thought to be motivated by a consideration of distributive justice, that is, minimizing unjustified inequalities in outcomes. In this sense, disparate impact roughly corresponds to the middle view of equality of opportunity that we discuss in chapter 4. Disparate impact tries to force decision makers to treat seemingly dissimilar people similarly on the belief that their current dissimilarity is a result of past injustice. It aims to compensate for at least some of the disadvantage suffered for unjust reasons. Indeed, in the case mentioned above, the Supreme Court pointed out that the racial performance disparity on aptitude tests could be explained by inequalities in the educational system. But disparate impact doctrine has evolved over the years and the extent to which it reflects distributive justice, as opposed to a device for illuminating well-concealed discrimination, is thought to have waned over time.

While we have discussed these two doctrines in the context of employment law, they are found in each of the six domains we discussed in the first section. Disparate impact has been so central to the legal understanding of discrimination that it was later incorporated into statutes, notably the ADA (disability law), but also into Title VII (equal employment law) itself through a 1991 amendment. But the Supreme Court has not extended the doctrine to situations where the laws or procedures of the state (rather than private actors) violate the Equal Protection Clause if they have a discriminatory purpose. In other words, there is no equivalent of disparate impact doctrine for state actors, only disparate treatment.

An important general observation about antidiscrimination law—especially for readers who may be accustomed to thinking about fairness in terms of the statistical properties of the outputs of decision-making processes—is that the law is primarily concerned with the processes themselves. Relatedly, the way that courts go about weighing evidence is also highly procedural, to the point where it may seem tangential to the substantive question of whether discrimination took place [244]. Even disparate impact, despite being motivated in part by distributive notions of justice, is treated in a formal and procedural manner. As an illustration of the centrality of the procedural element, the DOJ's legal manual for proving Title VI disparate impact claims is over 20,000 words long [245].

There are many possible reasons for the law's focus on process. One is historical: the statutes were primarily responding to blunt discrimination and formal denials of opportunity, as opposed to more subtle statistical phenomena. It is also a better match for how the law works: the definition of discrimination cannot be divorced from the procedure for proving it in court. A third reason is political: it is easier to achieve consensus about fair processes than what the right distributive outcome is. Finally, at a pragmatic level, it reflects the law's attention to the nuances of workplaces and other institutions, compared to which statistical fairness criteria seem crude and oversimplified.

Avoiding Excessive Burdens on Decision Makers

A recurring theme is how much burden on decision makers is justified in pursuit of fairness goals. For example, making accommodations for disabled employees results in some cost to a firm.

In general, the law gives substantial deference to the interests of the decision maker. This has been repeatedly made clear by lawmakers and the courts at various points in time. For example, the House Judiciary Committee said on the role of the EEOC at the moment of the agency's inception: "management prerogatives, and union freedoms are to be left undisturbed to the greatest extent possible" [246]. The Supreme Court clarified in 2015 that "the FHA (Fair Housing Act) is not an instrument to force housing authorities to reorder their priorities" [247].

One exception is the ADA, which imposes substantial compliance requirements on a large set of firms and governments. This shouldn't be a surprise since the law sought to create structural changes in society, especially to the built environment. The ADA does have an "undue hardship" defense to the requirement that employers provide "reasonable accommodations" to qualified employees with disabilities, but courts appear to tolerate a higher burden than under, say, Title VII. As an illustration, blind employees sued their employers for failing to provide paid readers for four hours of the workday; the court sided with the plaintiffs, implying that a roughly 50 percent increase in the cost of the employees to the employer did not constitute an undue hardship [248]. We hasten to add that undue hardship is a multifactor test and there is no clear or uniform cost threshold; cost is rarely the determinative factor.

There are a range of potential justifications of the burdens on decision makers in the academic scholarship and legislative history. Often the responsibilities of decision makers are justified by appeal to the human rights of those being harmed, rather than an economic analysis. For example, the ADA sought to intervene in discrimination against disabled people that often affected their livelihoods and sometimes cost lives [248]. Alternatively, burdens are sometimes justified because they pose only a "de minimis" cost. For example, Title VII requires employers to make accommodations for employees' religious beliefs, but not if it would pose more than a minimal cost.

In between these two types of cases lie a variety of others where a more careful balance between benefits and costs is necessary. We give a few brief examples here.

- *Positive externality*. A hope behind the ADA was that it would make it easier for disabled people to enter the workforce and contribute to the overall economy.

- *Regulation as collective action*. Title II of the Civil Rights Act of 1964 prohibits discrimination in places of public accommodation (e.g., restaurants). A major reason why such establishments discriminated against minorities was the prejudices of their White customers. Title II enabled them to stop discriminating, gaining business from minority customers without incurring lost business from their White customers; thus, the law did not impose a burden on them but rather created an opportunity [249]. Similarly, consider insurance. In the absence of regulation, if an insurer avoids calibrating premiums to risk in the interest of fairness, it might go out of business. But if all firms in the market have their behavior constrained by antidiscrimination law in the same way, they can no longer claim to be at a competitive disadvantage.

- *Cheapest cost avoider*. The cheapest cost avoider or least cost avoider principle assigns liability from a harm to the party that can avoid the harm at the lowest cost. It is the reason why firms, to some extent, bear liability for discrimination or harassment committed by their employees. If an employer is forced to internalize the costs of discriminatory harassment committed by its employees, it will, on standard economic theory, invest in precautions up to the point where they are no longer cost justified [250].

• *Correcting irrationality*. Some commentators suggest that the rampant discrimination against women before the passage of the ECOA was irrational behavior by creditors, and women were in fact good credit risks [251]. In this view, ECOA can be seen as correcting this irrationality rather than imposing a burden on creditors.

Limits of the Law in Curbing Discrimination

How effective has United States antidiscrimination law been? The best-case scenario is that the possibility of penalties has sufficiently deterred would-be discriminators that rates of discrimination have plummeted, and, in the few remaining cases of discrimination, victims manage to obtain redress through the courts. The worst-case scenario is that the laws have had virtually no effect, and any reductions in disparities since their passage can be attributed to other factors such as discriminators being less successful in the market.

The reality is somewhere in between. Rigorously evaluating the effect of laws is a tricky counterfactual problem and is subject to much uncertainty and debate. However, there is much evidence suggestive of a positive effect. For example, one study used a natural experiment to evaluate the impact of Title VII on job opportunities for African Americans relative to White Americans. It showed that the relative employment of African Americans increased more in industries and regions with a greater proportion of firms that were newly covered under Title VII by the Equal Employment Opportunity Act of 1972 [236, 252].

While the gains have been nonnegligible, the effectiveness of antidiscrimination law is blunted for many reasons which we now discuss. This motivates our view that work on algorithmic fairness should not treat the approach adopted in antidiscrimination law as a given, but should instead reconnect with the moral foundations of fairness.

Burdens on Victims of Discrimination

The law places an array of burdens on victims of discrimination if they wish to seek legal recourse. We will use the labor market as our running example, but our observations apply to other contexts as well.

To begin with, legal intervention is initiated by the victim, not the government, and cannot begin until after victims have already suffered discrimination. Regulators do not prospectively review employment practices, in contrast to other areas of law such as pharmaceutical regulation, where drugs must be thoroughly tested before being allowed on the market. Further, there is a fundamental information asymmetry between firms and employees (or job candidates). Victims may not even be aware that they have faced discrimination. After all, job candidates and employees have no direct visibility into employers' decision-making process and firms need not provide a justification for an adverse hiring or promotion decision. Only in the domain of credit does some form of transparency requirement exist.

Even if a victim becomes aware of discrimination, they face barriers that may deter them from suing. Litigation involves additional mental anguish. Victims may also be deterred by the high financial costs of litigation. Lawsuits typically take several years to reach a conclusion, by which time the victim's career may suffer a significant and irreparable setback. If the victim remains at the firm after filing suit, they face an uncomfortable situation in the best case and potentially retaliation from the employer (even though laws specifically prohibit retaliation, it remains a common result of discrimination lawsuits). And if the victim seeks employment elsewhere, future employers may weigh negatively the fact that the candidate sued a previous employer for discrimination.

Victims who decide to sue face a battery of procedural hurdles. If the employer has internal grievance procedures, the victim may be required to try those before suing (or risk losing her claims). Another prerequisite to filing suit is to file an administrative complaint with the EEOC promptly after the discrimination starts. The timeliness requirement often puts victims in a double bind because of the need to exhaust internal channels. It also makes it difficult to collect the evidence necessary to prevail in court [253, 254].

That brings us to the final and most serious difficulty that plaintiffs face, which is the burden of proof. To be sure, the standard of proof that the plaintiff must meet in discrimination cases is "preponderance of the evidence," which means more likely than not, which is lower than the standard in criminal cases. But even this standard has proved daunting. According to Katie Eyer, "anti-discrimination law is a highly rigid technical area of the law, in which any of a myriad of technical doctrines can lead to dismissal. Courts approach the question of discrimination as if it were a complex legal puzzle, in which any piece out of place must result in the dismissal of the plaintiffs' claims" [244].

Specifically, in disparate treatment cases, courts have created numerous defendant-friendly doctrines. Under the "stray remarks" doctrine, discriminatory comments made by the employer about the plaintiff do not constitute evidence of discriminatory intent unless there is a sufficiently clear causal nexus to the decision itself. Under the "same actor" defense, if the employer was willing to hire the plaintiff at a previous time, it is taken as evidence that the employer bears no discriminatory intent against the plaintiff. Under the "honest belief" rule, a case can be summarily dismissed if the employer "honestly believed" in the reasons for the decision, even if they can later be shown to be "mistaken, foolish, trivial, or baseless."

In disparate impact, an overlapping set of factors is arrayed against the plaintiff. While there is no need to establish intent, there is a new set of requirements: identifying a specific policy or practice that caused the adverse employment decision; compiling the requisite statistics to show that the policy has a disparate impact; and rebutting the employer's defense that the policy is justified by job-relatedness. The third prong is a particularly severe hurdle for plaintiffs as they are structurally poorly positioned to identify an alternative employment practice, lacking the knowledge of the internals of the business that the employer has [248, 255].

The net result of these barriers to plaintiffs is that their odds of success at trial are exceedingly low. Katie Eyer summarizes data from the Uncertain Justice project: "of every 100 discrimination plaintiffs who litigate their claims to conclusion (i.e., do not settle or voluntarily dismiss their claims), only 4 achieve any form (de minimis or not) of relief. ... These odds can properly be characterized as shockingly bad, and extend (with minor differences) to every category of discrimination plaintiff, including race, sex, age, and disability" [244].

We should note that there is a widespread view that employment discrimination lawsuits are too *easy* to file and too favorable to plaintiffs, a position we reject. Selmi critically examines this perception and notes that it is prevalent among judges; correcting this perceived imbalance may in fact be one reason for the creation of numerous hurdles for plaintiffs [256]. Whether or not one subscribes to the view that many "nuisance lawsuits" are filed by plaintiffs alleging discrimination, it is true that courts are highly strained and judges are wary of decisions that might open the "floodgates" to lawsuits. This suggests that the burdens we have discussed above are unlikely to go away [254, 257].

The Difficulty of Substantive and Structural Reform through Procedural Intervention

Even if compliance with antidiscrimination law is high, and legal remedies are readily attained, there may be even more fundamental limits to the effectiveness of the law. To what extent do the formal limits that the law imposes on individuals and organizations lead to a just society? How big is the gap between legal and moral notions of unfairness?

Stephen Halpern frames the issue thus [258]:

> In translating a social problem into the "language" of the law, lawyers must frame their analysis in terms of contrived concepts, issues, questions, and remedies that the legal system recognizes and deems legitimate. In that translation, as in any translation, there are constrictions and distortions. Framing a social problem as a legal issue produces a transformation of the issue itself—a

reconceptualization of the problem, yielding unique questions and concerns that first become the focus of the legal debate and subsequently tend to dominate public discussion. When racial problems are reformulated as questions of legal rights, the resulting dialogue does not capture the complexity and subtlety of those problems or permit consideration of the fullest range of remedies for them. Inevitably, the demands and limits of the legal process alter the public discourse about and understanding of vital racial issues.

Halpern's book is about racial inequality in education; his main example of his thesis is the effort that was put into ending the segregation of public schools without much attention paid to the quality of education received by Black students in integrated schools. Similarly, Title VI litigation focused on procedures for processing complaints of discrimination filed with the federal government, rather than mechanisms to vindicate substantive rights to an education of comparable quality. He argues that "few, if any, of the factors that have an impact on educational achievement are governed by 'legal rights' or are readily translatable into an issue of "racial discrimination.' " He gives two reasons why inequalities persist despite the law's formal remedies: the de facto segregation of American cities and the fact that academic differences often arise from instability in the home and other social, economic, and health disparities.

While the effect of school desegregation in the United States is a vast topic, the broader point is that limitations of the legal process restrict what is achievable—and even shape our understanding of the issues themselves. Another example of this comes from Richard Rothstein's book *Color of Law* [222].

> Although most African Americans have suffered under [historically racist government housing policies], they cannot identify, with the specificity a court case requires, the particular point at which they were victimized. For example, many African American World War II veterans did not apply for government guaranteed mortgages for suburban purchases because they knew that the Veterans Administration would reject them on account of their race, so applications were pointless. Those veterans then did not gain wealth from home equity appreciation as did white veterans, and their descendants could not then inherit that wealth as did white veterans' descendants. With less inherited wealth, African Americans today are generally less able than their white peers to afford to attend good colleges. If one of those African American descendants now learned that the reason his or her grandparents were forced to rent apartments in overcrowded urban areas was that the federal government unconstitutionally and unlawfully prohibited banks from lending to African Americans, the grandchild would not have the standing to file a lawsuit; nor would he or she be able to name a particular party from whom damages could be recovered.

Another impetus toward proceduralism comes from the interaction of the court system with the internal procedures of organizations. Under the theory of legal endogeneity, organizations enact procedural protections, such as diversity training programs, with the putative aim of curbing discrimination; over time, courts gradually come to mistake these procedural and symbolic compliance-oriented activities for substantive measures; but once these symbolic measures themselves attain legal significance, substantive concerns have been pushed outside the scope of legitimate debate [259].

Further, even substantive change at individual organizations may not imply structural change—that is, a change to the underlying factors in society that produce disparities in the first place. Even if an employer achieved statistical parity in hiring and promotion rates, the application rates might themselves reflect unequal opportunity in society and/or discrimination at previous levels or stages of the system, and there is little the law can do to compel individual decision makers to remedy these inequalitites.

Legal interventions whose effects are both substantive and structural are rare. One notable example is the impact of Title IX on women's athletics. The law has been interpreted not only to prohibit discrimination in a narrow sense but also to require equity in a number of areas such as scholarships, coaching, and facilities. Arguably, as a result of these interventions, women's athletics in the United States has gradually risen in prestige, weakening the gender hierarchy in athletics, leading to greater parity in athletics even outside the collegiate context. In general, however, these types of substantive interventions have so far proved less feasible than formal ones in part because of the funding they require.

Although we have contrasted procedural interventions with substantive and structural interventions above, the line between them can be murky, and the former can at least function as a toehold on the latter. To the extent that inequality persists because of entrenched policies that maintain an unequal distribution of resources, procedural interventions that allow members of historically oppressed groups to rise to positions of authority might allow them to more effectively alter these policies. Procedural interventions can help reduce the capacity of already advantaged groups from usurping full control over the policy-making process. Still, this is far from an ideal route to change, as it places the burden of advancing the interests of specific groups on individuals who belong to those groups.

Another seeming contrast is between discrimination law and redistributive policies, that is, the government directly taxing certain actors and reallocating those funds to the disadvantaged group. But discrimination law can be understood to be a mechanism that places the economic burden of rectifying past injustice to some extent on employers, lenders, and so on. In some ways, this might be similar to a policy of taxing employers and using those funds to support groups who have been subject to discrimination in the past.

Affirmative action policies, in particular, occupy a space that is squarely in between formal nondiscrimination and redistributive policies. An example of such a policy would be a job training program offered by an employer that favored groups with lower access to opportunities [260]. However, except in rare cases, the law does not *compel* affirmative action by private entities but merely *allows* it. More commonly seen are affirmative requirements for governments. The FHA, in addition to nondiscrimination mandates, requires HUD and recipients of federal funds from HUD to "affirmatively further" the policies and purposes of the act. This might enable, for instance, subsidized housing in high-income communities that opens up access to higher-quality schools and amenities. However, this part of the FHA has largely lain dormant. Thus, at least some of the limitation of the law in creating meaningful change can be attributed to the lack of political will to fully act on existing laws, rather than an inherent limitation of the legal system.

Regulating Machine Learning

Although US antidiscrimination law predates the widespread use of machine learning, it is just as applicable if a decision maker uses machine learning or other statistical techniques. That said, machine learning introduces many complications to the application of these laws. These complications are being vigorously debated in the legal scholarship and many scholars are concerned that existing law may be inadequate to address the types of discrimination that arise when machine learning is involved. At the same time, there is also an opportunity to exercise new regulatory tools to rein in algorithmic discrimination. There is little case law on this topic, so our discussion of these issues is based on legal scholarship. As before, our discussion focuses on the United States but we touch on the European Union's General Data Protection Regulation (GDPR) in a few places.

Disparate Treatment

Recall that the two main antidiscrimination doctrines are disparate treatment and disparate impact. Disparate treatment is principally concerned with the explicit intent to discriminate on the basis of legally protected characteristics; in contrast, disparate impact focuses on decision making where there is no explicit intent to discriminate but where even decisions made on the basis of seemingly benign characteristics nevertheless result in unjustified disparities along characteristics that are legally protected.

Most reports of discrimination in machine learning have been cases of unintentional rather than intentional discrimination. Besides, developers of machine

learning systems who intend to discriminate are unlikely to rely explicitly on protected attributes due to the easy availability of proxies. When this happens, it can
be hard to prove that there was an intent to mask discrimination. For these reasons,
disparate treatment is rarely invoked and disparate impact is seen as much more
relevant. We will return to disparate impact shortly, but one important question
involving disparate treatment relates to systems that explicitly rely on the protected attribute to correct data biases or mitigate the effects of past discrimination.
Does this constitute disparate treatment? In other words, does law impose limits on
algorithmic fairness interventions?

The answer is nuanced. One relatively bright line in the law is that selection
quotas are unconstitutional. In machine learning terms, this roughly maps to the difference between techniques that aim to enforce parity and those that merely penalize
disparity during the optimization step. The latter type of technique is analogous to a
process that is race conscious and values diversity but still allows the final distribution to vary depending on the set of candidates. It is helpful, as always, to remember
that technical distinctions rarely map cleanly on to legal determinations.

There is also a major difference between an individual decision made on the basis
of a protected attribute and an overall policy that takes the interests of protected
groups into account. Disparate treatment primarily applies to the former type of
decision. This is similar to the distinction between the use of a protected attribute at
training versus test time, although, again, this distinction by itself is far from legally
determinative.

A Supreme Court case that is often cited as an example of the tension between
disparate treatment and disparate impact (and the disparate-treatment pitfalls of
race-conscious decision making) is *Ricci v. DeStefano*. The case arose because the
New Haven fire department scrapped a promotional exam after finding that Black
firefighters had a lower passage rate than White firefighters. The department worried
that it would open itself to disparate impact liability. But it was then sued by the
White and Hispanic firefighters who would have qualified for promotion based on
the exam. The court agreed with the plaintiffs that the department had engaged in
disparate treatment against them.

Pauline Kim notes a crucial distinguishing feature of the *Ricci* case: the plaintiffs
had already invested considerable time and expense in studying for the exam, and
thus the department's actions resulted in concrete harm to specific individuals. In
Kim's view, the court's logic wouldn't apply when an employer *prospectively* makes
a change to its hiring practices in order to avoid the potential for disparate impact
[261].

Finally, even if a practice constitutes prima facie disparate treatment, it may
be legal if it is part of a valid affirmative action program, that is, one that aims
to remedy past discrimination. In employment, the Supreme Court has ruled

that race- or gender-based affirmative action programs are valid if they seek to eliminate "manifest imbalances" in "traditionally segregated job categories" and do not "unnecessarily trammel" the interests of other candidates. Some scholars have argued that this should hold for voluntary algorithmic affirmative action as well [262].

Disparate Impact

To understand how disparate impact applies to statistical decision making, we must unpack the legal doctrine. The burden-shifting framework established by the Supreme Court for Title VII employment discrimination claims works as follows [263]. First, the plaintiff must establish a prima facie case by showing a sufficient difference in selection rates between different groups. What constitutes a sufficient difference is unclear. The EEOC has established a threshold of four-fifths (i.e., a difference of 20 percent) as a guideline, but this is not a strict rule. In a big-data world, some commentators have argued that the criterion should be based on statistical significance of the difference rather than the magnitude.

If the plaintiff is successful at showing a sufficient difference, the burden shifts to the defendant, who must then establish that the challenged practice is "job related" and consistent with "business necessity." If the defendant can show this, the plaintiff can still win by showing that there is an "alternative employment practice" that would have achieved the employer's aims while being less discriminatory.

The critical step from the perspective of statistical decision making is the question of business necessity. One way the employer can show this is through "empirical data demonstrating that the selection procedure is predictive of or significantly correlated with important elements of job performance." Since machine learning is a technique for establishing predictive validity, commentators such as Solon Barocas and Andrew Selbst suggest that this represents an exceedingly low bar for employers [9]. As long as the target variable used in a predictive model is putatively job-related, the requirement is satisfied.

On the other hand, based on a close reading of the statute, Pauline Kim argues that Title VII can in fact effectively address discriminatory effects of machine learning [264]. However, the doctrine that has developed since its passage is a poor fit for addressing discriminatory machine learning. For example, the requirement for the plaintiff to identify a specific employment practice that caused the disparity developed in an era when written tests were the primary vehicle for disparate impact. But when a statistical model is at play, especially an uninterpretable one that uses a large number of features, it is not clear what the plaintiff is supposed to identify. Thus, the doctrine will have to evolve if Title VII is to address discriminatory machine learning.

Another issue that's specific to automated decision making arises from the fact that the software is usually not developed in-house by the decision maker but rather by specialized external firms. For example, companies such as Hirevue and Pymetrics offer tools to automate part of the hiring process, and Upstart provides a predictive model for loan underwriting. In such cases, who should bear liability? In employment law, employers, not vendors, bear legal liability [265]. But employers (and other clients of these tools) resist this since they usually lack the expertise to conduct statistical validation. Shifting some or all of the liability from clients to vendors would have pros and cons from an antidiscrimination perspective. It might mean that vendors become much more careful at testing their offerings. On the other hand, even if a tool has been broadly tested for disparate impact, it may perform differently in the context of a particular employer's applicant pool. Further, plaintiffs may face even more difficulty in showing an alternative employment practice.

While disparate treatment and disparate impact are the two main prongs of antidiscrimination law, when it comes to data-driven decisions the antidiscrimination toolbox is larger and includes privacy law, explanation, and potentially consumer protection law. We discuss these in turn.

Privacy

When we worry about privacy, the underlying concern is often that data about us could be used to discriminate or might result in adverse treatment. For example, if a job interviewer inquired about religion, it may be considered a privacy violation. The harm that animates this worry is the denial of a job. As another example, reports that the retailer Target uses shopping records to identify pregnant customers sparked outrage [266]. The potential for harm arises because pregnancy is a time when individuals are particularly susceptible to manipulation through marketing (which is the reason that marketers are interested in pregnancy in the first place).

Yet, data privacy law and antidiscrimination law have largely been separate in the United States. Returning to the above example, it is not privacy law that forbids interviewers from asking about religion. Rather, since employment law forbids discriminating on the basis of religion, interpretive guidance from the EEOC, and often from institutions themselves, discourages such questions during interviews.

Still, given the normative alignment, it is natural to wonder whether privacy law can be adapted to serve antidiscrimination ends. There is a lot of intuitive appeal to this idea, especially when it comes to machine learning. If a decision-making system relies on data, why not put restrictions on the flow of data to prevent unjustified discrimination?

But when we examine this argument in more detail, difficulties emerge. The most obvious is the issue of proxies. As we discussed in chapter 3, prohibiting access to sensitive attributes such as race or gender typically has a negligible impact on a classifier when rich datasets are available. It isn't just that the decision maker may train a model to predict the sensitive attribute from innocuous attributes, as in the Target example above. It may instead directly use the innocuous attributes to predict the outcome of interest, such as the susceptibility of a particular person to a particular marketing message. This is in fact exactly what has been shown to happen on Facebook-scale advertising platforms [267].

If proxies are the problem, another approach is to prohibit the collection of proxies. This is the idea behind "ban the box" laws in US states that prohibit employers from inquiring about criminal history. Ban-the-box has two motives. One is to make it easier for formerly incarcerated people to be rehabilitated into society through employment. In this view, criminal history itself can be seen as the sensitive attribute. The other motive is to combat the racially disparate impact of discrimination against fomerly incarcerated people. Here, criminal history can be seen as a proxy for race. It is this view that is of interest to us.

An influential study by Amanda Agan and Sonja Starr found that employers increased racial discrimination when they were subject to ban-the-box laws [268]. What does this mean for the prospect of preventing discrimination by prohibiting information flows? One view is that in the light of this finding, ban-the-box laws clearly harm more than they help. But another perspective is that racial discrimination is already unlawful, so what the study really reveals is the need to step up auditing and enforcement. If that were to happen, ban-the-box laws might be able to achieve their intended effects.

Going beyond protected attributes and their proxies, privacy law may make it harder to amass dossiers on individuals (for example, containing shopping or browsing records), and we might hope that this would make discrimination harder. While a full discussion is beyond our scope, US privacy law is often criticized for failing to accomplish this effectively, for several reasons. There is no general federal privacy law analogous to the GDPR in the EU. Only a few sectoral privacy laws exist, such as the Health Insurance Portability and Accountability Act (HIPAA). Privacy in most commercial transactions or interactions comes down to "notice and choice," which is ineffective for many reasons, including the power asymmetry and information asymmetry that exists between firms and individuals. In the machine learning context, the notice and choice approach to privacy is particularly ineffective as a barrier to firms building models that may infer sensitive attributes or make adverse decisions based on innocuous attributes. That's because of the "tyranny of the minority": it takes only a small number of individuals to consent to collection to be able to uncover the statistical patterns that make such inferences possible [269].

While privacy laws have not so far helped to address discrimination, discrimination law has sometimes helped to preserve privacy. The Genetic Information Nondiscrimination Act is an antidiscrimination law that mutated into a privacy law through expansive court decisions and EEOC interpretation [270]. Genetic information is an exception to the ubiquity of proxies, as it cannot readily be inferred with any degree of completeness or accuracy from observable characteristics.

Broader senses of the word "privacy" go beyond information flow and encompass transparency, explanation, and redress. We turn to those next.

Explanation

In the context of automated decision making, explanation could have one of two goals. The first is an explanation of the overall system. In a rule-based system this might be the set of decision rules. In a machine learning system it's less obvious what form this explanation should take, and it is a subject of active research in the field of interpretable machine learning.

An explanation of the overall system promotes fairness objectives because it allows regulators, users, and developers to check whether the system adheres to normative requirements. In many cases, explanation allows us to immediately spot potential unfairness. For example, if we know that a system used for detecting fake accounts on social media relies on an uncommon name as a signal of inauthenticity, it is easy to see why it may be more likely to incorrectly flag users who are from minority cultures, as we discussed in chapter 1.

The second goal of explanation is to describe how a particular decision was made given the characteristics of the decision subject. This goal can also promote fairness objectives. It satisfies a powerful innate need to understand how consequential decisions about us are made. The dread that arises when a decision system denies us such explanation is visceral enough that it has a name: Kafkaesque. Explanation of individual decisions also serves more instrumental purposes. It allows us to contest decisions that may have been made on the basis of erroneous information. Even if the decision was accurate, explanation allows recourse, that is, actions that decision subjects might take to alter the decision in the future. For example, a loan applicant who was denied because of a low credit score may make attempts to improve that score.

Taking a step back, decision systems can be analyzed at three levels. The highest level is that of values, goals, and normative constraints (for example, maximizing predictive accuracy while ensuring fairness). The second is the design of the system and its rules. The third is the level of individual decisions. Justification is needed at all three levels. In traditional decision-making systems, values and goals derive legitimacy through stakeholder participation, deliberation, and democratic debate.

Table 6.2
Comparison of the two flavors of model explanation

	Explanation of overall system	Explanation of specific decisions
Goal	Justify policy based on goals and values	Justify decision based on policy
Bureaucratic analogue	Rulemaking or policymaking	Adjudication
Technical tool	Global interpretability	Local explanation
Example legal requirement	GDPR: "meaningful information about the logic involved"	FCRA and ECOA: adverse action notice

It is often merged with the next step, rulemaking or policymaking, which is the process of going from the first level to the second—designing a decision system based on values and goals. If the first step was skipped, tensions between different values or between different stakeholders' objectives become apparent in this process. In administrative bureaucracies, they are resolved through a process of public participation [271]. In contrast, the process of *adjudication* bridges the second and the third levels [272]. For example, in the United States, bureaucrats periodically assess the value of homes and other real estate based on an elaborate policy in order to determine how much property tax should be levied. If the owner disagrees with the assessment, they can appeal, and they have the right to a hearing.

Automated systems erode the procedural protections involved in rulemaking and adjudication: public participation and appeals, respectively [273]. These issues are exacerbated when machine learning is involved, due to its inscrutability and nonintuitiveness [274]. The two goals of explanation might help mitigate these concerns, by allowing us to understand how the overall system and policy conform to normative constraints and how individual decisions conform to the policy. These roughly correspond to the distinction between "global" and "local" interpretability in the technical literature.

Requirements for both flavors of explanation can be seen in existing laws. The Fair Credit Reporting Act (FCRA) and ECOA contain an "adverse action notice" requirement. This is an example of the second goal, as it pertains only to the individual decision and does not require transparency about the overall model. In contrast, the GDPR requires "meaningful information about the logic involved" if an individual is subject to a consequential decision by an automated system. This is generally understood as requiring some degree of explanation of both the overall model and the specific decision (table 6.2).

Selbst and Barocas [9] describe several limitations to the usefulness of explanations. We highlight two main ones. The first is the difficulty of producing explanations that are simultaneously faithful to the model and understandable to a nonexpert. If a credit model combines dozens of variables in nonlinear ways, a reason such as "length of employment" or "insufficient income" might fall far short of fully explaining a decision; yet this is all that is required of adverse action notices. Conversely, an explanation of a decision that is fully faithful to a statistical model may be incomprehensible to most decision subjects.

There is an important distinction between explanations given willingly and those demanded by law of a decision maker who has no other incentive to provide them. So far, it has proven challenging for regulators to set legal requirements for what constitutes a good explanation and to assess whether they are working as intended. Empirical evidence supports the difficulty of compelling unwilling decision makers to provide meaningful explanations. For example, a 2018 study found that Facebook's "Why am I seeing this?" ad explanations are vague, incomplete, misleading, and generally useless [275]. The research literature shows that if Facebook *wanted* to provide good explanations, it is possible to do far better.

A more fundamental limitation described by Selbst and Barocas is that even explanations that are faithful and understandable may not enable normative assessment. If an employer uses a screening model that computes a score based on some keywords (a faithful and understandable explanation), it is normatively important to know whether those keywords represent job-related skills, act as proxies (for example, hobbies) that signal social class, or something else. We may be able to make such an assessment given the keywords, but it is not straightforward. Modern methods that provide explanations based on high-level concepts rather than low-level features hold promise in this regard, but the gap between explanations and a full normative justification is likely to remain.

Because of these limitations, there has been a gradual turn from explanations to *algorithmic impact assessments* (AIAs). A full discussion of AIAs is beyond our scope, but we point out how AIAs, at least in an idealized version, differ from explanations. First, AIAs go beyond explaining the model itself and focus on how it was created, how it will be used, and what impacts it is likely to have. Second, the primary consumers of AIAs are not decision subjects but rather regulators and other experts, which alleviates to some degree the faithfulness-comprehensibility trade-off. Third, AIAs must be performed before the model is deployed and must be updated periodically. Some visions of AIAs call for the involvement of impartial external parties in producing them.

The GDPR incorporates one version of AIAs, namely data protection impact assessments (DPIAs). DPIAs must include a description of the algorithm, and the purpose of the processing, an assessment of the necessity of processing in relation to the purpose, an assessment of the risks to individual rights and freedoms, and the measures a company will use to address these risks. It requires consultation "where appropriate" with impacted individuals. But DPIAs are not required to be released to the public. It is too early to tell how effective they will be in practice; much will rely on the behavior of regulators [276, 277].

AIAs are closely related to audits and the terms are sometimes used interchangeably. Nonetheless, there are several types that are worth distinguishing. A 2020 report classifies them into four categories [278]:

- *Bias audits* are conducted by researchers, journalists, or civil society organizations (inspired by social science audits, as we saw in the chapter on testing discrimination in practice).

- *Regulatory audits* are conducted by regulators with statutory powers to examine internal data and systems, modeled on financial audits.

- *Algorithmic risk assessments* are conducted by the developer or procurer of a tool, modeled on environmental impact assessments, to assess possible risks and mitigation strategies before deploying a system.

- *Algorithmic impact evaluations*, which are retrospective and modeled on policy evaluations, are conducted typically by public sector agencies with respect to algorithms which implement a policy.

Algorithmic impact assessments and audits are a burgeoning area [279], and their potential is still being explored. For example, Ifeoma Ajunwa ambitiously argues for reading in existing employment law a duty of care that would obligate employers to conduct audits of automated hiring systems [280]. Gianclaudio Malgieri and Frank Pasquale propose an ex-ante model of regulation where developers of consequential AI systems must perform a risk assessment before deployment and, in some cases, be required to get it approved by an authority [281]. These developments illustrate our point that the turn to machine learning, while indeed creating challenges for antidiscrimination law, creates opportunities alongside it. The software tool and data records involved in automated systems provide a leverage point for regulators.

Consumer Protection

Consumer protection law has completely different roots from either antidiscrimination law or privacy law. Consumer movements first gained ground in the United States in the early twentieth century, initially due to food safety issues [282]. The Federal Trade Commission was established in 1914. Although it initially focused on antitrust, consumer protection gradually became an equally important prong of its activities. It has been the primary agency responsible for consumer protection, and has the statutory authority to challenge "unfair or deceptive" practices in commerce. It is this authority that the agency uses to carry out the activities that it is well known for, such as policing false advertising and fraud, especially identity theft [283]. Many states have consumer protection laws with similar import, enforced by attorneys general.

Credit regulation is one area of consumer protection law that also serves fairness purposes, understood in a broad sense. The Fair Credit Reporting Act of 1974 narrows the permissible uses of credit reports so that they are not used for arbitrary purposes. It gives consumers ways to contest inaccuracies in the data, considering that they are used to make consequential decisions. And it requires notifying the consumer when adverse action is taken against them. FCRA does not address discrimination in the sense of disparate treatment or disparate impact; that would come later, in the Equal Credit Opportunity Act of 1976. In other areas such as employment law, consumer protection does not currently play a role, although scholars have speculatively advocated for treating job candidates as consumers [255].

Outside the traditional sectors of antidiscrimination law, there is a vast swath of everyday digital products in which machine learning biases manifest, and this is where consumer protection law is potentially highly relevant. For example, if a face unlock feature on a smartphone is substantially less accurate for some groups of users, this is not a violation of any of the sector-specific statutes we've discussed so far, but it may fall under FTC authority. Even in domains such as employment discrimination, there are peculiar gaps such as the fact that vendors of algorithmic screening tools are not covered entities, and consumer protection law can potentially help fill this gap.

As of this writing, this is all speculative. So far, the FTC hasn't gone after discriminatory practices except when the company also violates an antidiscrimination statute such as ECOA, which the FTC has authority to enforce [284]. The term "unfairness" in the FTC act has traditionally meant something quite different: taking unjustified advantage of consumers that they cannot avoid. The prototypical example of an unfair commercial practice would be selling snake oil, and a more modern example would be lax data security leading to a data breach. But note that, unlike the antidiscrimination statutes, the FTC has substantially more power to determine what constitutes deceptive and unfair. It is quite possible that the agency will take a broad view of unfairness and that the courts will permit it. The statute allows the FTC to look to "established public policies" in determining what is unfair. It has been suggested that the FTC can thus look to antidiscrimination statutes and rules as scaffolding to build a framework for making determinations about algorithmic discrimination [285].

The FTC's deception authority offers a clearer but more circumscribed option. Companies often make affirmative claims about their products being unbiased. If those claims turn out to be false, that's deception. The same goes for false claims about products being *effective*. This is relevant since many predictive decision-making tools on the market lack evidence of predictive validity, which means that they may subject people to arbitrary decisions. Yet, unless those arbitrary decisions are also systematically biased, they are difficult to challenge under antidiscrimination law. Finally, a lack of transparency may also constitute a deceptive practice. Indeed, the FTC took action against a company that trained a face recognition model on its users' photos while falsely telling them that the feature was an opt-in [286].

Historically, the FTC has had a roller coaster ride in terms of how broadly it treats its authority and how much it flexes its muscles. After being ineffective in the 1960s and reinvigorated in the 1970s [287], Congress rebuked it in the early 1980s and limited its authority, due to lobbying by powerful business interests [283]. It has remained cautious since, and was further caught off guard in the technology era due to limitations of in-house technical expertise. This led to withering criticism for failures such as allowing Cambridge Analytica's exfiltration of Facebook users' data, despite the FTC's long being aware of similar previous events and supposedly closely monitoring Facebook under a "consent decree." In the 2020s, the agency has shown some signs of being invigorated. Specifically on algorithmic discrimination, it published a blog post containing surprisingly strong language [288]. A whitepaper coauthored by a sitting commissioner also lays out an ambitious agenda [289]. All this is to say: the relevance of consumer protection law to algorithmic discrimination remains a wild card.

Beyond the traditional conception of consumer protection, there are emerging ideas such as a duty of loyalty for companies who are entrusted with customers' data [290]. Such companies would be obligated to act in the best interests of people exposing their data and online experiences. The duty of loyalty is a common obligation in fiduciary relationships (for example, a lawyer owes such a duty to her client). But its application to the holders of personal data is a relatively new idea. Although it has been proposed mainly with the aim of improving privacy and minimizing manipulative practices such as "dark patterns," it would have some implications for nondiscrimination as well.

Concluding Thoughts

We've covered a lot of ground in this chapter. We reviewed how the various civil rights movements together gave rise to a relatively robust body of antidiscrimination law in the United States. Generally, this law aims to strike a balance between preventing (and remedying) discrimination, on the one hand, and avoiding excessive burdens on decision makers, on the other. It has been refined, contested, and implemented over decades by the push and pull of court decisions, regulatory agencies, institutional bureaucrats, continued civil rights activism, and shifts in public opinion. It has important limitations: practically, private plaintiffs find it difficult to find legal recourse; more fundamentally, the law itself is far from an ideal route to bring about structural changes.

Turning to the novel challenges raised by automated decision making, there is a risk that discriminatory machine learning might slip through the gaps in how the law conceives of discrimination. In our view, this risk is counterbalanced by the expanded legal toolkit available: privacy law, requirements regarding explanation and impact assessment, and consumer protection law. So far, this potential has lain mostly dormant for various reasons: a narrow conception of privacy, a lack of broad legislation in the United States requiring explanation of consequential decisions, and the timidity of consumer protection agencies. This could yet change; it is possible that the law and enforcement agencies could be reformed to effectively address the new problems. At a minimum, even if not enshrined into law, the tools for intervention that we've discussed offer a blueprint for public interest advocates seeking to hold companies accountable.

7

Testing Discrimination in Practice

In previous chapters, we have seen statistical, causal, legal, and normative fairness criteria. This chapter is about the complexities that arise when we want to apply them in practice.

A running theme of this book is that there is no single test for fairness, that is, there is no single criterion that is both necessary and sufficient for fairness. Rather, there are many criteria that can be used to diagnose potential unfairness or discrimination.

There's often a gap between moral notions of fairness and what is measurable by available experimental or observational methods. This does not mean that we can select and apply a fairness test based on convenience. Far from it: we need moral reasoning and domain-specific considerations to determine which test(s) are appropriate, how to apply them, determine whether the findings indicate wrongful discrimination, and whether an intervention is called for. We will see examples of such reasoning throughout this chapter. Conversely, if a system passes a fairness test, we should not interpret it as a certificate that the system is fair.

In this chapter, our primary objects of study will be real systems rather than models of systems. We must bear in mind that there are many necessary assumptions in creating a model which may not hold in practice. For example, so-called automated decision-making systems rarely operate without any human judgment. Or, we may assume that a machine learning system is trained on a sample drawn from the same population as the one on which it makes decisions, which is also almost never true in practice. Further, decision making in real life is rarely a single decision point, but rather a cumulative series of small decisions. For example, hiring includes sourcing, screening, interviewing, selection, and evaluation, and those steps themselves include many components [291].

An important source of difficulty for testing discrimination in practice is that researchers have a limited ability to observe—much less manipulate—many of the steps in a real-world system. In fact, we'll see that even the decision maker faces limitations in its ability to study the system.

Despite these limitations and difficulties, empirically testing fairness is vital. The studies that we discuss serve as an existence proof of discrimination and provide a lower bound of its prevalence. They enable tracking trends in discrimination over time. When the findings are sufficiently blatant, they justify the need for intervention regardless of any differences in interpretation. And when we do apply a fairness intervention, they help us measure its effectiveness. Finally, empirical research can also help uncover the mechanisms by which discrimination takes place, which enables more targeted and effective interventions. This requires carefully formulating and testing hypotheses using domain knowledge.

The first half of this chapter surveys classic tests for discrimination that were developed in the context of human decision-making systems. The underlying concepts are just as applicable to the study of fairness in automated systems. Much of the first half builds on the discussion of causality in chapter 5 and explains concrete techniques including experiments, difference-in-differences, and regression discontinuity. While these are standard tools in the causal inference toolkit, we will learn about the specific ways in which they can be applied to fairness questions. Then we turn to the application of the observational criteria from chapter 3. Table 7.1 lists, for each test, the fairness criterion that it probes, the type of access to the system that is required, and other nuances and limitations. The second half of the chapter is about testing fairness in algorithmic decision making, focusing on issues specific to algorithmic systems.

Two quick points of terminology: We use the terms *unfairness* and *discrimination* roughly synonymously. There is no overarching definition of either term, but we will make our discussion precise by referring to a specific criterion whenever possible. We use *system* as a shorthand for a decision-making system, such as hiring at a company. It may or may not involve any automation or machine learning.

Part 1: Traditional Tests for Discrimination

Audit Studies

The audit study is a popular technique for diagnosing discrimination. It involves a study design called a field experiment. "Field" refers to the fact that it is an experiment on the actual decision-making system of interest (in the "field," as opposed to a lab simulation of decision making). Experiments on real systems are hard to pull off. For example, we usually have to keep participants unaware that they are in an experiment. But field experiments allow us to study decision making as it actually happens rather than worrying that what we're discovering is an artifact of a lab setting. At the same time, the experiment, by carefully manipulating and controlling variables, allows us to observe a treatment effect, rather than merely observing a correlation.

Table 7.1
Summary of traditional tests and methods, highlighting the relationship to fairness, the observational and experimental access required by the researcher, and limitations

	Test / study design	Fairness notion / application	Access	Notes / limitations
1	Audit study	Blindness	$A\text{-exp} :=, X :=, R$	Difficult to interpret
2	Natural experiment especially diff-in-diff	Impact of blinding	$A\text{-exp} \sim, R$	Confounding; SUTVA violations; other
3	Natural experiment	Arbitrariness	$W \sim, R$	Unobserved confounders
4	Natural experiment especially regr. disc.	Impact of decision	R, Y or Y'	Sample size; confounding; other technical difficulties
5	Regression analysis	Blindness	X, A, R	Unreliable due to proxies
6	Regression analysis	Cond. demographic parity	X, A, R	Weak moral justification
7	Outcome test	Predictive parity	$A, Y \mid \widehat{Y} = 1$	Infra-marginality
8	Threshold test	Sufficiency	$X', A, Y \mid \widehat{Y} = 1$	Model-specific
9	Experiment	Separation/error rate parity	$A, R, \widehat{Y} :=, Y$	Often unethical or impractical
10	Observational test	Demographic parity	A, R	See chapter 3
11	Mediation analysis	"Relevant" mechanism	X, A, R	See chapter 5

Notes:
- $:=$ indicates intervention on some variable (that is, $X :=$ does not represent a new random variable but is simply an annotation describing how X is used in the test)
- $\sim$ represents natural variation in some variable exploited by the researcher
- A-exp represents exposure of a signal of the sensitive attribute to the decision maker
- W represents a feature that is considered irrelevant to the decision
- X' represents a set of features which may not coincide with those observed by the decision maker|
- Y' represents an outcome that may or may not be the one that is the target of prediction|

How to interpret such a treatment effect is a trickier question. In our view, most audit studies, including the ones we describe, are best seen as attempts to test blindness: whether a decision maker directly uses a sensitive attribute. Recall that this notion of discrimination is not necessarily a counterfactual in a valid causal model (chapter 5). Even as tests of blindness, there is debate about precisely what it is that they measure, since the researcher can at best signal race, gender, or another sensitive attribute. This will become clear when we discuss specific studies.

Audit studies were pioneered by the US Department of Housing and Urban Development (HUD) in the 1970s for the purpose of studying the adverse treatment faced by minority home buyers and renters [292]. They have since been successfully applied to many other domains.

In one landmark study by Ayres and Siegelman, the researchers recruited thirty-eight testers to visit about 150 car dealerships to bargain for cars, and record the price they were offered at the end of bargaining [293]. Testers visited dealerships in pairs; testers in a pair differed in terms of race or gender. Both testers in a pair bargained for the same model of car, at the same dealership, usually within a few days of each other.

Pulling off an experiment such as this in a convincing way requires careful attention to detail; here we describe just a few of the many details in the paper. Most significantly, the researchers went to great lengths to minimize any differences between the testers that might correlate with race or gender. In particular, all testers were between twenty-eight and thirty-two years old, had three our four years of post-secondary education, and "were subjectively chosen to have average attractiveness." Further, to minimize the risk of testers' interaction with dealers being correlated with race or gender, every aspect of their verbal or nonverbal behavior was governed by a script. For example, all testers "wore similar 'yuppie' sportswear and drove to the dealership in similar rented cars." They also had to memorize responses to a long list of questions they were likely to encounter. All of this required extensive training and regular debriefs.

The paper's main finding was a large and statistically significant price penalty in the offers received by Black testers. For example, Black males received final offers that were about $1,100 more than White males, which represents a threefold difference in dealer profits based on data about dealer costs. The analysis in the paper has alternative target variables (initial offers instead of final offers; percentage markup instead of dollar offers), alternate model specifications (e.g., to account the two audits in each pair having correlated noise), and additional controls (e.g., bargaining strategy). Thus, there are a number of different estimates, but the core findings remain robust.

A tempting interpretation of this study is that if two people were identical except for race, with one being White and the other being Black, then the offers they should expect to receive would differ by about $1,100. But what does it mean for two people to be identical except for race? Which attributes about them would be the same, and which would be different?

With the benefit of the discussion of ontological instability in chapter 5, we can understand the authors' implicit framework for making these decisions. In our view, they treat race as a stable source node in a causal graph, attempt to hold constant all of its descendants, such as attire and behavior, in order to estimate the direct effect of race on the outcome. But what if one of the mechanisms of what we understand as "racial discrimination" is based on attire and behavior differences? The social construction of race suggests that this is plausible [294].

Note that the authors did not attempt to eliminate differences in accent between testers. Why not? From a practical standpoint, accent is difficult to manipulate. But a more principled defense of the authors' choice is that accent is a part of how we understand race; a part of what it means to *be* Black, White, and so on, so that even if the testers could manipulate their accents, they shouldn't. Accent is subsumed into the "race" node in the causal graph.

To take an informed stance on questions such as this, we need a deep understanding of cultural context and history. They are the subject of vigorous debate in sociology and critical race theory. Our point is this: the design and interpretation of audit studies requires taking positions on contested social questions. It may be futile to search for a single "correct" way to test even the seemingly straightforward fairness notion of whether the decision maker treats similar individuals similarly regardless of race. Controlling for a plethora of attributes is one approach. Arguably, it yields *lower bounds* on the amount of discrimination since it incorporates a thin conception of race. Another is to simply recruit Black testers and White testers, have them behave and bargain as would be their natural inclination, and measure the demographic disparity. Each approach tells us something valuable, and neither is "better."

Another famous audit study by Marianne Bertrand and Sendhil Mullainathan tested discrimination in the labor market [295]. Instead of sending testers in person, the researchers sent in fictitious résumés in response to job ads. Their goal was to test if an applicant's race had an impact on the likelihood of an employer inviting them for an interview. They signaled race in the résumés by using White-sounding names (Emily, Greg) or Black-sounding names (Lakisha, Jamal). By creating pairs of résumés that were identical except for the name, they found that White names were 50 percent more likely to result in a callback than Black names. The magnitude of the effect was equivalent to an additional eight years of experience on a résumé.

Despite the study's careful design, debates over interpretation have inevitably arisen, primarily due to the use of candidate names as a way to signal race to employers. Did employers even notice the names in all cases, and might the effect have been even stronger if they had? Or, can the observed disparities be better explained based on factors correlated with race, such as a preference for more common and familiar names, or an inference of higher socioeconomic status for the candidates with White-sounding names? (Of course, the alternative explanations don't make the observed behavior morally acceptable, but they are important to consider.) Although the authors provide evidence against these interpretations, debate has persisted. For a discussion of critiques of the validity of audit studies, see Devah Pager's survey [296].

In any event, like other audit studies, this experiment tests fairness as blindness. Even simple proxies for race, such as residential neighborhood, were held constant between matched pairs of résumés. Thus, the design likely underestimates the extent to which morally irrelevant characteristics affect callback rates in practice. This is just another way to say that attribute flipping does not generally produce counterfactuals that we care about, and it is unclear if the effect sizes measured have any meaningful interpretation that generalizes beyond the context of the experiment.

Rather, as Issa Kohler-Hausmann argues, audit studies are valuable because they trigger a strong and valid moral intuition [297]. They also serve a practical purpose: when designed well, they illuminate the mechanisms that produce disparities and help guide interventions. For example, the car-bargaining study concluded that the preferences of owners of dealerships don't explain the observed discrimination, that the preferences of other customers may explain some of it, and it offered strong evidence that dealers themselves (rather than owners or customers) are the primary source of the observed discrimination.

Résumé-based audit studies, also known as correspondence studies, have been widely replicated. We briefly present some major findings, with the caveat that there may be publication biases. For example, studies finding no evidence of an effect are in general less likely to be published. Alternately, published null findings might reflect poor experiment design or might simply indicate that discrimination is expressed only in certain contexts.

A 2016 review by Marianne Bertrand and Esther Duflo lists thirty studies from fifteen countries covering nearly all continents revealing pervasive discrimination against racial and ethnic minorities [298]. As the review suggests, the method has also been used to study discrimination based on gender, sexual orientation, and physical appearance. It has also been used outside the labor market, in retail and academia [298]. Finally, trends over time have been studied: a meta-analysis found no change in racial discrimination in hiring against African Americans from 1989 to 2015. There was some indication of declining discrimination against Latin Americans, although the data on this question were sparse [299].

Collectively, audit studies have helped nudge the academic and policy debate away from the naïve view that discrimination is a concern of a bygone era. From a methodological perspective, our main takeaway from the discussion of audit studies is the complexity of defining and testing blindness.

Testing the Impact of Blinding

In some situations, it is not possible to test blindness by randomizing the decision maker's perception of race, gender, or other sensitive attribute. For example, suppose we want to test whether there's gender discrimination in peer review in a particular research field. Submitting real papers with fictitious author identities may result in the reviewer attempting to look up the author and realizing the deception. A design in which the researcher changes author names to those of real people is even more problematic.

There is a slightly different strategy that's more viable: an editor of a scholarly journal in the research field could conduct an experiment in which each paper received is randomly assigned to be reviewed in either a single-blind fashion (in which the author identities are known to the referees) or double-blind fashion (in which author identities are withheld from referees). Indeed, such experiments have been conducted [300], but in general even this strategy can be impractical.

At any rate, suppose that a researcher has access to only observational data on journal review policies and statistics on published papers. Among ten journals in the research field, some introduced double-blind review, and did so in different years. The researcher observes that in each case, right after the switch, the fraction of female-authored papers rose, whereas there was no change for the journals that stuck with single-blind review. Under certain assumptions, this enables the researcher to estimate the impact of double-blind reviewing on the fraction of accepted papers that are female-authored. This hypothetical example illustrates the idea of a "natural experiment," so called because experiment-like conditions arise due to natural variation. Specifically, the study design in this case is called "differences in differences." The first "difference" is between single-blind and double-blind reviewing, and the second "difference" is between journals (row 2 in table 7.1).

Differences-in-differences is methodologically nuanced, and a full treatment is beyond our scope [301]. We briefly note some pitfalls. There may be unobserved confounders: perhaps the switch to double-blind reviewing at each journal happened as a result of a change in editorship, and the new editors also instituted policies that encouraged female authors to submit strong papers. There may also be spillover effects (which violates the SUTVA): a change in policy at one journal can cause a change in the set of papers submitted to other journals. Outcomes are serially correlated (if there is a random fluctuation in the gender composition of the research field due to an entry or exodus of some researchers, the effect will last many years). This complicates the computation of the standard error of the estimate [302]. Finally, the effect of double-blinding on the probability of acceptance of female-authored papers (rather than on the fraction of accepted papers that are female authored) is not identifiable using this technique without additional assumptions or controls.

Even though testing the impact of blinding sounds similar to testing blindness, there is a crucial conceptual and practical difference. Since we are not asking a question about the impact of race, gender, or another sensitive attribute, we avoid running into ontological instability. The researcher doesn't need to intervene on the observable features by constructing fictitious résumés or training testers to use a bargaining script. Instead, the natural variation in features is left unchanged; the study involves real decision subjects. The researcher intervenes only on the decision-making procedure (or exploits natural variation) and evaluates the impact of that intervention on groups of candidates defined by the sensitive attribute A. Thus, A is not a node in a causal graph but merely a way to split the units into groups for analysis. Questions of whether the decision maker actually inferred the sensitive attribute or merely a feature correlated with it are irrelevant to the interpretation of the study. Further, the effect sizes measured do have a meaning that generalizes to scenarios beyond the experiment. For example, a study tested the effect of "résumé whitening," in which minority applicants deliberately concealed cues of their racial or ethnic identity in job application materials to improve their chances of getting a callback [303]. The effects reported in the study are meaningful to job seekers who engage in this practice.

Revealing Extraneous Factors in Decisions

Sometimes natural experiments can be used to show the arbitrariness of decision making rather than unfairness in the sense of nonblindness (row 3 in table 7.1). Recall that arbitrariness is one type of unfairness that we are concerned about in this book (chapter 2). Arbitrariness may refer to the lack of a uniform decision-making procedure or to the incursion of irrelevant factors into the procedure.

For example, a study looked at decisions made by judges in Louisiana juvenile courts, including sentence lengths [304]. It found that in the week following an upset loss suffered by the Louisiana State University (LSU) football team, judges imposed sentences that were 7 percent longer on average. The impact was greater for Black defendants. The effect was driven entirely by judges who got their undergraduate degrees at LSU, suggesting that the effect is due to the emotional impact of the loss. For readers unfamiliar with the culture of college football in the United States, the paper helpfully notes that "Describing LSU football just as an event would be a huge understatement for the residents of the state of Louisiana."

Another well-known study by Shai Danziger, Jonathan Levav, and Liora Avnaim-Pesso on the supposed unreliability of judicial decisions is in fact a poster child for the danger of confounding variables in natural experiments. The study tested the relationship between the order in which parole cases are heard by judges and the outcomes of those cases [305]. It found that the percentage of favorable rulings started out at about 65 percent early in the day before gradually dropping to nearly zero right before the judges' food break, returned to around 65 percent after the break, with the same pattern repeated for the following food break! The authors suggested that judges' mental resources are depleted over the course of a session, leading to poorer decisions. It quickly became known as the "hungry judges" study and has been widely cited as an example of the fallibility of human decision makers (figure 7.1).

The finding would be extraordinary if the order of cases was truly random. In fact, it would be so extraordinary that it has been argued that the study should be dismissed simply based on the fact that the effect size observed is far too large to be caused by psychological phenomena such as judges' attention [306].

The authors were well aware that the order wasn't random, and performed a few tests to see if the effect was associated with factors pertinent to the case (since

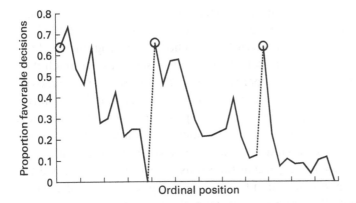

Figure 7.1
Fraction of favorable rulings over the course of a day. The dotted lines indicate food breaks. From Shai Danziger, Jonathan Levav, and Liora Avnaim-Pesso. 2011.

those factors might also impact the probability of a favorable outcome in a legitimate way). They did not find such factors. But it turned out they didn't look hard enough. A follow-up investigation revealed multiple confounders and potential confounders, including the fact that prisoners without an attorney are presented last within each session, and tend to prevail at a much lower rate [307]. This invalidates the conclusion of the original study.

Testing the Impact of Decisions and Interventions

An underappreciated aspect of fairness in decision making is the impact of the decision on the decision subject. In our prediction framework, the target variable (Y) is not impacted by the score or prediction (R). But this is not true in practice. Banks set interest rates for loans based on the predicted risk of default, but setting a higher interest rate makes a borrower more likely to default. The impact of the decision on the outcome is a question of causal inference.

There are other important questions we can ask about the impact of decisions. What is the utility or cost of a positive or negative decision to different decision subjects (and groups)? For example, admission to a college may have a different utility to different applicants based on the *other* colleges where they were or weren't admitted. Decisions may also have effects on people who are not decision subjects. For instance, incarceration impacts not just individuals but communities [169]. Measuring these costs allows us to be more scientific about setting decision thresholds and adjusting the trade-off between false positives and negatives in decision systems.

One way to measure the impact of decisions is via experiments, but again, they can be infeasible for legal, ethical, and technical reasons. Instead, we highlight a natural experiment design for testing the impact of a decision—or a fairness intervention—on the candidates, called *regression discontinuity* (row 4 in table 7.1).

Suppose we would like to test if a merit-based scholarship program for first-generation college students has lasting beneficial effects—say, on how much they earn after college. We cannot simply compare the average salary of students who did and did not win the scholarship, as those two variables may be confounded by intrinsic ability or other factors. But suppose the scholarships were awarded based on test scores, with a cutoff of 85 percent. Then we can compare the salary of students with scores of 85 to 86 percent (and who thus were awarded the scholarship) with those of students with scores of 84 to 85 percent (and thus who were not awarded the scholarship). We may assume that within this narrow range of test scores, scholarships are awarded essentially randomly. For example, if the variation (standard error) in test scores for students of identical ability is 5 percentage points, then the difference between 84 and 86 percent is of minimal significance. Thus we can estimate the impact of the scholarship as if we did a randomized controlled trial.

We need to be careful, though. If we consider too narrow a band of test scores around the threshold, we may end up with insufficient data points for inference. If we consider a wider band of test scores, the students in this band may no longer be exchangeable units for the analysis.

Another pitfall arises because we assumed that the set of students who receive the scholarship is precisely the set that is above the threshold. If this assumption fails, it immediately introduces the possibility of confounders. Perhaps the test score is not the only scholarship criterion, and income is used as a secondary criterion. Or, some students offered the scholarship may decline it because they already received another scholarship. Other students may not avail themselves of the offer because the paperwork required to claim it is cumbersome. If it is possible to take the test multiple times, wealthier students may be more likely to do so until they meet the eligibility threshold.

Purely Observational Tests

The final category of quantitative tests for discrimination is purely observational. When we are not able to do experiments on the system of interest, nor do we have the conditions that enable quasi-experimental studies, there are still many questions we can answer with purely observational data.

One question that is often studied using observational data is whether the decision maker used the sensitive attribute; this can be seen as a loose analogue of audit studies. This type of analysis is often used in the legal analysis of disparate treatment, although there is a deep and long-standing legal debate on whether and when explicit consideration of the sensitive attribute is necessarily unlawful [308].

The most common way to do this is to use regression analysis to see if attributes other than the protected attributes can collectively "explain" the observed decisions [309] (row 5 in table 7.1). If they don't, then the decision maker must have used the sensitive attribute. However, this is a brittle test. As discussed in chapter 3, given a sufficiently rich dataset, the sensitive attribute can be reconstructed using the other attributes. It is no surprise that attempts to apply this test in a legal context can turn into dueling expert reports, as seen in the *SFFA v. Harvard* case discussed in chapter 5.

We can of course try to go deeper with observational data and regression analysis. To illustrate, consider the gender pay gap. A study might reveal that there is a gap between genders in wage per hour worked for equivalent positions in a company. A rebuttal might claim that the gap disappears after controlling for college GPA and performance review scores. Such studies can be seen as tests for *conditional demographic parity* (row 6 in table 7.1). Note that this requires strong assumptions about the functional form of the relationship between the independent variables and the target variable.

It can be hard to make sense of competing claims based on regression analysis. Which variables should we control for, and why? There are two ways in which we can put these observational claims on a more rigorous footing. The first is to use a causal framework to make our claims more precise. In this case, causal modeling might alert us to unresolved questions: Why do performance review scores differ by gender? What about the gender composition of different roles and levels of seniority? Exploring these questions may reveal unfair practices. Of course, in this instance, the questions we raised are intuitively obvious, but other cases may be more intricate.

The second way to go deeper is to apply our normative understanding of fairness to determine which paths from gender to wage are morally problematic. If the pay gap is caused by the (well-known) gender differences in negotiating for pay raises, does the employer bear the moral responsibility to mitigate it? This is, of course, a normative and not a technical question.

Outcome-Based Tests
So far in this chapter we've presented many scenarios—screening job candidates, peer review, parole hearings—that have one thing in common: while they all aim to predict some outcome (job performance, paper quality, recidivism), the researcher does not have access to data on the true outcomes.

Lacking ground truth, the focus shifts to the observable characteristics at decision time, such as job qualifications. A persistent source of difficulty in these settings is for the researcher to construct two sets of samples that differ only in the sensitive attribute and not in any of the relevant characteristics. This is often an untestable assumption. Even in an experimental setting such as a résumé audit study, there is substantial room for different interpretations: Did employers infer race from names or socioeconomic status? And in observational studies, the findings might turn out to be invalid because of unobserved confounders (such as in the hungry judges study).

But if outcome data are available, then we can do at least one test of fairness without needing any of the observable features (other than the sensitive attribute): specifically, we can test for sufficiency, which requires that the true outcome be conditionally independent of the sensitive attribute given the prediction ($Y \perp A | R$). For example, in the context of lending, if the bank's decisions satisfy sufficiency, then among applicants in any narrow interval of predicted probability of default (R), we should find the same rate of default (Y) for applicants of any group (A).

Typically, the decision maker (the bank) can test for sufficiency, but an external researcher cannot, since the researcher gets to observe only $\widehat{Y}$ (i.e., whether or not the loan was approved) and not R. Such a researcher can test predictive parity rather than sufficiency. Predictive parity requires that the rate of default (Y) for favorably classified applicants ($\widehat{Y} = 1$) of any group (A) be the same. This observational test is called the *outcome test* (row 7 in table 7.1).

Here is a tempting argument based on the outcome test: if one group (say, women) who receive loans have a *lower* rate of default than another (men), it suggests that the bank applies a *higher* bar for loan qualification for women. Indeed, this type of argument was the original motivation behind the outcome test. But it is a logical fallacy; sufficiency does not imply predictive parity (or vice versa). To see why, consider a thought experiment involving the Bayes optimal predictor. In the hypothetical figure 7.2, applicants to the left of the vertical line qualify for the loan.

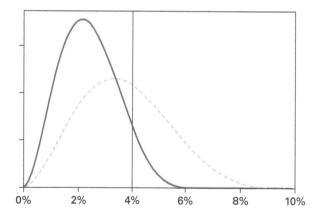

Figure 7.2
Hypothetical probability density of loan default for two groups, women (solid line) and men (dashed line).

Since the area under the curve to the left of the line is concentrated further to the right for men than for women, men who receive loans are more likely to default than women. Thus, the outcome test would reveal that predictive parity is violated, whereas it is clear from the construction that sufficiency is satisfied, and the bank applies the same bar to all groups.

This phenomenon is called *inframarginality*, that is, the measurement is aggregated over samples that are far from the decision threshold (margin). If we are indeed interested in testing sufficiency (equivalently, whether the bank applied the same threshold to all groups), rather than predictive parity, this is a problem. To address it, we can somehow try to narrow our attention to samples that are close to the threshold. This is not possible with $(\widehat{Y}, A, Y)$ alone: without knowing R, we don't know which instances are close to the threshold. However, if we also had access to some set of features X' (which need not coincide with the set of features X observed by the decision maker), it becomes possible to test for violations of sufficiency. The *threshold test* is a way to do this (row 8 in table 7.1). A full description is beyond our scope [310]. One limitation is that it requires a model of the joint distribution of (X', A, Y), whose parameters can be inferred from the data, whereas the outcome test is model-free.

While we described inframarginality as a limitation of the outcome test, it can also be seen as a benefit. When using a marginal test, we treat the distribution of applicant characteristics as a given, and miss the opportunity to ask: *Why* are some individuals so far from the margin? Ideally, we can use causal inference to answer this question, but when the data at hand don't allow this, nonmarginal tests might be a useful starting point for diagnosing unfairness that originates "upstream" of the decision maker. Similarly, error rate disparity, to which we will now turn, while crude by comparison to more sophisticated tests for discrimination, attempts to capture some of our moral intuitions for why certain disparities are problematic.

Separation and Selective Labels

Recall that separation is defined as $R \perp A | Y$. At first glance, it seems that there is a simple observational test analogous to our test for sufficiency ($Y \perp A | R$). However, this is not straightforward, even for the decision maker, because outcome labels can be observed for only some of the applicants (i.e., the ones who received favorable decisions). Trying to test separation using this sample suffers from selection bias. This is an instance of what is called the *selective labels problem*. The issue also affects the computation of false positive and false negative rate parity, which are binary versions of separation.

More generally, the selective labels problem is the issue of selection bias in evaluating decision-making systems due to the fact that the very selection process we wish to study determines the sample of instances that are observed. It is not specific to the issue of testing separation or error rates: it affects the measurement of other fundamental metrics such as accuracy as well. It is a serious and often overlooked issue that has been the subject of some study [311].

One way to get around this barrier is for the decision maker to employ an experiment in which some sample of decision subjects receive positive decisions regardless of the prediction (row 9 in table 7.1). However, such experiments raise ethical concerns and are rarely done in practice. In machine learning, some experimentation is necessary in settings where there does not exist offline data for training the classifier, which must instead simultaneously learn and make decisions [312].

One scenario where it is straightforward to test separation is when the "prediction" is not actually a prediction of a future event, but rather when machine learning is used for automating human judgment, such as harassment detection in online comments. In these applications, it is indeed possible and important to test error rate parity.

Taste-Based and Statistical Discrimination

We have reviewed several methods of detecting discrimination but we have not addressed the question of why discrimination happens. A long-standing way to try to answer this question from an economic perspective is to classify discrimination as *taste-based* or *statistical*. A taste-based discriminator is motivated by an irrational animus or prejudice against a group. As a result, they are willing to make suboptimal decisions by passing up opportunities to select candidates from that group, even though they will incur a financial penalty for doing so. This is the classic model of discrimination in labor markets introduced by Gary Becker in 1957 [313].

A statistical discriminator, in contrast, aims to make optimal predictions about the target variable using all available information, including the protected attribute. This theory was developed in the early 1970s by Edmund Phelps and Kenneth Arrow among others [314, 315]. In the simplest model of statistical discrimination, two conditions hold. First, the distribution of the target variable differs by group. The usual example is of gender discrimination in the workplace, involving an employer who believes that women are more likely to take time off due to pregnancy (resulting in lower job performance). The second condition is that the observable characteristics do not allow a perfect prediction of the target variable, which is essentially always the case in practice. Under these two conditions, the optimal prediction will differ by group even when the relevant characteristics are identical. In this example, the employer would be less likely to hire a woman than an equally qualified man. There's a nuance here: from a moral perspective we would say that the employer above discriminates against all female candidates. But under the definition of statistical discrimination, the employer discriminates against only the female candiates who would not have taken time off if hired (and in fact discriminates in favor of the female candidates who would take time off if hired).

While some authors put much weight understanding discrimination based on the taste-based versus statistical categorization, we deemphasize it in this book. Several reasons motivate our choice. First, since we are interested in extracting lessons for statistical decision-making systems, the distinction is not that helpful: such systems will not exhibit taste-based discrimination unless prejudice is explicitly programmed into them (while that is certainly a possibility, it is not a primary concern of this book).

Second, there are practical difficulties in distinguishing between taste-based and statistical discrimination. Often, what might seem to be a "taste" for discrimination is simply the result of an imperfect understanding of the decision maker's information and beliefs. For example, at first sight, the findings of the car-bargaining study may look like a clear-cut case of taste-based discrimination. But maybe the dealer knows that different customers have different access to competing offers and therefore have different willingness to pay for the same item. Then, the dealer uses race as a proxy for this amount (correctly or not). In fact, the paper provides tentative evidence toward this interpretation. The reverse is also possible: if the researcher does not know the full set of features observed by the decision maker, taste-based discrimination might be mischaracterized as statistical discrimination.

Third, many of the fairness questions of interest to us, such as structural discrimination, don't map on to either of these criteria (as they consider only causes that are relatively proximate to the decision point). We discuss structural discrimination in chapter 8.

Finally, it's also worth noting that thinking about discrimination in terms of the dichotomy of taste-based and statistical is associated with the policy position that fairness interventions are unnecessary. In this view, firms that practice taste-based discrimination will go out of business. As for statistical discrimination, it is argued to be either justified, or futile to proscribe because firms will find workarounds. For example, laws restricting employers from asking about applicants' criminal history resulted in employers using race as a proxy for it [316].

Of course, that's not necessarily a reason to avoid discussing taste-based and statistical discrimination, as the policy position in no way follows from the technical definitions and models themselves; it's just a relevant caveat for the reader who might encounter these dubious arguments in other sources.

Although we deemphasize this distinction, we consider it critical to study the sources and mechanisms of discrimination. This helps us design effective and well-targeted interventions. For example, several studies (including the car-bargaining study) test whether the source of discrimination lies in the owner, employees, or customers.

An example of a study that can be difficult to interpret without understanding the mechanism is a 2015 résumé-based audit study that revealed a 2:1 faculty preference for women for STEM tenure-track positions [317]. Consider the range of possible explanations: animus against men; a desire to compensate for past disadvantage suffered by women in STEM fields; a preference for a more diverse faculty (assuming that the faculties in question are currently male dominated); a response to financial incentives for diversification frequently provided by universities to STEM departments; and an assumption by decision makers that due to prior descrimination, a female candidate with a CV equivalent to that of a male candidate is of greater intrinsic ability. Note that if this assumption is correct, then a preference for female candidates is accuracy maximizing (as a predictor of career success). It is also required by some fairness criteria, such as counterfactual fairness.

To summarize, rather than a one-size-fits-all approach to understanding mechanisms such as taste-based versus statistical discrimination, more useful is a nuanced and domain-specific approach where we formulate hypotheses in part by studying decision-making processes and organizations, especially in a qualitative way. Let us now turn to those studies.

Studies of Decision-Making Processes and Organizations

One way to study decision-making processes is through surveys of decision makers or organizations. Sometimes such studies reveal blatant discrimination, such as strong racial preferences by employers [318]. Over the decades, however, such overt attitudes have become less common, or at least less likely to be expressed [319]. Discrimination tends to operate in more subtle, indirect, and covert ways.

Ethnographic studies excel at helping us understand covert discrimination. Ethnography is one of the main research methods in the social sciences and is based on the idea of the researcher being embedded among the research subjects for an extended period of time as they go about their daily activities. It uses a set of qualitative methods that are complementary to and symbiotic with quantitative ones. Ethnography allows us to ask questions that are deeper than quantitative methods permit and to produce richly detailed accounts of culture. It also helps formulate hypotheses that can be tested quantitatively.

A good illustration is the book *Pedigree* by Lauren Rivera, which examines hiring practices in a set of elite consulting, banking, and law firms [320]. These firms together constitute the majority of the highest-paying and most desirable entry-level jobs for college graduates. The author used two standard ethnographic research methods. The first is a set of 120 interviews in which she presented as a graduate student interested in internship opportunities. The second method is called *participant observation*: she worked in an unpaid human resources position at one of the firms for nine months, after obtaining consent to use her observations for research. There are several benefits to the researcher becoming a participant in the culture: it provides a greater level of access, allows the researcher to ask more nuanced questions, and makes it more likely that the research subjects would behave as they would when not being observed.

Several insights from the book are relevant to us. First, the hiring process has about nine stages, including outreach, recruitment events, screening, multiple rounds of interviews and deliberations, and "sell" events. This highlights why any quantitative study that focuses on a single slice of the process (say, evaluation of résumés) is limited in scope. Second, the process bears little resemblance to the ideal of predicting job performance based on a standardized set of attributes, albeit noisy ones, that we describe in chapter 1. Interviewers pay a surprising amount of attention to attributes that should be irrelevant or minimally relevant, such as leisure activities, but that instead serve as markers of class. Applicants from privileged backgrounds are more likely to be viewed favorably, both because they are able to spare more time for such activities and because they have the insider knowledge that these seemingly irrelevant attributes matter in recruitment. The signals that firms do use as predictors of job performance, such as admission to elite universities—the *pedigree* in the book's title—are also highly correlated with socioeconomic status. The authors argue that these hiring practices help explain why elite status is perpetuated in society along hereditary lines. In our view, the careful use of statistical methods in hiring, despite their limits, may mitigate the strong preferences based on social class exposed in the book.

Another book, *Inside Graduate Admissions* by Julie Posselt, focuses on education rather than labor market [321]. It resulted from the author's observations of decision making by graduate admissions committees in nine academic disciplines over two years. A striking theme that pervades this book is the tension between formalized and holistic decision making. For instance, committees arguably overrely on GRE scores, despite stating that they consider their predictive power to be limited. As it turns out, one reason for the preference for GRE scores and other quantitative criteria is that they avoid the difficulties of subjective interpretation associated with signals such as reference letters. This is considered valuable because it *minimizes tensions between faculty members* in the admissions process. On the other hand, decision makers are implicitly aware (and occasionally explicitly articulate) that if admissions criteria are too formal, then some groups of applicants—notably, applicants from China—would be successful at a far greater rate, and this is considered undesirable. This motivates a more holistic set of criteria, which often include idiosyncratic factors such as an applicant's hobby being considered "cool" by a faculty member. The author argues that admissions committees use a facially

neutral set of criteria, characterized by an almost complete absence of explicit, substantive discussion of applicants' race, gender, or socioeconomic status, but which
nonetheless perpetuates inequities. For example, there is a reluctance to take on students from underrepresented backgrounds whose profiles suggest that they would
benefit from more intensive mentoring.

This concludes the first part of the chapter. Now let us turn to algorithmic systems. The background we've built up so far will prove useful. In fact, the traditional
tests of discrimination are just as applicable to algorithmic systems. But we will also
encounter many novel issues.

Part 2: Testing Discrimination in Algorithmic Systems

An early example of discrimination in an algorithmic system is from the 1950s.
In the United States, applicants for medical residency programs provide a ranked
list of their preferred hospital programs to a centralized system, and hospitals likewise rank applicants. A matching algorithm takes these preferences as input and
produces an assignment of applicants to hospitals that optimizes mutual desirability.

Early versions of the system discriminated against couples who wished to stay
geographically close, because couples could not accurately express their joint preferences: for example, each partner might prefer a hospital over all others but only
if the other partner also matched to the same hospital [43, 322]. This is a noncomparative notion of discrimination: the system does injustice to an applicant (or
a couple) when it does not allow them to express their preferences, regardless of
how other applicants are treated. Note that none of the tests for fairness that we
have discussed are capable of detecting this instance of discrimination, as it arises
because of dependencies between pairs of units, which is not something we have
modeled.

There was a crude attempt in the residency matching system to capture joint preferences, involving designating one partner in each couple as the "leading member";
the algorithm would match the leading member without constraints and then match
the other member to a proximate hospital if possible. Given the prevailing gender
norms at that time, it is likely that this method had a further discriminatory impact
on women in heterosexual couples.

Despite this and other early examples, it was in the 2010s that testing unfairness
in real-world algorithmic systems became a pressing concern and a distinct area of
research. This work has much in common with the social science research that we
reviewed, but the targets of research have expanded considerably. In the rest of this
chapter, we review and attempt to systematize the research methods in several areas
of algorithmic decision making: various applications of natural language processing
and computer vision, ad targeting platforms, search and information retrieval tools,

and online markets (ride hailing, vacation rentals, etc.). Much of this research has focused on drawing attention to the discriminatory effects of specific, widely used tools and platforms at specific points in time. While that is a valuable goal, we aim to highlight broader, generalizable themes in our review. We close the chapter by identifying common principles and methods behind this body of research.

Fairness Considerations in Applications of Natural Language Processing

One of the central tasks in natural language processing (NLP) is language identification: determining the language that a given text is written in. It is a precursor to virtually every other NLP operation on the text, such as translation to the user's preferred language on social media platforms. It is considered a more or less solved problem, with relatively simple models based on n-grams of characters achieving high accuracies on standard benchmarks, even for short texts that are a few words long.

However, a 2016 study showed that a widely used tool, langid.py, which incorporates a pretrained model, had substantially more false negatives for tweets written in African American English (AAE) compared to those written in more common dialectal forms: 13.2 percent of AAE tweets were classified as non-English compared to 7.6 percent of "White-aligned" English tweets. AAE is a set of English dialects commonly spoken by Black people in the United States (of course, there is no implication that all Black people in the United States primarily speak AAE or even speak it at all). The authors' construction of the AAE and White-aligned corpora themselves involved machine learning as well as validation based on linguistic expertise; we defer a full discussion to chapter 9, on datasets. The observed error rate disparity is likely a classic case of underrepresentation in the training data.

Unlike the audit studies of car sales or labor markets discussed earlier, here it is not necessary (or justifiable) to control for any features of the texts, such as the level of formality. While it may certainly be possible to "explain" disparate error rates based on such features, that is irrelevant to the questions of interest in this context, such as whether NLP tools will perform less well for one group of users compared to another.

NLP tools range in their application from aids to online interaction to components of decisions with major career consequences. In particular, NLP is used in predictive tools for screening of résumés in the hiring process. There is some evidence of potential discriminatory impacts of such tools, both from employers themselves [323] and from applicants [324], but it is limited to anecdotes. There is also evidence from the lab experiments on the task of predicting occupation from online biographies [325].

Table 7.2
Four types of NLP tasks and the types of unfairness that can result. Note that the traditional tests discussed in part 1 operate in the context of predicting outcomes (row 3 in this table)

Type of task	Examples	Sources of disparity	Harm
Perception	Language ID, speech-to-text	Underrepresentation in training corpus	Degraded service
Automating judgment	Toxicity detection, essay grading	Human labels, underrepresentation in training corpus	Adverse decisions
Predicting outcomes	Résumé filtering	Various, including human labels	Adverse decisions
Sequence prediction	Language generation, translation	Cultural stereotypes, historical prejudices	Representational harm

We briefly survey other findings. Automated essay grading software tends to assign systematically lower scores to some demographic groups [326] compared to human graders, whose scores may themselves be discriminatory [327]. Hate speech detection models use markers of dialect as predictors of toxicity, according to a lab study [328], resulting in discrimination against minority speakers. Many sentiment analysis tools assign systematically different scores to text based on race-aligned or gender-aligned names of people mentioned in the text [329]. Speech-to-text systems perform worse for speakers with certain accents [330]. In all these cases, the author or speaker of the text is potentially harmed. In other NLP systems, such as those involving natural language generation or translation, there is a different type of fairness concern, namely the generation of text reflecting cultural prejudices resulting in representational harm to a group of people [331]. Table 7.2 summarizes this discussion.

There is a line of research on cultural stereotypes reflected in word embeddings. Word embeddings are representations of linguistic units; they do not correspond to any linguistic or decision-making task. As such, lacking any notion of ground truth or harms to people, it is not meaningful to ask fairness questions about word embeddings without reference to specific downstream tasks for which they might be used. More generally, it is meaningless to ascribe fairness as an attribute of models as opposed to actions, outputs, or decision processes.

Demographic Disparities and Questionable Applications of Computer Vision

Like NLP, computer vision technology has made major headway in the 2010s due to the availability of large-scale training corpora and improvements in hardware for training neural networks. Today, many types of classifiers are used in commercial products to analyze images and videos of people. Unsurprisingly, they often exhibit disparities in performance based on gender, race, skin tone, and other attributes, as well as deeper ethical problems.

A prominent demonstration of error rate disparity comes from Joy Buolamwini and Timnit Gebru's analysis of three commercial tools designed to classify a person's gender as female or male based on an image, developed by Microsoft, IBM, and Face++, respectively [332]. The study found that all three classifiers perform better on male faces than female faces (8.1–20.6 percent difference in error rate). Further, all perform better on lighter faces than darker faces (11.8–19.2 percent difference in error rate), and worst on darker female faces (20.8–34.7 percent error rate). Finally, since all classifiers treat gender as binary, the error rate for people of nonbinary gender can be considered to be 100 percent.

If we treat the classifier's target variable as gender and the sensitive attribute as skin tone, we can decompose the observed disparities into two separate issues: first, female faces are classified as male more often than male faces are classified as female. This can be addressed relatively easily by recalibrating the classification threshold without changing the training process. The second and deeper issue is that darker faces are misclassified more often than lighter faces.

Image classification tools have found it particularly challenging to achieve geographic equity due to the skew in training datasets. A 2019 study evaluated five popular object recognition services on images of household objects from fifty-four countries [333]. It found significant accuracy disparities between countries, with images from lower-income countries being less accurately classified. The authors point out that household objects such as dish soap or spice containers tend to look very different in different countries. These issues are exacerbated when images of people are being classified. A 2017 analysis found that models trained on ImageNet and Open Images, two prominent datasets for object recognition, performed dramatically worse at recognizing images of bridegrooms from countries such as Pakistan and India compared to those from North American and European countries (the wedding attire of the former were often classified as chain mail, a type of armor) [334].

Several other types of unfairness are known through anecdotal evidence in image classification and face recognition systems. At least two different image classification systems are known to have applied demeaning and insulting labels to photos of people [335, 336]. Face recognition systems have been anecdotally reported to exhibit the cross-race effect, wherein they are more likely to confuse faces of two people who are from a racial group that is underrepresented in the training data [337]. This possibility was shown in a simple linear model of face recognition as early as 1991 [338]. Many commercial products have had difficulty detecting faces of darker-skinned people [339, 340]. Similar results are known from lab studies of publicly available object detection models [341].

More broadly, computer vision techniques seem to be particularly susceptible to being misused in ways that are fundamentally ethically questionable regardless of accuracy. Consider gender classification: while Microsoft, IBM, and Face++ have worked to mitigate the accuracy disparities discussed above, a more important question is: Why build a gender classification tool in the first place? By far the most common application appears to be displaying targeted advertisements based on inferred gender (and many other inferred characteristics, including age, race, and current mood) in public spaces, such as billboards, stores, or screens in the back seats of taxis. We won't recap the objections to targeted advertising here, but it is an extensively discussed topic and the practice is strongly opposed by the public, at least in the United States [342].

Morally dubious computer vision technology goes well beyond this example, and includes apps that "beautify" images of users' faces, that is, edit them to better conform to mainstream notions of attractiveness; apps that "recognize" emotion, which has been alleged to be a pseudoscience; and apps that analyze video footage for cues such as body language for screening job applicants [343].

Search and Recommendation Systems: Three Types of Harms

Search engines, social media platforms, and recommendation systems have different goals and underlying algorithms, but they do have many things in common from a fairness perspective. They are not decision systems and don't provide or deny people opportunities, at least not directly. Instead, there are (at least) three types of disparities and attendant harms that may arise in these systems. First, they may serve the informational needs of some consumers (searchers or users) better than others. Second, they may create inequities among *producers* (content creators) by privileging certain content over others. Third, they may create representational harms by amplifying and perpetuating cultural stereotypes. There are a plethora of other ethical concerns about information platforms, such as the potential to contribute to the political polarization of society. However, we limit our attention to harms that can be considered to be forms of discrimination.

Unfairness to Consumers

An illustration of unfairness to consumers comes from a study of collaborative filtering recommender systems that used theoretical and simulation methods (rather than a field study of a deployed system) [344]. Collaborative filtering is an approach to recommendations that is based on the explicit or implicit feedback (e.g., ratings and consumption, respectively) provided by other users of the system. The intuition behind it is seen in the "users who liked this item also liked . . ." feature on many services. The study found that such systems can underperform for minority groups in the sense of being worse at recommending content that those users would like. A related but distinct reason for underperformance occurs when users from one group are less observable, that is, less likely to provide ratings. The underlying assumption is that different groups have different preferences, so that what the system learns about one group doesn't generalize to other groups.

In general, this type of unfairness is hard to study in real systems (not just by external researchers but also by system operators themselves). The main difficulty is accurately measuring the target variable. The relevant target construct from a fairness perspective is users' satisfaction with the results or how well the results served the users' needs. Metrics such as clicks and ratings serve as crude proxies for the target, and are themselves subject to demographic measurement biases. Companies do expend significant resources on A/B testing or other experimental methods for optimizing search and recommendation systems, and frequently measure demographic differences as well. But to reiterate, such tests almost always emphasize metrics of interest to the firm rather than the benefit or payoff for the user.

A rare attempt to transcend this limitation comes from an (internal) audit study of the Bing search engine by Rishabh Mehrotra, Ashton Anderson, Fernando Diaz et al. [345]. The authors devised methods to disentangle user satisfaction from other demographic-specific variation by controlling for the effects of demographic factors on behavioral metrics. They combined it with a method for inferring latent differences directly instead of estimating user satisfaction for each demographic group and then comparing these estimates. This method infers which impression, among a randomly selected pair of impressions, led to greater user satisfaction. They did this using proxies for satisfaction such as reformulation rate. Reformulating a search query is a strong indicator of dissatisfaction with the results. Based on these methods, they found no gender differences in satisfaction but mild age differences.

Unfairness to Producers

In 2019, a group of content creators sued YouTube alleging that YouTube's algorithms as well as human moderators suppressed the reach of LGBT-focused videos and the ability to earn ad revenue from them. This is a distinct type of issue from that discussed above, as the claim is about a harm to producers rather than consumers (although, of course, YouTube viewers interested in LGBT content are also presumably harmed). There are many other ongoing allegations and controversies that fall into this category: partisan bias in search results and social media platforms, search engines favoring results from their own properties over competitors, fact-checking of online political ads, and inadequate (or, conversely, overaggressive) policing of purported copyright violations. It is difficult to meaningfully discuss and address these issues through the lens of fairness and discrimination rather than a broader perspective of power and accountability. The core issue is that when information platforms have control over public discourse, they become the arbiters of conflicts between competing interests and viewpoints. From a legal perspective, these issues fall primarily under antitrust law and telecommunication regulation rather than antidiscrimination law.

Representational Harms

The book *Algorithms of Oppression* drew attention to the ways in which search engines reinforce harmful racial, gender, and intersectional stereotypes [48]. There have also been quantitative studies of some aspects of these harms. In keeping with our quantitative focus, let's discuss a study that measured how well the gender skew in Google image search results for forty-five occupations (*author, bartender, construction worker, ...*) corresponded to the real-world gender skew of the respective occupations [37]. This can be seen as a test for calibration: instances are occupations and the fraction of women in the search results is viewed as a predictor of the fraction of women in the occupation in the real world. The study found weak evidence for stereotype exaggeration, that is, imbalances in occupational statistics are exaggerated in image search results. However, the deviations were minor.

Consider a thought experiment: suppose the study had found no evidence of miscalibration. Is the resulting system fair? It would be simplistic to answer in the affirmative for at least two reasons. First, the study tested calibration between image search results and occupational statistics *in the United States*. Gender stereotypes of occupations as well as occupational statistics differ substantially between countries and cultures. Second, accurately reflecting real-world statistics may still constitute a representational harm when those statistics are skewed and themselves reflect a history of prejudice. Such a system contributes to the lack of visible role models for underrepresented groups. To what extent information platforms should bear responsibility for minimizing these imbalances, and what types of interventions are justified, remain matters of debate.

Understanding Unfairness in Ad Targeting

Ads have long been targeted in relatively crude ways. For example, a health magazine might have ads for beauty products, exploiting a coarse correlation. In contrast to previous methods, online targeting offers several key advantages to advertisers: granular data collection about individuals, the ability to reach niche audiences (in theory, the audience size can be one, since ad content can be programmatically generated and customized with user attributes as inputs), and the ability to measure conversion (conversion is when someone who views the ad clicks on it, and then takes another action such as a purchase). To date, ad targeting has been one of the most commercially impactful applications of machine learning.

The complexity of modern ad targeting results in many avenues for disparities in the demographics of ad views, which we will study. But it is not obvious how to connect these disparities to fairness. After all, many types of demographic targeting such as clothing ads by gender are considered innocuous.

There are two frameworks for understanding potential harms from ad targeting. The first framework sees ads as unlocking opportunities for their recipients, because they provide information that the viewer might not have. This is why targeting employment or housing ads based on protected categories may be unfair and unlawful. The domains where targeting is legally prohibited broadly correspond to those that impact civil rights, and reflect the complex histories of discrimination in those domains, as discussed in chapter 6.

The second framework views ads as tools of persuasion rather than information dissemination. In this framework, harms arise from ads being manipulative—that is, exerting covert influence instead of making forthright appeals—or exploiting stereotypes [346]. Users are harmed by being targeted with ads that provide them negative utility, as opposed to the first framework, in which the harm comes from missing out on ads with positive utility. The two frameworks don't necessarily contradict each other. Rather, individual ads or ad campaigns can be seen as either primarily informational or primarily persuasive, and accordingly, one or the other framework might be appropriate for analysis.

There is a vast literature on how race and gender are portrayed in ads; we consider this literature to fall under the persuasion framework [347]. However, this line of inquiry has yet to turn its attention to online targeted advertising, which has the potential for accentuating the harms of manipulation and stereotyping by targeting specific people and groups. Thus, the empirical research that we highlight falls under the informational framework.

There are roughly three mechanisms by which the same targeted ad may reach one group more often than another. The most obvious is the use of explicit targeting criteria by advertisers: either the sensitive attribute itself or a proxy for it (such as ZIP code as a proxy for race). For example, Facebook allows thousands of targeting categories, including categories that are automatically constructed by the system based on users' free-form text descriptions of their interests. Investigations by ProPublica found that these categories included "Jew haters" and many other antisemitic terms [348]. The company has had difficulty eliminating even direct proxies for sensitive categories, resulting in repeated exposés.

The second disparity-producing mechanism is optimization of click rate (or another measure of effectiveness), which is one of the core goals of algorithmic targeting. Unlike the first category, this does not require explicit intent by the advertiser or the platform. The algorithmic system may predict a user's probability of engaging with an ad based on her past behavior, her expressed interests, and other factors (including, potentially, explicitly expressed sensitive attributes).

The third mechanism is market effects: delivering an ad to different users may cost the advertiser different amounts. For example, some researchers have observed that women cost more to advertise to than men and hypothesized that this is because women clicked on ads more often, leading to a higher measure of effectiveness [267, 349]. Thus if the advertiser simply specifies a total budget and leaves the delivery up to the platform (which is a common practice), then the audience composition will vary depending on the budget: smaller budgets will result in the less expensive group being overrepresented.

In terms of methods to detect these disparities, researchers and journalists have used broadly two approaches: interact with the system either as a user or as an advertiser. Amit Datta, Michael Carl Tschantz, and Anupam Datta created simulated users that had the "gender" attribute in Google's ad settings page set to female or male, and found that Google showed the simulated male users ads from a certain career coaching agency that promised large salaries more frequently than the simulated female users [350]. While this type of study establishes that employment ads through Google's ad system are not blind to gender (as expressed in the ad settings page), it cannot uncover the mechanism, that is, distinguish between explicit targeting by the advertiser and platform effects of various kinds.

Interacting with ad platforms as an advertiser has proved to be a more fruitful approach so far, especially to analyzing Facebook's advertising system. This is because Facebook exposes vastly more details about its advertising system to advertisers than to users. In fact, it allows advertisers to learn more information it has inferred or purchased about a user than it will allow the users themselves to access [351]. The existence of antisemitic autogenerated targeting categories, mentioned above, was uncovered using the advertiser interface. Ad delivery on Facebook has been found to introduce demographic disparities due to both market effects and optimization effects [267]. To reiterate, this means that even if the advertiser does not explicitly target an ad by, say, gender, there may be a systematic gender skew in the ad's audience. The optimization effects are enabled by Facebook's analysis of the contents of ads. Interestingly, this includes image analysis, which researchers revealed using the clever technique of serving ads with transparent content that is invisible to humans but nonetheless can be extracted by content analysis algorithms. The researchers found that the contents of transparent images had an effect on ad delivery, showing that Facebook does perform automated image analysis [267].

Fairness Considerations in the Design of Online Marketplaces

Online platforms for ride hailing, short-term housing, and freelance (gig) work rose to prominence in the 2010s: notable examples are Uber, Lyft, Airbnb, and TaskRabbit. They are important targets for the study of fairness because they directly impact people's livelihoods and opportunities. We will set aside some types of markets from our discussion. Online dating apps share some similarities with these markets, but they require an entirely separate analysis because the norms governing romance are different from those governing commerce and employment [352]. Then there are marketplaces for goods such as Amazon and eBay. In these markets the characteristics of the participants are less salient than the attributes of the product, so discrimination is less of a concern (which is not to say that it is nonexistent [353]).

Unlike the domains studied so far, machine learning is not a core component of the algorithms in online marketplaces. (Nonetheless, we consider it in scope because of our broad interest in decision making and fairness, rather than just machine learning.) Therefore fairness concerns are less about training data or algorithms; the far more serious issue is discrimination by buyers and sellers. For example, one study found that Uber drivers turned off the app in areas where they did not want to pick up passengers [354].

Methods to detect discrimination in online marketplaces are fairly similar to traditional settings such as housing and employment; a combination of audit studies and observational methods have been used. A notable example is a field experiment targeting Airbnb by Benjamin Edelman, Michael Luca, and Dan Svirsky [355]. The authors created fake guest accounts whose names signaled race (African American or White) and gender (female or male), but were otherwise identical. Twenty different names were used: five in each combination of race and gender. They then contacted the hosts of 6,400 listings in five cities through these accounts to inquire about availability. They found a 50 percent probability of acceptance of inquiries from guests with White-sounding names, compared to 42 percent for guests with names that sounded African American. The effect was persistent regardless of the host's race, gender, and experience on the platform, as well as listing type (high or low priced; entire property or shared) and diversity of the neighborhood. Note that the accounts did not have profile pictures; if inference of race by hosts happens in part based on appearance, a study design that varied the accounts' profile pictures might find a greater effect.

Compared to traditional settings, some types of observational data are readily available on online platforms, which can be useful to the researcher. In the above study, the public availability of reviews of listed properties proved useful. It was not essential to the design of the study, but it allowed an interesting validity check. When the analysis was restricted to the 29 percent of hosts in the sample who had received at least one review from an African American guest, the racial disparity in responses declined sharply. If the study's findings were a result of a quirk of the experimental design, rather than actual racial discrimination by Airbnb hosts, it would be difficult to explain why the effect would disappear for this subset of hosts. This supports the study's external validity.

In addition to discrimination by participants, another fairness issue that many online marketplaces must contend with is geographic differences in effectiveness. One study of TaskRabbit and Uber found that neighborhoods with high population density and high-income neighborhoods receive the largest benefits from the sharing economy [356]. Due to the pervasive correlation between poverty and race/ethnicity, these also translate to racial disparities. In the Chicago area, where this study was conducted, Black and Latin American neighborhoods have a lower population density, further exacerbating this effect.

Of course, geographic and structural disparities in these markets are not caused by online platforms, and no doubt they exist in offline analogues such as word-of-mouth gig work. In fact, the magnitude of racial discrimination is much larger in scenarios such as hailing taxis on the street [357] compared to technologically mediated interactions. However, in comparison to markets regulated by antidiscrimination law, such as hotels, discrimination in online markets is more severe. In any case, the formalized nature of online platforms makes audits easier. As well, the centralized nature of these platforms is a powerful opportunity for fairness interventions.

There are many ways in which platforms can use design to minimize users' ability to discriminate (such as by withholding information about counterparties) and the impetus to discriminate (such as by making participant characteristics less salient compared to product characteristics in the interface) [358]. There is no way for platforms to take a neutral stance toward discrimination by participants: even choices made without explicit regard for discrimination can affect the extent to which users' prejudicial attitudes translate into discriminatory behavior.

As a concrete example of design decisions to mitigate discrimination, the authors of the Airbnb study recommend that the platform withhold guest information from hosts prior to booking. (Note that ride-hailing services do withhold customer information. Carpooling services, on the other hand, allow users to view names when selecting matches; unsurprisingly, this enables discrimination against ethnic minorities [359].) The authors of the study on geographic inequalities suggest, among other interventions, that ride-hailing services provide a "geographic reputation" score to drivers to combat the fact that drivers often incorrectly perceive neighborhoods to be more dangerous than they are.

Mechanisms of Discrimination

We've looked at a number of studies on detecting unfairness in algorithmic systems. Let's take stock.

In the introductory chapter we discussed, at a high level, different ways in which unfairness could arise in machine learning systems. Here, we see that the specific sources and mechanisms of unfairness can be intricate and domain-specific. Researchers need an understanding of the domain to effectively formulate and test hypotheses about sources and mechanisms of unfairness.

For example, consider the study of gender classification systems discussed above. It is easy to guess that unrepresentative training datasets contributed to the observed accuracy disparities, but unrepresentative in what way? A follow-up paper by Vidya Muthukumar, Pedapati Tejaswini, Ratha Nalini et al. considered this question [360]. It analyzed several state-of-the-art gender classifiers (in a lab setting, as opposed to field tests of commercial APIs [application programming interfaces] in the original paper) and argued that underrepresentation of darker skin tones in the training data is *not* a reason for the observed disparity. Instead, one mechanism suggested by the authors is based on the fact that many training datasets of human faces comprise photos of celebrities. They found that photos of female celebrities have more prominent makeup compared to photos of women in general. This led to classifiers using makeup as a proxy for gender in a way that didn't generalize to the rest of the population.

Slightly different hypotheses can produce vastly different conclusions, especially in the presence of complex interactions between content producers, consumers, and platforms. For example, one study by Ronald Robertson, Shan Jiang, Kenneth Joseph et al. tested claims of partisan bias by search engines, as well as related claims that search engines return results that reinforce searchers' existing views (the "filter bubble" hypothesis) [361]. The researchers recruited participants with different political views, collected Google search results on a political topic in both standard and incognito windows from those participants' computers, and found that standard (personalized) search results were no more partisan than incognito (nonpersonalized) ones, seemingly finding evidence against the claim that online search reinforces users' existing beliefs.

This finding is consistent with the fact that Google doesn't personalize search results except based on searcher location and immediate (ten-minute) history of searches. This is known from Google's own admission [362] and prior research [363].

However, a more plausible hypothesis for the filter-bubble effect in searches comes from a qualitative study by Francesca Tripodi [364]. Simplified somewhat for our purposes, it goes as follows: when an event with political significance unfolds, key influencers (politicians, partisan news outlets, interest groups, political message boards) quickly craft their own narratives of the event. Those narratives selectively reach their respective partisan audiences through partisan information networks. Those people then turn to search engines to learn more or to "verify the facts." Crucially, however, they use different search terms to refer to the same event, reflecting the different narratives to which they have been exposed. The results for these different search terms are often starkly different, because the producers of news and commentary selectively and strategically cater to partisans using these same narratives. Thus, searchers' beliefs are reinforced. Note that this filter-bubble-producing mechanism operates effectively even though the search algorithm itself is arguably neutral.

A final example to reinforce the fact that disparity-producing mechanisms can be subtle and that domain expertise is required to formulate the right hypothesis: an investigation by journalists found that staples.com showed discounted prices to individuals in some ZIP codes; these ZIP codes were, on average, wealthier [365]. However, the actual pricing rule that explained most of the variation, as they reported, was that if there was a competitor's physical store located within twenty miles or so of the customer's inferred location, then the customer would see a discount! Presumably this strategy is intended to infer the customer's reservation price or willingness to pay. Incidentally, this is a similar kind of "statistical discrimination" as seen in the car sales discrimination study discussed at the beginning of this chapter.

Fairness Criteria in Algorithmic Audits

While the mechanisms of unfairness are different in algorithmic systems, the applicable fairness criteria are the same for algorithmic decision making as for other kinds of decision making. That said, some fairness notions are more often relevant, and others less so, in algorithmic decision making compared to human decision making. We offer a few selected observations on this point.

Fairness as Blindness

This is seen less often in audit studies of algorithmic systems; such systems are generally designed to be blind to sensitive attributes. Besides, fairness concerns often arise precisely from the fact that blindness is generally not an effective fairness intervention in machine learning. Two exceptions are ad targeting and online marketplaces (where the nonblind decisions are in fact being made by users and not the platform).

Unfairness as Arbitrariness

There are roughly two senses in which decision making could be considered arbitrary and hence unfair. The first is when decisions are made on a whim rather than a uniform procedure. Since automated decision making results in procedural uniformity, this type of concern is generally not salient.

The second sense of arbitrariness applies even when there is a uniform procedure, if that procedure relies on a consideration of factors that are thought to be irrelevant, either statistically or morally. Since machine learning excels at finding correlations, it commonly identifies factors that seem puzzling or blatantly unacceptable. For example, in aptitude tests such as the Graduate Record Examination (GRE), essays are graded automatically. Although e-rater and other tools used for this purpose are subject to validation checks, and are found to perform similarly to human raters on samples of actual essays, they are able to be fooled into giving perfect scores to machine-generated gibberish. Recall that there is no straightforward criterion that allows us to assess if a feature is morally valid (chapter 2), and this question must be debated on a case-by-case basis.

More serious issues arise when classifiers are not even subjected to proper validity checks. For example, there are a number of companies that claim to predict candidates' suitability for jobs based on personality tests or body language and other characteristics in videos [343]. There is no peer-reviewed evidence that job performance is predictable using these factors, and no basis for such a belief. Thus, even if these systems don't produce demographic disparities, they are unfair in the sense of being arbitrary: candidates receiving an adverse decision lack due process to understand the basis for the decision, contest it, or determine how to improve their chances of success.

Observational Fairness Criteria

Criteria including demographic parity, error rate parity, and calibration have received much attention in algorithmic fairness studies. Convenience has probably played a big role in this choice: these metrics are easy to gather and straightforward to report without necessarily connecting them to moral notions of fairness. We reiterate our caution about the overuse of parity-based notions; parity should rarely be made a goal by itself. At a minimum, it is important to understand the sources and mechanisms that produce disparities as well as the harms that result from them before deciding on appropriate interventions.

Representational Harms

Traditionally, allocative and representational harms were studied in separate literatures, reflecting the fact that they are mostly seen in separate spheres of life (for instance, housing discrimination versus stereotypes in advertisements). Many algorithmic systems, on the other hand, are capable of generating both types of harms. A failure of face recognition for darker-skinned people is demeaning, but it could also prevent someone from being able to access a digital device or enter a building that uses biometric security.

Information Flow, Fairness, Privacy

A notion called "information flow" is seen frequently in algorithmic audits. This criterion requires that sensitive information about subjects not flow from one information system to another or from one part of a system to another. For example, a health website may promise that user activity, such as searches and clicks, are not shared with third parties such as insurance companies (since that may lead to potentially discriminatory effects on insurance premiums). It can be seen as a generalization of blindness: whereas blindness is about not acting on available sensitive information, restraining information flow ensures that the sensitive information is not available to act upon in the first place.

There is a powerful test for violations of information flow constraints, which we call the adversarial test [350]. It does not directly detect information flow, but rather decisions that are made on the basis of that information. It is powerful because it does not require specifying a target variable, which minimizes the domain knowledge required of the researcher. To illustrate, let's revisit the example of the health website. The adversarial test operates as follows:

1. Create two groups of simulated users (A and B), that is, bots, that are identical except for the fact that users in group A, but not group B, browse the sensitive website in question.

2. Have both groups of users browse *other* websites that are thought to serve ads from insurance companies, or personalize content based on users' interests, or somehow tailor content to users based on health information. This is the key point: the researcher does not need to hypothesize a mechanism by which potentially unfair outcomes result—for example, which websites (or third parties) might receive sensitive data, whether the personalization might take the form of ads, prices, or some other aspect of content.

3. Record the contents of the web pages seen by all users in the previous step.

4. Train a binary classifier to distinguish between web pages encountered by users in group A and those encountered by users in group B. Use cross-validation to measure its accuracy.

5. If the information flow constraint is satisfied (i.e., the health website did not share any user information with any third parties), then the websites browsed in step 2 are blind to user activities in step 1; thus the two groups of users look identical, and there is no way to systematically distinguish the content seen by group A from that seen by group B. The classifier's test accuracy should not significantly exceed $\frac{1}{2}$. The permutation test can be used to quantify the probability that the classifier's observed accuracy (or better) could have arisen by chance if there is in fact no systematic difference between the two groups [366].

There are additional nuances relating to proper randomization and controls, for which we refer the reader to the study by Datta, Tschantz, and Datta [350]. Note that if the adversarial test fails to detect an effect, it does not mean that the information flow constraint is satisfied. Also note that the adversarial test is not capable of measuring an effect size. Such a measurement would be meaningless anyway, since the goal is to detect information flow, and any effect on observable behavior of the system is merely a proxy for it.

This view of information flow as a generalization of blindness reveals an important connection between privacy and fairness. Many studies based on this principle can be seen as either privacy or fairness investigations. For example, a study found that Facebook solicits phone numbers from users with the stated purpose of improving account security, but uses those numbers for ad targeting [367]. This is an example of undisclosed information flow from one part of the system to another. Another study used ad retargeting—in which actions taken on one website, such as searching for a product, result in ads for that product on another website—to infer the exchange of user data between advertising companies [368]. Neither study used the adversarial test.

Comparison of Research Methods

For auditing user fairness on online platforms, there are two main approaches: creating fake profiles and recruiting real users as testers. Each has its pros and cons. Both approaches have the advantage, compared to traditional audit studies, of allowing a potentially greater scale due to the ease of creating fake accounts or recruiting testers online (e.g., through crowd-sourcing).

Scaling is especially relevant for testing geographic differences, given the global reach of many online platforms. It is generally possible to simulate geographically dispersed users by manipulating testing devices to report faked locations. For example, the above-mentioned investigation of regional price differences on staples.com actually included a measurement from each of the 42,000 ZIP codes in the United States [369]. They accomplished this by observing that the website stored the user's inferred location in a cookie and programmatically changing the value stored in the cookie to each possible value.

That said, practical obstacles commonly arise in the fake-profile approach. In one study, the number of test units was practically limited by the requirement for each account to have a distinct credit card associated with it [370]. Another issue is bot detection. For example, the Airbnb study was limited to five cities, even though the researchers originally planned to test more, because the platform's bot-detection algorithms kicked in during the course of the study to detect and shut down the anomalous pattern of activity. It's easy to imagine an even worse outcome where accounts detected as bots are somehow treated differently by the platform (e.g., messages from those accounts are more likely to be hidden from intended recipients), compromising the validity of the study.

As this example illustrates, the relationship between audit researchers and the platforms being audited is often adversarial. Platforms' efforts to hinder researchers can be technical but also legal. Many platforms, notably Facebook, prohibit both fake-account creation and automated interaction in their terms of service. The ethics of terms-of-service violation in audit studies is a matter of ongoing debate, paralleling some of the ethical discussions during the formative period of traditional audit studies. In addition to ethical questions, researchers incur a legal risk when they violate terms of service. In fact, under laws such as the US Computer Fraud and Abuse Act, it is possible that they may face criminal as opposed to just civil penalties.

Compared to the fake-profile approach, recruiting real users allows less control over profiles but is better able to capture the natural variation in attributes and behavior between demographic groups. Thus, neither design is always preferable, and they are attuned to different fairness notions. When testers are recruited via crowdsourcing, the result is generally a convenience sample (i.e., the sample is biased toward people who are easy to contact), resulting in a nonprobability (nonrepresentative) sample. It is generally infeasible to train such a group of testers to carry out an experimental protocol; instead, such studies typically handle the interaction between testers and the platform via software tools (e.g., browser extensions) created by the researcher and installed by the tester. For more on the difficulties of research using nonprobability samples, see the book *Bit by Bit* by Matthew Salganik [371].

Due to the serious limitations of both approaches, lab studies of algorithmic systems are commonly seen. The reason that lab studies have value at all is that since automated systems are fully specified using code, the researcher can hope to simulate them relatively faithfully. Of course, there are limitations: the researcher typically doesn't have access to training data, user interaction data, or configuration settings. But simulation is a valuable way for developers of algorithmic systems to test their *own* systems, and this is a common approach in the industry. Companies often go so far as to make deidentified user interaction data publicly available so that external researchers can conduct lab studies to develop and test algorithms. The Netflix Prize is a prominent example of such a data release [372]. So far, these efforts have almost always been about improving the accuracy rather than the fairness of algorithmic systems.

Lab studies are especially useful for getting a handle on questions that cannot be studied by other empirical methods, notably the *dynamics* of algorithmic systems, that is, their evolution over time. One prominent result from this type of study is the quantification of feedback loops in predictive policing [30, 31]. Another insight is the increasing homogeneity of users' consumption patterns over time in recommender systems [373].

Observational studies and observational fairness criteria continue to be impor-tant. Such studies are typically carried out by algorithm developers or decision makers, often in collaboration with external researchers [374, 375]. It is relatively rare for observational data to be made publicly available. A rare exception, the COMPAS dataset, involved a Freedom of Information Act request.

Finally, it is worth reiterating that quantitative studies are narrow in what they can conceptualize and measure [376]. Qualitative and ethnographic studies of deci-sion makers thus provide an invaluable complementary perspective. To illustrate, we'll discuss a study by Samir Passi and Solon Barocas that reports on six months of ethnographic fieldwork in a corporate data science team [170]. The team worked on a project in the domain of car financing that aimed to "improve the quality" of leads (leads are potential car buyers in need of financing who might be converted to actual buyers through marketing). Given such an amorphous high-level goal, for-mulating a concrete and tractable data science problem is a necessary and nontrivial task—a task that is further complicated by the limitations of the data available. The paper documents how there is substantial latitude in problem formulation, and spot-lights the iterative process that was used, which resulted in a series of proxies for lead quality. The authors show that different proxies have different fairness implica-tions: one proxy would maximize people's lending opportunities and another would alleviate dealers' existing biases, both potentially valuable fairness goals. However, the data scientists were not aware of the normative implications of their decisions and did not explicitly deliberate them.

Looking Ahead

In this chapter, we cover traditional tests for discrimination as well as fairness stud-ies of various algorithmic systems. Together, these methods constitute a powerful toolbox for interrogating a single decision system at a single point in time. But there are other types of fairness questions we can ask: What is the cumulative effect of the discrimination faced by a person over the course of a lifetime? What structural aspects of society result in unfairness? We cannot answer such questions by looking at individual systems. The next chapter is all about broadening our view of discrim-ination and then using that broader perspective to study a range of possible fairness interventions.

Chapter Notes

To understand social science audit studies in more depth, see the paper by Devah Pager [296]. S. Michael Gaddis provides a more recent introduction and survey [377].

Auditing of algorithmic systems is a young, quickly evolving field: a 2014 paper issued a call to action toward this type of research [378]. Most of the studies that we cite postdate that piece. For a more recent practitioner-focused overview, Sasha Costanza-Chock, Inioluwa Deborah Raji, and Joy Buolamwini compile best practices for auditors and provide recommendations for policymakers based on interviews with over 150 auditors [279]. Briana Vecchione, Karen Levy, and Solon Barocas draw lessons for algorithm audits and justice from the history of audits in the social sciences [379]. Miles Brundage, Shahar Avin, Jasmine Wang et al. put third-party audits in the context of many other ways of supporting verifiable claims about the impacts of AI systems [380].

For in-depth treatments of the history and politics of information platforms, see *The Master Switch* by Tim Wu [381], *The Politics of "Platforms"* and *Custodians of the Internet* by Tarleton Gillespie [382, 383], and *The New Governors* by Kate Klonick [384].

8

A Broader View of Discrimination

Machine learning systems don't operate in a vacuum; they are adopted in societies that already have many types of discrimination intertwined with systems of oppression such as racism. This is at the root of fairness concerns in machine learning. In this chapter we take a systematic look at discrimination in society. This gives us a more complete picture of the potential harmful impacts of machine learning. We see that while a wide variety of fairness interventions are possible—and necessary—only a small fraction of them translate to technical fixes.

Case Study: The Gender Earnings Gap on Uber

We'll use a paper that analyzes the gender earnings gap on Uber [385] as a way to apply some of the lessons from the previous two chapters while setting up some of the themes of this chapter. The study was coauthored by current and former Uber employees.

The authors start with the observation that female drivers earn 7 percent less on Uber per active hour than male drivers. They conclude that this gap can be explained by three factors: gender differences in drivers' choices of where to drive, men's greater experience on the platform, and men's tendency to drive faster. They find that customer discrimination and algorithmic discrimination do not contribute to the gap. We take the paper's technical claims at face value, but use the critical framework we've introduced to interpret the findings quite differently from the authors.

First, let's understand the findings in more detail.

The paper analyzes observational data on trips in the United States, primarily in Chicago. Figure 8.1 is a causal graph showing what we consider to be the core of the causal model studied in the paper (the authors do not draw such a graph and do not pose their questions in a causal framework; we have chosen to do so for pedagogical purposes). A full graph would be much larger than the figure; for example, we've omitted a number of additional controls, such as race, that are presented in the appendix.

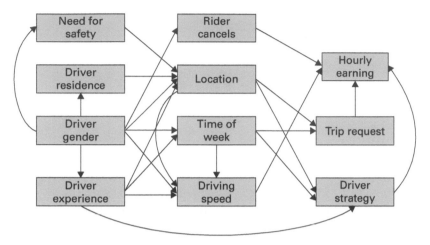

Figure 8.1
Our understanding of the causal model implicit in the Uber study.

We'll use this graph to describe the findings. At a high level, the graph describes a joint distribution whose samples are trips. To illustrate, different trips corresponding to the same driver will have the same *residence* (unless the driver moved during their tenure on the platform), but different *experience* (measured as number of prior trips).

Drivers' hourly earnings are primarily determined by the algorithm that allocates trip requests from riders to drivers. The allocation depends on demand, which in turn varies by location and time of the week (the week-to-week variation is considered noise). Uber's algorithm ignores driver attributes including experience and gender, hence there are no arrows from those nodes to *trip request*. In addition, a few other factors might affect earnings. Drivers who drive faster complete more trips, drivers may strategically accept or cancel trips, and riders might discriminate by canceling trips after the driver accepts.

The paper uses a technique called *Gelbach decomposition* to identify the effect of each of several variables on the hourly earnings. Decomposition is a set of techniques used in economics for quantifying the contribution of various sources to an observed difference in outcomes. Although the authors don't perform causal inference, we will continue to talk about their findings in causal terms for pedagogical purposes. The difference is not salient to the high-level points we wish to make.

The authors find that the earnings gap (i.e., the effect of *driver gender* on *hourly earning*) can be entirely explained by paths involving *driver experience*, *location*, and *driving speed*. Paths through *rider cancellation* and *time of week* don't have significant effects.

The authors further interrogate the effect of gender on location (i.e., the choice of where to drive), and find that women are less likely to drive in less safe areas that also turn out to be more lucrative. They then dig deeper and argue that this effect operates almost entirely by women *residing* in safer areas and choosing to drive based on where they live.

The returns to experience could operate in several ways. The authors don't decompose the effect but suggest several possibilities: the choice of where and when to drive and other elements of strategy including which rides to accept. A key finding of the paper is the effect of gender on experience. Men are less likely to leave the platform and they drive more hours during each week that they stay on the platform, resulting in a large experience differential. There are no gender differences in *learning* from experience: male and female drivers' behavior changes at the same rate for a given number of trips.

The paper highlights questions that can be studied using observational data but not necessarily with field experiments (audit studies). An audit study of the Uber gender pay gap (along the lines of those discussed in chapter 7) may have involved varying the driver's name to test the effect on rider cancellation and ratings. Such an experiment would have no way to uncover the numerous other paths by which gender affects earnings. An audit study would be more suited for studying discrimination *by drivers against riders*, in part because drivers in these systems exercise more choice in the matching process than riders do. Indeed, a study found that UberX and Lyft drivers discriminate against Black and female riders [357].

Causal diagrams in realistic scenarios are more complex than typical textbook examples. We reiterate that the graph above is much simplified compared to the (implicit) graph in the paper. The estimation in the paper proceeds as a series of regressions focusing iteratively on small parts of the graph, rather than an analysis of the entire graph at once. In any messy exercise such as this, there is always the possibility of unobserved confounders.

Despite the number of possible effects considered in the study, it leaves out many others. For example, some drivers may move to take advantage of the earning potential. This would introduce a cycle into our causal graph (*location* $\rightarrow$ *residence*). This type of behavior might seem unlikely for an individual driver, which justifies ignoring such effects in the analysis. Over time, however, the introduction of transportation systems has the potential to reshape communities [386, 387]. Today's empirical methods have limitations in understanding these types of long-term phenomena that involve feedback loops.

A more notable omission from the paper is the effect of driver gender on experience. Why do women drop off the platform far more frequently? Could one reason be that they face more harassment from riders? The authors don't seem to consider this question.

This leads to our most salient observation about this study: the narrow definition of discrimination. First, as noted, the study doesn't consider that differential dropout rates might be due to discrimination. This is especially pertinent since the gender gap in hourly earnings is merely 7 percent whereas the gap in participation rate is a factor of 2.7! One would think that if there is rider discrimination, it would be most apparent in its effect on dropout rates. In contrast, the only avenue of discrimination considered in the paper involves a (presumably misogynistic) rider who cancels a ride, incurring delays and potentially algorithmic penalties, based solely on the driver's gender.

Further, the authors take an essentialist view of the gender difference in average speed (e.g., "men are more risk tolerant and aggressive than women"). We may question how innate these differences are, given that in contemporary US society women may face social penalties when they are perceived as aggressive. If this is true of driver-rider interactions, then women who drive as fast as men will receive lower ratings with attendant negative consequences. This is a form of discrimination by riders.

Another possible view of the speed difference, also not considered by the authors, is that male drivers on average provide a lower quality of service due to an increase in accident risk resulting from greater speed (which also creates negative externalities for others on the road). In this view, Uber's matching algorithm discriminates against female drivers by *not* accounting for this difference.

Finally, the paper doesn't consider structural discrimination. It finds that women reside in less lucrative neighborhoods and that their driving behavior is shaped by safety considerations. However, a deeper understanding of the reasons for these differences is outside the scope of the paper. In fact, gender differences in safety risks and the affordability of residential neighborhoods can be seen as an example of the greater burden that society places on women. In other words, Uber operates in a society in which women face discrimination and have unequal access to opportunities, and the platform perpetuates those differences in the form of a pay gap.

Let us generalize a bit. There is a large set of studies that seek to explain the reasons for observed disparities in wages or other outcomes. Generally these studies find that the direct effect of gender, race, or another sensitive attribute is much smaller than the indirect effect. Frequently this leads to a vigorous debate on whether or not the findings constitute evidence of discrimination or unfairness. There is room for different views on this question. The authors of the Uber study interpreted none of the three paths by which gender impacts earnings—experience, speed, and location—as discrimination; we've argued that all three can plausibly be interpreted as discrimination. Different moral frameworks will lead to different answers. Views on these questions are also politically split. As well, scholars in different fields often tend to answer these questions differently (including, famously, social science and economics [388]).

Certainly these definitional questions are important. However, perhaps the greatest value of studies on mechanisms of discrimination is that they suggest avenues for intervention *without* having to resolve definitional questions. Looking at the Uber study through this lens, several interventions are apparent. Recall that there is a massive gender disparity in the rate at which drivers drop out of the platform. Uber could more actively solicit and listen to feedback from female drivers and use that feedback to inform the design of the app. This may lead to interventions such as making it easier for drivers (and riders) to report harassment and taking stronger action in response to such reports.

As for the speed difference, Uber could issue warnings to drivers who exceed the speed limit or whose speed results in a predicted accident risk that crosses some threshold (such a prediction is presumably possible given Uber's access to data). In addition, Uber could use its predictive tools to educate drivers about strategy, decreasing the advantage that experienced drivers have over inexperienced ones, regardless of gender. Finally, the findings also give greater urgency to structural efforts to make neighborhoods safe for women. None of these interventions requires a consensus on whether or not female drivers on Uber are discriminated against.

Three Levels of Discrimination

Sociologists organize discrimination into three levels: structural, organizational, and interpersonal [319, 388]. Structural discrimination arises from the ways in which society is organized, both through relatively hard constraints such as discriminatory laws and through softer ones such as norms and customs. Organizational factors operate at the level of decision-making units, such as a company making hiring decisions. Interpersonal factors refer to the attitudes and beliefs that result in discriminatory behavior by individuals.

A separate way to classify discrimination is as direct or indirect. By direct discrimination we mean actions or decision processes that make explicit reference to a sensitive attribute. Indirect discrimination refers to actions or decision processes that make no such reference, yet disadvantage one or more groups. The line between direct and indirect discrimination is hazy and it is better to think of it as a spectrum rather than a binary category.

Structural Factors

Structural factors refers to ways in which society is organized. A law that overtly limits opportunities for certain groups is an example of a direct structural factor. Due to various rights revolutions around the world, there are fewer of these laws

Table 8.1
Examples of discrimination organized into three levels and on a spectrum of directness

Level	More direct	More indirect
Structural	Laws against same-sex marriage	Better-funded schools in wealthier, more segregated areas
Organizational	Lack of disability accommodations	Networked hiring
Interpersonal	Overt animus	Belief in need for innate brilliance (combined with gender stereotypes)

today than there used to be. Yet, discriminatory laws are far from a thing of the past. For example, as of 2021, a mere twenty-nine countries recognize marriage equality [389]. Further, discriminatory laws of the past have created structural effects which persist today [390].

Indirect structural discrimination is pervasive in virtually every society. Here are two well-known examples affecting the United States. Drug laws and drug policies, despite being facially neutral, have the effect of disproportionately affecting minority groups, especially Black people [391]. Schools in high-income neighborhoods tend to be better funded (since public schools are funded primarily through property taxes) and attract more qualified teachers, transmitting an educational advantage to children of higher-income parents.

Other factors are even less tangible yet no less serious in terms of their effects, such as cultural norms and stereotypes. In the case study of gender bias in Berkeley graduate admissions in chapter 5, we encountered the hypothesis that societal stereotypes influence people's career choices in a way that reproduces gender inequalities in income and status [175]:

> The bias in the aggregated data stems ... apparently from prior screening at earlier levels of the educational system. Women are shunted by their socialization and education toward fields of graduate study that are generally more crowded, less productive of completed degrees, and less well funded, and that frequently offer poorer professional employment prospects.

Organizational Factors
Organizational factors operate at the level of organizations or decision-making units: how they are structured, the decision-making rules and processes they put in place, and the context in which individual actors operate. Again, these lie on a spectrum between direct and indirect.

The most direct form of discrimination—excluding people from participation explicitly based on group membership—is mostly unlawful in liberal democracies. However, practices such as lack of disability accommodations and failure to combat sexual harassment are rampant. A more indirectly discriminatory policy is the use of employees' social networks in hiring, an extremely common practice. One observational study found that the use of employee referrals in predominantly White firms reduced the probability of a Black hire by nearly 75 percent relative to the use of newspaper ads [392]. The study controlled for spatial segregation, occupational segregation, city, and firm size.

Organizational discrimination can be revealed and addressed at the level of a single organization, unlike structural factors (e.g., no individual school is responsible for teachers being attracted to schools in high-income neighborhoods).

Interpersonal Factors

Interpersonal factors comprise the attitudes and beliefs that result in discriminatory behavior by individuals. Sometimes people may discriminate because of an overt animus for a certain group, in the sense that the discriminator does not attempt to justify it by any appeal to rationality.

More often, the mechanisms involved are relatively indirect. A 2015 study by Sarah-Jane Leslie, Andrei Cimpian, Meredith Meyer et al. found that academic fields in which achievement is believed to be driven by innate brilliance exhibit a greater gender disparity, that is, they have fewer women [393]. The authors propose that the disparity is caused by the combination of the belief in the importance of innate brilliance together with stereotypes about lower innate brilliance in women. This combination could then impact women in brilliance-emphasizing disciplines in two ways: either by practitioners of those disciplines exhibiting biases against women or by women internalizing those stereotypes and self-selecting out of those disciplines (or performing more poorly than they otherwise would). The authors don't design tests to distinguish between these competing mechanisms. However, they do test whether the observed disparities could alternatively be caused by actual innate differences (rather than beliefs in innate differences) in ability or aptitude, or willingness to work long hours. Using various proxies (such as GRE score for innate ability), they argue that such competing explanations cannot account for the observed differences.

One may wonder: Can we not test for innate differences more rigorously, such as by examining young children? A follow-up study showed that children as young as six tend to internalize gendered stereotypes about innate brilliance, and these stereotypes influence their selection of activities [394]. These difficulties hint at the underlying complexity of the concept of gender, which is produced and reinforced in part through these very stereotypes [395].

To recap, we've discussed structural, organizational, and interpersonal discrimination, and the fact that these are often indirect and pervasive. The three levels are interconnected: for example, in the Uber case study, structural inequalities don't

perpetuate themselves, but rather they are perpetuated through organizational decisions; those decisions at Uber are made by individuals whose worldviews are shaped
by culture. In other words, even structural discrimination is actively perpetuated,
and we collectively have the power to mitigate it and to reverse course. It would be
a mistake to resign ourselves to viewing structural discrimination as simply the way
the world is.

Notice that adopting statistical decision making is not automatically a way out of
any of these factors, which operate for the most part in the background and not at a
single, discrete moment of decision making.

The Persistence and Magnitude of Inequality

Formal equality under the law primarily addresses direct discrimination and has
relatively little effect on indirect discrimination, whether structural, organizational,
or interpersonal. This is one reason why inequality can be persistent in societies
that seemingly promise equal opportunity. Here are two stark examples of how long
inequalities can sustain themselves.

Beginning in 1609, Jesuit missions were established in the Guaraní region of
South America that overlaps modern day Argentina, Paraguay, and Brazil. In addition to practicing religious conversion, the missionaries undertook educational
efforts among the indigenous people. However, due to political upheaval in Spain
and Portugal, the missions abruptly ended in 1767–1768 and the missionaries were
expelled. How long after this date would we expect the geographic inequalities
introduced by Jesuit presence to persist? Perhaps a generation or two? Remarkably, the Jesuit effect on educational attainment has been found to persist 250 years
later: areas closer to a former mission have 10–15 percent higher literacy rates as
well as 10 percent higher incomes. The study, by Felipe Valencia Caicedo, makes
use of a clever idea to argue that the mission locations were essentially random,
making this a natural experiment [396]. Another study of the long-run persistence
of inequality shows the present-day effects of a system of colonial forced labor in
Peru and Bolivia between 1573 and 1812 [397].

More evidence for the long-run persistence of inequality comes from the city of
Florence, based on a unique dataset containing tax-related data for all individuals
from the year 1427. A working paper finds that surnames associated with wealthier individuals in the dataset are associated with wealthier individuals today, six
hundred years later [398].

While these are just a few examples, research shows that persistence of inequality
over generations along social and geographic lines is the norm. Yet it is not widely
appreciated. For example, Americans believe that an individual born into the bottom quintile of the income distribution has a one-in-six chance of rising to the top
quintile, but the observed likelihood is one-in-twenty [399]. Mobility in the United

States has decreased since the 1980s, and is lower for Black Americans than White Americans [400].

These inequalities are significant because of their magnitude in addition to their persistence. Median income of Black Americans is about 65 percent that of White Americans [401]. Wealth inequality is much more severe: the median wealth of Black households is about 11 percent that of White households. A data analysis combined with simulations suggests that the gap may never close without interventions such as reparations [402]. Most Americans are not aware of this gap: on average, survey respondents estimated the wealth of a typical Black family to be about 90 percent of that of a typical White family [403].

Turning to gender: full-time, year-round working women earned 80 percent of what their male counterparts earned [404]. Geographic inequalities also exist. For example, the richest and poorest census tracts in the United States differ in average income by a factor of about thirty [405].

Machine Learning and Structural Discrimination

For a book about machine learning, we've covered a lot of ground on discrimination and inequality in society. There's a reason. To understand fairness, it isn't enough to think about the moment of decision making. We also need to ask: What impact does the adoption of machine learning by decision makers have in long-lasting cycles of structural inequality in society? Does it help us make progress toward enabling equality of opportunity, or other normative ideals, over the course of people's lives? Here are some observations that can help answer those questions.

Predictive Systems Tend to Preserve Structural Advantages and Disadvantages

Predictive systems tend to operate within existing institutions. When such institutions perpetuate inequality due to structural factors, predictive systems will only reify those effects, absent explicit intervention. Predictive systems tend to inherit structural discrimination because the objective functions used in predictive models usually reflect the incentives of the organizations deploying them. As an example, consider a 2019 study that found strong racial bias in a system used to identify patients with a high risk of adverse health outcomes, in the sense that Black patients were assigned lower scores compared to equally at-risk White patients [374]. The authors found that this happened because the model was designed to predict health care costs instead of needs, and the health care system spends less caring for Black patients than White patients even when they have the same health conditions.

Suppose a firm makes hiring decisions based on a model that predicts job performance based on educational attainment. Imagine a society where students from higher-income families, on average, have had better educational opportunities that translate to greater job skills. This is not a measurement bias in the data that can be corrected away: education level genuinely predicts job performance. Thus, an accurate predictive system will rank higher-income candidates higher on average.

The structural effect of such systems becomes clear when we imagine every employer applying similar considerations. Candidates with greater educational opportunities end up with more desirable jobs and higher incomes. In other words, predictive systems have the effect of transferring advantages from one phase of life to the next, and one generation to the next.

This phenomenon shows up in less obvious ways. For instance, online ad targeting is based on the assumption that differences in past behavior between users reflect differences in preferences. But they might also result from differences in structural *circumstances*, and there is no way for targeting engines to tell the difference. This helps explain why ads, including job ads, may be targeted in ways that reinforce stereotypes and structural discrimination [406].

This aspect of predictive systems is amplified by compounding injustice [407, 408]. That is, individuals are subject to a series of decisions over the course of their lives, and the effects of these decisions both accumulate and compound over time. When a person receives (or is denied) one opportunity, they are likely to appear more (or less) qualified at their next encounter with a predictive system.

Machine Learning Systems May Make Self-Fulfilling Predictions

Suppose we find that chess skill is correlated with productivity among software engineers. Here are a few possible explanations:

1. Chess skill makes one a better software engineer.
2. There are underlying cognitive skills that make one better at both.
3. College professors hold stereotypes about chess skills and software engineering, and steered students good at chess into computer science courses.
4. People with more leisure time were both able to pursue chess as a hobby and devote time to improving their software engineering skills.

Standard supervised learning does not distinguish between these causal paths. Regardless of the correct causal explanation, once a large swath of employers start using chess skill as a hiring criterion, they contribute to the perpetuation of the observed correlation. That is because applicants who are better at chess will have better opportunities for software engineering positions in this world, and these opportunities will allow them to develop their software engineering skills.

Machine learning automates the discovery of correlations such as the above. When we deploy those correlations as decision criteria, we alter the very phenomena that we are supposedly measuring. In other words, using noncausal variables as decision criteria may give them causal powers over time. This is not limited to machine learning: sociologists have long recognized that stereotypes that are used to justify discrimination may in fact be produced by that discrimination [409].

Algorithmic Recommendation Systems May Contribute to Segregation

Even small preferences for homogeneous neighborhoods can lead to dramatic large-scale effects. In the appendix, we discuss a toy model of residential segregation showing such effects. But what about the online world, for example, online social networks? The phenomenon of people making friends with similar others (online or offline) is called *homophily*.

In the early days of social media, there was a hope—now seen as naïve—that in the online sphere there would be no segregation, due to the ease with which people can connect with each other. Instead, we observe similar patterns of homophily and segregation online as offline. This is partly because real-world relationships are reflected online, but in part it is because segregation emerges through our online preferences and behaviors [410].

As social media has matured, concerns arising from homophily have expanded from demographic segregation to ideological echo chambers. The causal mechanisms behind polarized online discourse and the role of recommendation algorithms are being researched and debated (see chapter 7 on testing), but there is no doubt that online media can have structural effects.

Machine Learning May Lead to Homogeneity of Decision Making

A company that hires only people whose names begin with certain letters of the alphabet may seem absurd but it's not necessarily a cause for alarm. One reason behind this intuition is that we expect that the effect of any such idiosyncratic policies will cancel out, given that job candidates have many firms to apply to. If, on the other hand, every employer adopted such a policy, then the experience of job seekers becomes radically different.

Machine learning results in more homogeneous decision making compared to the vagaries of individual decisions. Studies of human behavior show that human decisions have a lot of "noise." Removing the noise is one of the main attractions of statistical decision making. But there are also risks. If statistical decision making results in similar decisions being made by many decision makers, otherwise-idiosyncratic biases could become amplified and reified to the point where they create structural impediments [54].

Homogeneity can happen in many ways. At a high level, if many machine learning systems use the same training data and the same target variable, they will make roughly the same classifications, even if the learning algorithms are very different. Intuitively, if this *weren't* the case, one could make more accurate classifications by ensembling their predictions. For a stark illustration of homogeneous predictions from the domain of predicting life outcomes, see the Fragile Families Challenge [90].

Alternatively, many decision makers could use the same underlying system. Kleinberg and Raghavan call this situation *algorithmic monoculture* [411]. There are anecdotes of job seekers being repeatedly screened out of jobs on the basis of personality tests, all offered by the same vendor [412].

Even *individual* algorithmic systems may have such an outsized influence in society that their policies may have structural effects. The most obvious example is a system adopted by the state, such as a predictive policing system that leads to the overpolicing of low-income neighborhoods.

But it is private platforms, especially those with a global scale, where this effect has been most prominent. Take content moderation: a small number of social media companies together determine which types of speech can be a part of mainstream online discourse and which communities are able to mobilize online. Platform companies have faced criticism for allowing content that incites violence and, conversely, for being overzealous in deplatforming individuals or groups.

In some cases, platform policies are shaped by the capabilities and limitations of machine learning [413]. For example, algorithms are relatively good at detecting nudity but relatively poor at detecting context. Companies such as Facebook have had broad bans on nudity without much attention to context, often taking down artwork and iconic historical images.

Machine Learning Shifts Power

Like all technologies, machine learning shifts power. To make this more precise, we analyze the adoption of machine learning by a bureaucracy. We don't mean the term "bureaucracy" in its colloquial, pejorative sense of an inefficient, rule-bound government agency. We rather use the term as social scientists do: a bureaucracy is a public or private entity in which highly trained workers called bureaucrats, operating in a hierarchical structure, make decisions in a way that is constrained by rules and policies but also requires expert judgment. Firms, universities, hospitals, police forces, and public assistance programs are all bureaucracies to various degrees. Most of the decision making scenarios that motivate this book are situated in bureaucracies.

To understand the effect of adopting machine learning, we consider five types of stakeholders: decision subjects, the people who provide the training data, domain experts, machine learning experts, and policymakers. Our analysis builds on a talk by Pratyusha Kalluri [414].

Machine learning as generally implemented today shifts power away from the first three categories. By representing decision subjects as standardized feature vectors, statistical decision making removes their agency and ability to advocate for themselves. In many domains, notably the justice system, this ability is central to the rights of decision subjects. Even in a relatively less consequential domain, such as college admissions, the personal statement provides this ability and is a key component of the evaluation.

People who provide training data may have *knowledge* about the task at hand, but provide only their *behavior* as input to the system (think of email recipients clicking the "Report spam" button). Machine learning instead constructs a form of knowledge in a centralized way. In contrast, domain experts learn in part from the knowledge and lived experience of the individuals they interact with. Admittedly, experts such as physicians are often criticized for devaluing the knowledge and experience of decision subjects (patients) [415]. But the fact that such a debate is happening at all is evidence of the fact that patients have at least some power in the traditional system [416].

The role of domain experts is also more limited compared to traditional decision making, where the discretion and judgment of such experts holds sway. In supervised machine learning, domain expertise is primarily needed in two of the steps: formulating the problem and task, and labeling training examples. In practice, domain expertise is often not valued by tool developers, and hence experts' roles are even more circumscribed. For example, one study found that based on sixty-eight interviews, "developers conceived of [domain experts] as corrupt, lazy, non-compliant, and datasets themselves, pursuing surveillance and gamification to discipline workers to collect better quality data" [417].

The fairness implications of this shift in power are complex. In government bureaucracies, the power wielded by "street-level bureaucrats" such as police officers and social service caseworkers—the people who translate policy into individual decisions—can be abused, and removing their discretion is often seen as a fairness intervention. Yet the discretion and human intelligence of these decision makers can also be a vital fairness-promoting element due to the existence of extenuating factors or novel circumstances not seen in the training data or covered in existing policies [59, 418]. And when the system itself is unjust, the humans tasked with implementing it can be an important source of resistance through noncompliance or whistle-blowing.

In contrast to street-level bureaucrats, machine learning empowers policy makers or centralized decision makers—those at the top of the bureaucracy. Consider a risk prediction tool used by a child protection agency to screen calls. Depending on the agency's budget and other factors, the decision maker may want to screen in a higher or lower proportion of calls. With a statistical tool, such a policy change can be implemented instantly, and it is enormously easier than the alternative of retraining hundreds of case workers to adjust their mental heuristics. This is just one example that illustrates why such tools have proven so attractive to those who make the decision to deploy them.

Machine learning experts, of course, tend to have a central role. Stakeholders' requirements have to be translated into implementation by these experts; whether intentionally or unintentionally, there are often substantial gaps between the desired policy and the policy that's realized in practice [273]. In every automated system, there is something lost in the translation of policy from human language to computer code. For example, there have been cases where software miscalculated prison inmates' eligibility for early release, with harrowing consequences including being held in prison too long and being returned to prison after being released [419, 420]. But in those classic automated systems, these gaps tend to be mistakes that are generally obvious on manual inspection (not that it is of any comfort to those who are harmed). But when machine learning is involved, the involvement of the expert is often necessary even to recognize that something has gone wrong. This is because the policy tends to be more ambiguous (what does "high risk" mean?) and because deviations from the policy become apparent only in aggregate.

In addition, decision makers often abdicate their power to tool developers, making them even more powerful. Deirdre Mulligan and Kenneth Bamberger explain how government agencies acquire machine learning systems through procurement processes—the same processes used to secure a contractor to build a bridge [421]. The procurement mindset ignores the fact that the resulting products are used to make consequential decisions, that is, effectively to make policy. Procurement emphasizes factors such as price and risk avoidance rather than transparency or oversight of decision making.

Structural Interventions for Fair Machine Learning

The fact that machine learning may contribute to structural discrimination motivates the need for interventions that are similarly broad in scope. We call these "structural interventions": changing the way machine learning gets built and deployed. The changes we have in mind go beyond the purview of any single organization and require collective action. This could take the form of a broad social movement, or other collectives including communities, workers, researchers, and users.

Reforming the Underlying Institutions
One approach is to focus on the underlying institution rather than the technology, and change it so that it is less prone to adopt harmful machine learning tools in the first place. For example, shifting the focus of the criminal justice system from incapacitation to rehabilitation could decrease the demand for risk prediction tools [422]. Many scholars and activists distinguish between reform and abolition

(sometimes called nonreformist reform), abolition being a more radical and transformative approach [193, 423, 424]. For our purposes, however, they both have the effect of centering the intervention on the institution rather than the technology.

In many domains, the very purposes and aims of our institutions remain contested. For example, what are the goals of policing? Commonly accepted goals include deterrence and prevention of crime, ensuring public safety and minimizing disorder, and bringing offenders to justice; they might also include broader efforts to improve the health and vitality of communities. The relative importance of these goals varies between communities and over time. Thus, formulating police allocation decisions as an optimization problem, as predictive policing systems do, involves taking positions on these deeply contested issues.

History shows us that many institutions that may feel like fixtures of modern society, such as higher education, have in fact repeatedly redefined their goals and purposes to adapt to a changing world. In fact, sometimes the impetus for such shifts was to *more effectively discriminate*. In the early twentieth century, elite American universities morphed from treating size (in terms of enrollment) as a source of prestige to favoring selectivity. A major reason for this change was to curtail the rising proportion of Jewish students without having to introduce explicit quotas; the newfound mission of being selective enabled them to emphasize traits like character and personality in admissions, which in turn allowed much leeway for discretion. In fact, the system that Harvard adopted in 1926 was the origin of the holistic approach to admissions that continues to be contentious today, as Jerome Karabel explains in the book *The Chosen* [425].

Some scholars have gone beyond the position that intervention to address algorithmic harms should focus on the underlying institution, and have argued that the adoption of automated decision making actually enables resistant institutions to stave off necessary reform. Virginia Eubanks examines four public assistance programs for poor people in the United States—food assistance, Medicaid, support for the homeless, and care for at-risk children [47]. In each case there are eligibility criteria administered automatically, some of which use statistical techniques. Virginia Eubanks's book documents the harmful effects of these systems, including the punitive effects on those deemed ineligible; the disproportionate impact of those burdens on low-income people of color, especially women; the lack of transparency and seeming arbitrariness of the decisions; and the tracking and surveillance of the lives of poor that is necessary for these systems to operate.

These problems may be fixable to some extent, but Eubanks has a deeper critique: that these systems distract from the more fundamental goal of eradicating poverty ("We manage the individual poor in order to escape our shared responsibility for eradicating poverty"). In theory, the two approaches may coexist. In practice, Eubanks argues, these systems legitimize the idea that there is something wrong with some people, hide the underlying structural problem, and foster inaction. They also incur a high monetary cost that could otherwise be put toward more fundamental reform.

Community Rights

Harmful technologies are often legally justified under a notice-and-consent framework, which rests on an individualistic conception of rights and is ill-equipped to address collective harms. For example, police departments obtain footage en masse from residential security cameras with the consent of residents through centralized platforms like Amazon Ring [426]. However, consent is not a meaningful check in this scenario, because the people who stand to be harmed by police abuse of surveillance footage—such as protesters or members of racial minorities who had the police called on them for "acting suspiciously"—are not the ones whose consent is sought or obtained.

This gap is especially salient in machine learning applications: even if a classifier is trained on data provided with consent, it may be applied to nonconsenting decision subjects. An alternative is to allow groups, such as geographic communities, the right to collectively consent to or reject the adoption of technology tools. In response to the police use of facial recognition, civil liberties activists advocated for a community right to reject such tools; the success of this advocacy has led to various local bans and moratoriums [427]. In contrast, consider online targeted advertising, another technology that has faced widespread dissent. In this case, there are no analogous collectives who can organize effective resistance, and hence attempts to reject the technology have been much less successful [428].

Beyond collective consent, another goal of community action is to obtain a seat at the table in the design of machine learning systems as stakeholders and participants whose expertise and lived experience shapes the conception and implementation of the system rather than as mere data providers and decision subjects. Among other benefits, this approach would make it easier to foresee and mitigate representational harms—issues such as demeaning categories in computer vision datasets or image search results that represent offensive stereotypes. But there are also potential risks to participatory design: it may create further burdens for members of underrepresented communities and it may act as a smokescreen for organizations resisting meaningful change. It is essential that participation be recognized as labor and be fairly compensated [429].

Regulation

Regulation that promotes fair machine learning can take the form of applying existing laws to decision systems that incorporate machine learning, or laws that specifically address the use of technology and its attendant harms. Examples of the latter include the above-mentioned bans on facial recognition, and restrictions on automated decision making under the EU's GDPR. Both flavors of regulation are evolving in response to the rapid adoption of machine learning in decision-making systems. Regulation is a major opportunity for structural intervention for fair machine learning. Yet, because of the tendency of law to conceptualize discrimination in narrow terms, its practical effect on curbing harmful machine learning largely remains to be seen [430].

The gap between the pace of adoption of machine learning and the pace of law's evolution has led to attempts at self-regulation: a 2019 study found eighty-four AI ethics guidelines around the world [431]. Such documents don't have the force of law but attempt to shape norms for organizations and/or individual practitioners. While self-regulation has been effective in some fields, such as medicine, it is doubtful if AI self-regulation can address the thorny problems we have identified in this chapter. Indeed, industry self-regulation generally aims to forestall actual regulation and the structural shifts it may necessitate.

Workforce Interventions

Machine learning shifts power to machine learning experts, which makes the machine learning workforce an important locus of interventions. One set of efforts is aimed at enabling more people to benefit from valuable job opportunities in the industry [432] and to fight imbalances of power within the workforce—notably, between technology experts and those who perform other roles such as annotation [433]. Another set of efforts seeks to align the uses of machine learning with ethical values of the workforce. The nascent unionization movement in technology companies seems to have both objectives.

While a more diverse workforce is morally valuable for its own sake, it is interesting to ask what effect it has on the fairness of the resulting products. One experimental study of programmers found that the gender or race of programmers did not impact whether they produced biased code [434]. However, this is a lab study and should not be seen as a guide to the effects of structural interventions. For example, one causal path by which workforce diversity could impact products (not captured in the study's design) is that a team with a diversity of perspectives may be more willing to ask critical questions about whether a product should be built or deployed.

Other workforce interventions are education and training. Ethics education for computer science students is on the rise, and a 2018 compilation included over 200 such courses [435]. A long-standing debate is about the relative merits of stand-alone courses and the integration of ethics into existing computer science courses [436]. Professional organizations such as the Association for Computing Machinery

(ACM) have had codes of ethics for several decades, but it's unclear if these codes have had a meaningful impact on practitioners.

In many professional fields, including some in engineering, ethical responsibilities are enforced in part through licensing of practitioners. Professionals such as doctors and lawyers must master a body of professional knowledge, including ethical codes, are required by law to pass standardized exams before being licensed to practice, and may have that license revoked if they commit ethical transgressions. This is not the case for software engineering. At any rate, the software engineering certification standards that do exist [437] have virtually no overlap with the topics in this book.

The Research Community

The machine learning research community is another important locus for reform and transformation. The most significant push for change has been the ongoing fight to treat research topics such as fairness, ethics, and justice as legitimate and first-rate. Traditionally, a few topics in machine learning such as optimization algorithms have been considered "core" or "real" machine learning, and other topics—even dataset construction—have been seen as peripheral and less intellectually serious. Abeba Birhane, Pratyusha Kalluri, Dallas Card et al. performed a text analysis of papers at premier machine learning conferences, ICML and NeurIPS, and found that most papers justify themselves by appealing to values such as performance and generalization, and only 1 percent mentioned potential negative effects [438].

A few other key debates: Should all machine learning researchers be required to reflect on the ethics of their research [439]? Is there too much of a focus on fixing bias as opposed to deeper questions about power and justice [440]? How should the perspectives of people and communities affected by machine learning systems be centered? What is the role of industry research on fair machine learning given the conflicts of interest?

Organizational Interventions for Fairer Decision Making

The structural interventions we've discussed above require social movements or other collective action and have been evolving on a timescale of years to decades. This is not to say that an organization should throw up its hands and wait for structural shifts. A plethora of interventions are available to most types of decision makers. This section is an overview of the most important ones.

As you read, observe that the majority of interventions attempt to improve outcomes for all decision subjects rather than viewing fairness as an inescapable trade-off. One reason this is possible is that many of them don't operate at the moment

Table 8.2
A summary of major types of organizational interventions

Type	Intervention	Example
Modifying the outputs	Reallocation	Group-specific decision thresholds
Modifying the decision process	Combatting interpersonal discrimination	Implicit bias training
	Formalization	Adopting statistical decision making
	Procedural protections	Explanation and recourse
Before the decision	Outreach	Sending mailers about scholarships
	Intervening on causal factors	Job training, preventive health
After the decision	Modifying the environment	Helping defendants show up to court

of decision. Note, also, that evaluating the effects of interventions—whether with respect to fairness or other metrics—generally requires causal inference. Finally, only a small subset of potential fairness interventions can be implemented in the framework of machine learning. The others focus on organizational or human practices rather than the technical subsystem involved in decision making.

Redistribution or Reallocation

Redistribution and reallocation are terms that refer to interventions that modify a decision-making process to introduce an explicit preference for one or more groups, usually groups considered to be disadvantaged. When we talk about fairness interventions, this might be the kind that most readily comes to mind.

When applied to selection problems where there is a relatively static number of slots, as is typical in hiring or college admissions, a plethora of algorithmic fairness interventions reduce to different forms of reallocation. This includes techniques such as adding a fairness constraint to the optimization step or a postprocessing adjustment to improve the scores of the members of the disadvantaged groups. This is true regardless of whether the goal is demographic parity or any other statistical criterion.

Reallocation is appealing because it doesn't require a causal understanding of why the disparity arose in the first place. By the same token, reallocation is a crude intervention. It is designed to benefit a group—and it has the advantage of providing a measure of transparency by allowing a quantification of the group benefit—but most reallocation procedures don't incorporate a notion of deservingness of members within that group. Often, reallocation is accomplished by a uniform preference for members of the disadvantaged group. Alternatively, it may be accomplished by tinkering with the optimization objective to incorporate a group preference. In this approach, distributing the fruits of reallocation within the group is delegated to the model, which may end up learning a nonintuitive and unintended allocation (for example, an intersectional subgroup may end up further disadvantaged compared to a no-intervention condition). At best, reallocation methods will aim to ensure that relative ranking within groups is left unchanged.

As crude as reallocation is, another intervention with an even worse tradeoff is to omit features correlated with group identity from consideration. To be clear, if the feature is statistically, causally, or morally irrelevant, that may be a good reason for omitting it (chapter 2). But what if the feature is in fact relevant to the outcome? For example, suppose that people who contribute to open-source software projects tend to be better software engineers. This effect acts through a morally relevant causal path because programmers obtain useful software engineering skills through open-source participation. Unfortunately, many open-source communities are hostile and discriminatory to women and minorities (this is perhaps because they lack the formal organizational structures that firms use to keep interpersonal discrimination in check to some degree). Recognizing this, a software company could either explicitly account for it in hiring decisions or simply omit consideration of open-source contributions as a criterion. If it does the latter, it ends up with less qualified hires on average; it also disadvantages the people who braved discrimination to develop their skills, arguably the most deserving group.

Omitting features based on statistical considerations without a moral or causal justification is extremely popular in practice because it is simple to implement, politically palatable, and avoids the legal risk of disparate treatment.

Combatting Interpersonal Discrimination

Rather than intervene directly on the outputs, organizations can try to improve the process of decision making. In many cases, discriminators are surprisingly candid about their prejudices in surveys and interviews [318]. Can they perhaps be trained out of their implicit or overt biases? This is the idea behind prejudice reduction, often called "diversity training."

But does diversity training work? Elizabeth Paluck and Donald Green conducted a massive review of nearly a thousand such interventions in 2009 [441]. The interventions include promoting contact with members of different groups, recategorization of social identity, explicit instruction, consciousness raising, targeting emotions, targeting value consistency and self-worth, cooperative learning, entertainment (reading, media), discussion, and peer influence. Unfortunately, only a small fraction of the published studies reported on field experiments; Paluck and Green are dubious about both observational field studies and lab experiments. Overall, the field experiments don't provide much support for the effectiveness of diversity interventions. That said, there were many promising lab methods that hadn't yet been tested in the field. A more recent review summarizes the research progress from 2007 to 2019 [442].

Minimizing the Role of Human Judgment via Formalization

Approaches like implicit bias training seek to improve the judgment of human decision makers, but ultimately defer to that judgment. In contrast, formalization aims to curb judgment and discretion.

The simplest formalization technique is to withhold the decision subject's identity (or other characteristics considered irrelevant) from the decision maker. Although this idea dates to antiquity, in many domains the adoption of anonymous evaluation is a recent phenomenon and has been made easier by technology [443]. Two major limitations of this approach are the ubiquitous availability of proxies and the fact that anonymization is not feasible in many contexts such as in-person hiring interviews.

A more ambitious approach is rule-based or statistical decision making that removes human discretion entirely. For example, removing lender discretion in loan underwriting was associated with a nearly 30 percent increase in the approval rates of minority and low-income applicants, while at the same time increasing predictive accuracy (of the risk of default) [1]. Human decision makers tend to selectively ignore credit history irregularities of White applicants [444].

In some ways, machine learning can be seen as a natural progression of the shift from human judgment to rule-based decision making. In machine learning, the discovery of the rule—and not just its application—is deferred to the data and is implemented by an automated system. Based on this, one might naively hope that machine learning will be even more effective at minimizing discrimination.

However, there are several counterarguments. First, claims of the superiority of statistical formulas over human judgment, at least in some domains, have been questioned as being based on apples-to-oranges comparisons, because the human experts did not view their role as pure prediction. For example, judges making sentencing decisions may consider the wishes of victims, and may treat youth as a morally exculpatory factor deserving of leniency [2]. Second, there has been a recognition of all the ways in which machine learning can be discriminatory, which is of course a central theme of this book. Third, there are numerous potential drawbacks, such as a loss of explainability and structural effects, that are not captured by the human-machine comparisons.

Perhaps most significantly, incomplete formalization can simply shift the abuse of discretion elsewhere. In Kentucky, the introduction of pretrial risk assessment *increased* racial disparities for defendants with the same predicted risk. The effect appears to arise partly because of differential adoption of risk assessment in counties with different racial demographics and partly because even the same judges are more likely to override the recommended decision for Black defendants compared to White defendants [445, 446]. In Ontario, social service caseworkers describe how they manipulated the inputs to the automated system to get the outcomes they want [447]. In Los Angeles, police officers used many strategies to resist being managed by predictive policing algorithms [448].

The most pernicious effect of formalization as a fairness intervention is that it may shift discretion to earlier stages of the process, making discrimination *harder* to mitigate. Examples abound. Mandatory minimum sentencing guidelines for drug possession in the United States in the 1980s were justified in part as a way to combat judges' prejudices and arbitrariness [449], but are now widely recognized as overly punitive and structurally racist. One way in which such laws can encode race is the 100-to-1 sentencing disparity between crack and powder cocaine, the popularity of the two forms of the same drug differing by income and socioeconomic status [450]. A very different kind of example comes from Google, which has had a vaunted, highly formalized process for recruiting in order to combat unconscious bias and enhance the quality of decisions [451]. But recruiters have argued that this process in fact bakes in racial discrimination because it incorporates a ranking of colleges in which HBCUs are not ranked at all [452].

The Harvard admissions lawsuit from chapter 5 is another case study of formalization versus holistic decision making. Plaintiffs point out that the admissions criteria include subjective assessments of personality traits such as likability, integrity, helpfulness, kindness, and courage. Harvard scored Asian American applicants on average far lower on these traits than any other racial group. Harvard, on the other hand, argues that evaluating the "whole person" is important to identify those with unique life experiences who would contribute to campus diversity, and that a consideration of subjective traits is a necessary component of this evaluation.

Procedural Protections

Diversity training and formalization are examples of procedural fairness interventions. There are many other procedural protections: notably, making the process transparent, providing explanations of decisions, and allowing decision subjects to contest decisions that may have been made in error. As we discussed above, procedural protections are more important when machine learning is involved than for other types of automated systems.

United States law emphasizes procedural fairness over outcomes. This is one reason for the great popularity of diversity training despite its questionable effectiveness [453]. When the decision maker is the government, the legal conception of fairness is even more focused on procedure. For example, there is no notion of disparate impact under United States constitutional law.

While some procedural interventions such as diversity training have been widely adopted, many others remain rare despite their obvious fairness benefits. For example, few employers offer candid explanations for job rejection. Decision makers turning to automated systems are often looking to cut costs, and may hence be especially loath to adopt procedural protections. An illustrative scenario comes from Amazon, which uses an automated system to manage contract delivery drivers, including contract termination: insiders reported that "it was cheaper to trust the algorithms than pay people to investigate mistaken firings so long as the drivers could be replaced easily" [454].

There are many examples of fairness concerns with automated systems for which *only* procedural protections can be an effective remedy (other than scrapping the system altogether). For example, Google's policy is to suspend users across its entire suite of services if they violate its terms of service. There are many anecdotal reports from users who have lost years' worth of personal and professional data, insist that Google's decision was made in error, and that Google's appeal process did not result in a meaningful human review of the decision.

Outreach

The rest of the interventions are not about changing the decision-making process (or outcomes). Instead, they change something about the decision subjects or the organizational environment.

A 2018 study by Susan Dynarski, C. J. Libassi, Katherine Michelmore et al. sought to address the puzzling phenomenon that low-income students tend not to attend highly selective colleges, even when their strong academic credentials qualify them for admission and despite the availability of financial aid that would make it *cheaper* to attend a selective institution [455]. The authors designed an intervention in which they sent flyers to low-income high school students informing them about a new scholarship at the University of Michigan, and found that compared to a control group, these students were more than twice as likely to apply as well as enroll at the university. The effect was entirely due to students who would have otherwise attended other selective colleges or not attended college at all. The targets of outreach were highly qualified students identified based on standardized test scores (ACT and SAT), which allowed the university to guarantee financial aid conditional on admission. It is worth reiterating that this was a purely informational intervention: the scholarship was equally available to students in the control group, who received only postcards listing University of Michigan application deadlines.

To the extent that disparities are due to disadvantaged groups lacking knowledge of opportunities, informational interventions should decrease those disparities, but this point doesn't appear to be well researched. For example, the Michigan study targeted the intervention at low-income students, so it doesn't address the question of whether informing *all* students would close the income gap.

Intervening on Causal Factors

If we understand the causal factors that lead to underperformance of some individuals or groups, we can intervene to mitigate them. Like informational interventions, this approach seeks to help all individuals rather than simply minimize disparities. This type of intervention is extremely common. Some examples: job training programs for formerly incarcerated people to improve welfare and decrease the chances of recidivism; efforts to bolster math and science education to address an alleged labor shortage of engineers (a so-called pipeline problem); and essentially all of public health and preventive health care. The use of randomized controlled trials to identify and intervene on the causes of poverty has been so influential in development economics that it led to the 2019 Nobel Prize to Abhijit Banerjee, Esther Duflo, and Michael Kremer.

In a competitive market, such as an employer competing for workers, this intervention may not pay off for an individual decision maker from an economic perspective: job seekers who have benefited from the intervention may choose to join other firms instead. Many approaches have been used to overcome this misalignment of incentives. Firms may act collectively, or the state may fund the intervention. If a firm is large enough, the overall payoffs could be so high relative to the cost of the intervention that the reputational benefit to the firm may be sufficient to justify it.

Modifying the Organizational Environment

If decision makers have many opportunities to intervene before the point of decision (e.g., hiring), they also have opportunities to intervene after that point to ensure that individuals fulfill their potential. If a firm finds that few minority employees are successful, it may be because the workplace is hostile and discriminatory.

In other cases, some individuals or groups may need additional accommodations to remedy past disadvantages or because of morally irrelevant differences. A few examples: remedial courses for disadvantaged students, a peer group for first-time college students, need-based scholarships, a nursing mother's room in a workplace, and disability accommodations.

Accommodation isn't simply redistribution in disguise: it does not (or need not) involve an explicit preference for the disadvantaged group. Even if the accommodation is made available to everyone, the disadvantaged group will preferentially benefit from it. This is obvious in the case of, say, disability accommodations. In other cases it is less obvious, but no less true. Even if financial aid were available to all students at a university, it would differentially benefit low-income students.

However, the actual effects of accommodations can be hard to predict and must be carefully measured empirically. A notable example comes from a study showing that men benefit from gender-neutral clock-stopping policies [456]. Such policies in universities allow both men and women to add time to the tenure clock with the birth of a child. While they are often adopted in the interest of fairness, the study shows that they increase men's tenure rates and lower women's; this is presumably because men are able to be more productive during their extended time due to differences in childcare responsibilities or the impact of the birth itself. That said, note that the policy has two fairness goals: to mitigate the adverse career impact of childbirth and to decrease gender disparities in said impacts. Presumably the policy still meets the first goal even if it fails the second.

Here's a stark example of how organizational policies can cause people to fail and how easily they can be remedied. In New York City, there are approximately 300,000 cases of low-level offenses every year. The defendants are required to appear in court (except for offenses of the lowest severity, which may be resolved by mail). If they fail to appear, arrest warrants are automatically issued. Historically, a remarkable 40 percent of defendants fail to appear in court. The resulting negative consequences of Failure to Appear (FTA) are both severe and unequally distributed: for instance, members of groups that are subject to overpolicing are more likely to be arrested. Remarkably, a study by Alissa Fishbane, Aurelie Ouss, and Anuj K. Shah found that FTA rates decreased from 41 to 26 percent when they simply redesigned the summons form to be less confusing and sending defendants text messages shortly before their court dates [457]!

Concluding Thoughts

We looked at seven broad types of fairness interventions that organizations can deploy. The majority of these interventions potentially improve opportunity for all decision subjects as they are motivated by some underlying injustice rather than merely mitigating some disparity. In fact, interventions that aim to address an underlying injustice might sometimes increase certain disparities between groups—a possibility that would be morally justified under a noncomparative notion of fairness that calls for treating each subject as they ought to be treated [458].

It is appealing to focus on comparative notions of fairness because they are easy to quantify, but we shouldn't forget the deeper questions. A domain where this seems to have happened is algorithmic hiring. Tools used in algorithmic hiring utilize situational judgment tests, personality tests, and sometimes much more dubious techniques—increasingly involving machine learning—for screening and selecting candidates. Firms adopt such tools to cut recruitment costs, especially for low-wage positions, where the cost of hiring a worker through the traditional process can be seen as significant in relation to a worker's contribution to the firm's revenue over the course of the period of their employment.

These tools are problematic for many reasons. While they aim to formalize the hiring process, they often use attributes that are morally and causally irrelevant to job performance. HireVue, for example, previously relied on facial expressions and intonations in a person's voice as part of its automated assessment. They also fail to take a broad view of discrimination. Focusing narrowly on minimizing disparities in hiring rates across groups leaves unaddressed what kind of environment employees will encounter once hired. If job applicants from certain groups were previously predicted to perform poorly in a certain workplace, the employer should strive to understand the reasons for this difference in success, rather than simply try to find members of these groups that might be able to succeed under such unfavorable, unwelcoming, or hostile conditions. Parity-promoting interventions change the selection process, but preserve the organizational status quo, endorsing the idea that the candidates who have been selected should be able to deal with these conditions sufficiently well to be as productive as their peers who don't face similar challenges. Other productive—and potentially less harmful—forms of intervention include on-the-job training (which might be understood as a way of intervening on causal factors), meaningful feedback for rejected applicants (which would provide some degree of procedural protection, but also help guide applicants' future investment in their own development), and a strategic approach to sourcing candidates who firms with more accurate tools might now be better able to assess.

The narrow focus on disparities can mean that there is little consideration of the quality of decisions made by the tools. Tools that simply lack validity raise a host of normative concerns. Notably, assessments that achieve approximate demographic parity but continue to suffer from accuracy disparity (also called *differential validity*) can set up members of certain groups for failure by expecting them to be able to perform better than they would be currently prepared to.

To reiterate, we do not advocate for treating statistical fairness criteria as constraints, at least in the first instance. That approach assumes that reallocation is the only available intervention. Instead, if we treat statistical fairness criteria as diagnostics, we are likely to uncover deeper problems that require remedying. Unfortunately, these deeper remedies are also harder. They require both causal inference and normative depth. That is of course why they are often ignored, and foundational questions remain unaddressed.

A case in point: a 2021 paper by Megan Stevenson and Sandra Mayson analyzes the fairness of pretrial detention in a noncomparative sense [459]. How risky does a defendant have to be that the expected benefit to public safety justifies the harm to the defendant from detention? Using the clever approach of asking survey recipients to choose between being detained and becoming victims of certain crimes, the authors conclude that pretrial detention is essentially never justified.

The study's method is sure to be debated, but the point remains that there have been relatively few principled, quantitative attempts to justify the risk thresholds used in pretrial detention. There have been many other calls to end pretrial detention based on different moral and legal arguments. When such foundational questions continue to be debated, it would be exceedingly premature to declare a risk-based pretrial detention system to be "fair" because it satisfied some statistical criterion.

Chapter Notes

The first part of the chapter draws heavily from the sociology of discrimination. A review of racial discrimination by Devah Pager and Hana Shepherd is a good entry point into this literature [319]. Mario Small and Devah Pager distill six lessons from the sociology of discrimination [388].

The complex ways in which discrimination operates—feedback loops that sustain persistent inequalities, multiple interlocking systems of discrimination that together structure society—mean that the quantitative tests for discrimination discussed in the previous chapter frame the question narrowly and are inherently limited in what they can reveal. For more on the limits of the quantitative approach, see Arvind Narayanan's talk [376] or the discussion in the final section of Kevin Lang and Ariella Kahn-Lang Spitzer's paper [461].

The chapter then turns to the practice of machine learning. To reinforce the idea that developers of machine learning systems must make many choices requiring normative judgment throughout the development process, see Jessica Eaglin's case study of recidivism risk prediction [463]. Turning to the question of what technologists should do differently in their everyday work, we recommend the book *Human-Centered Data Science* by Cecilia Aragon, Shion Guha, Marina Kogan et al., and Ben Green's paper urging data scientists to recognize that their work is political [463].

The final part of the chapter argues that most fairness interventions should target organizational culture and processes rather than tweaking decision criteria. This means that designing effective fairness interventions requires understanding the organizations that are meant to adopt them. This is a vast area of sociology; we give a few samples. Michael Lipsky's classic text *Street-Level Bureaucracy* discusses the complex relationship between individual decision makers and the government agency in which they are embedded in [418]. Rebecca Johnson and Simone Zhang deconstruct the process by which social service bureaucracies make and implement policies, and advocate for the benefits of formalizing the process through algorithmic decision making [72].

As for texts about specific bureaucracies or organizations, *Misdemeanorland* by Issa Kohler-Hausmann dives into New York City's lower criminal courts, and shows that the system's purpose and operation is almost diametrically different from how most people and most textbooks (including this one) conceive of it [464]. The books *Pedigree* [320] and *Inside Graduate Admissions* [321] shed light on the hiring processes at elite firms and the admissions processes at graduate schools, respectively. *Uberland* by Alex Rosenblat describes the stories and working conditions of Uber drivers in the United States and Canada [467].

Appendix: A Deeper Look At Structural Factors

Let us briefly discuss two phenomena that help explain the long-run persistence of inequality: segregation and feedback loops.

The Role of Segregation

A structural factor that exacerbates all of the mechanisms of discrimination we discussed is the segregation of society along the lines of group identity. Segregation arguably enables interpersonal discrimination because increased contact among groups decreases prejudice toward outgroups—the controversial contact hypothesis [465].

At a structural level, segregation sustains inequality because an individual's opportunities for economically productive activities depend on her social capital, including the home, community, and educational environment. A strand of the economics literature has built mathematical models and simulations to understand how group inequalities—especially racial inequalities—arise and persist indefinitely even in the absence of interpersonal discrimination, and despite no intrinsic differences between groups. In the extreme case, if we imagine two or more groups belonging to noninteracting economies that grow at the same rate, it is intuitively clear that differences can persist indefinitely. If segregation is imperfect, do gaps eventually close? This is sensitive to the assumptions in the model. In Lundberg and Startz's model the gaps close eventually, although extremely slowly [466]. In Bowles and Sethi's model, they don't under some conditions [468]; one reason is that the disadvantaged group might face higher costs of labor-market skill acquisition due to lower social capital [469].

In the United States, after the civil rights legislation of the 1960s and 1970s, residential segregation by race has been decreasing, albeit slowly. On the other hand, residential segregation by income appears to be increasing [470].

The Role of Feedback Loops

There is a classic economic model of feedback loops in the context of a labor market [315]. There are two groups of workers and two types of jobs: high- and low- skilled, with high-skilled jobs requiring certain qualifications to be performed effectively. Under suitable assumptions (especially, employers cannot perfectly observe worker qualifications before hiring them, but only after providing costly on-the-job skills training) there exists an economic equilibrium in which the following feedback loop sustains itself:

1. The employer practices wage discrimination between the two groups.

2. As a result, the disadvantaged group achieves lower returns to investment in qualifications.

3. Workers, assumed to be rational, respond to such a differential by invest-ing differently in acquiring qualifications, with one group acquiring more qualifications.

4. The employer—again, under certain rationality assumptions—wage discrimi-nates because of the observed difference in qualifications.

The significance of this model is that it can explain the persistence of inequality (and discrimination) without assuming intrinsic differences between the groups, and without employers discriminating between equally qualified workers. It should be viewed as showing only the possibility of such feedback loops. Like any theoretical model, a claim that such a feedback loop explains some actually observed disparity would require careful empirical validation.

9

Datasets

It has become commonplace to point out that machine learning models are only as good as the data they're trained on. The old slogan "garbage in, garbage out" no doubt applies to machine learning practice, as does the related catchphrase "bias in, bias out." Yet, these proverbs still understate—and somewhat misrepresent—the significance of data for machine learning.

It's not only the output of a learning algorithm that may suffer with poor input data. A dataset serves many other vital functions in the machine learning ecosystem. The dataset itself is an integral part of the problem formulation. It implicitly sorts out and operationalizes what the problem is that practitioners end up solving. Datasets have also shaped the course of entire scientific communities in their capacity to measure and benchmark progress, support competitions, and interface between researchers in academia and practitioners in industry.

If so much hinges on data in machine learning, it might come as a surprise that there is no simple answer to the question of what makes data good for what purpose. The collection of data for machine learning applications has not followed any established theoretical framework, certainly not one that was recognized a priori.

In this chapter, we take a closer look at popular datasets in the field of machine learning and the benchmarks that they support. We use this to tease apart the different roles datasets play in scientific and engineering contexts. Then we review the harms associated with data and discuss how they can be mitigated based on the dataset's role. We conclude with several broad directions for improving data practices.

We limit the scope of this chapter in some important ways. Our focus will be largely on publicly available datasets that support training and testing purposes in machine learning research and applications. Our focus excludes large swaths of industrial data collection, surveillance, and data-mining practices. It also excludes data purposely collected to test specific scientific hypotheses, such as experimental data gathered in a medical trial.

A Tour of Datasets in Different Domains

The creation of datasets in machine learning does not follow a clear theoretical framework. Datasets aren't collected to test a specific scientific hypothesis. In fact, we will see that there are many different roles data play in machine learning. As a result, it makes sense to start by looking at a few influential datasets from different domains to get a better feeling for what they are, what motivated their creation, how they organized communities, and what impact they had.

TIMIT

Automatic speech recognition is a machine learning problem of significant commercial interest. Its roots date back to the early twentieth century [471].

Interestingly, speech recognition also features one of the oldest benchmarks datasets, the TIMIT (Texas Instruments/Massachusetts Institute for Technology) data. The creation of the dataset was funded through a 1986 program on speech recognition by the Defense Advanced Research Projects Agency (DARPA). In the mid-eighties, artificial intelligence was in the middle of a "funding winter," where many governmental and industrial agencies were hesitant to sponsor AI research because it often promised more than it could deliver. DARPA program manager Charles Wayne proposed that a way round this problem was to establish more rigorous evaluation methods. Wayne enlisted the National Institute of Standards and Technology to create and curate shared datasets for speech, and he graded success in his program based on performance on recognition tasks on these datasets.

Many now credit Wayne's program with kick-starting a revolution of progress in speech recognition [472–474]. According to Kenneth Ward Church,

> It enabled funding to start because the project was glamour-and-deceit-proof, and to continue because funders could measure progress over time. Wayne's idea makes it easy to produce plots which help sell the research program to potential sponsors. A less obvious benefit of Wayne's idea is that it enabled hill climbing. Researchers who had initially objected to being tested twice a year began to evaluate themselves every hour [473].

A first prototype of the TIMIT dataset was released in December 1988 on a CD-ROM. An improved release followed in October 1990. TIMIT already featured the training/test split typical for modern machine learning benchmarks. There's a fair bit we know about the creation of the data due to its thorough documentation [475].

TIMIT features a total of about five hours of speech, composed of 6,300 utterances, specifically, ten sentences spoken by each of 630 speakers. The sentences were drawn from a corpus of 2,342 sentences such as the following.

Table 9.1
Demographic information about the TIMIT speakers

	Male	Female	Total (%)
White	402	176	578 (91.7)
Black	15	11	26 (4.1)
American Indian	2	0	2 (0.3)
Spanish American	2	0	2 (0.3)
Oriental	3	0	3 (0.5)
Unknown	12	5	17 (2.6)

```
She had your dark suit in greasy wash water all year.
 (sa1)
Don't ask me to carry an oily rag like that. (sa2)
This was easy for us. (sx3)
Jane may earn more money by working hard. (sx4)
She is thinner than I am. (sx5)
Bright sunshine shimmers on the ocean. (sx6)
Nothing is as offensive as innocence. (sx7)
```

The TIMIT documentation distinguishes between eight major dialect regions in the United States, documented as *New England, Northern, North Midland, South Midland, Southern, New York City, Western, Army Brat (moved around)*. Of the speakers, 70 percent are male and 30 percent are female. All native speakers of American English, the subjects were primarily employees of Texas Instruments at the time. Many of them were new to the Dallas area where they worked.

Racial information was supplied with the distribution of the data and coded as "White," "Black," "American Indian," "Spanish American," "Oriental," and "Unknown." Of the 630 speakers, 578 were identified as White, twenty-six as Black, two as American Indian, two as Spanish-American, three as Oriental, and seventeen as unknown (table 9.1).

The documentation notes:

> In addition to these 630 speakers, a small number of speakers with foreign accents or other extreme speech and/or hearing abnormalities were recorded as "auxiliary" subjects, but they are not included on the CD-ROM.

It comes as no surprise that early speech-recognition models had significant demographic and racial biases in their performance.

Today, several major companies, including Amazon, Apple, Google, and Microsoft, all use speech-recognition models in a variety of products, from cell phone apps to voice assistants. There is no longer a major open benchmark that would support training models competitive with the industrial counterparts. Industrial speech recognition pipelines are generally complex and use proprietary data sources that we don't know a lot about. Nevertheless, today's speech-recognition systems continue to exhibit performance disparities along racial lines [476].

UCI Machine Learning Repository

The UCI Machine Learning Repository currently hosts more than 500 datasets, mostly used for different classification and regression tasks. Most datasets are relatively small, consisting of a few hundred or a few thousand instances. The majority are structured tabular data sets with a handful or a few tens of attributes.

The UCI Machine Learning Repository contributed to the adoption of the train-test paradigm in machine learning in the late 1980s. Pat Langley recalls:

> The experimental movement was aided by another development. David Aha, then a PhD student at UCI, began to collect data sets for use in empirical studies of machine learning. This grew into the UCI Machine Learning Repository (http://archive.ics.uci.edu/ml/), which he made available to the community by FTP in 1987. This was rapidly adopted by many researchers because it was easy to use and because it let them compare their results to previous findings on the same tasks [477].

The most popular dataset in the repository is the Iris dataset, containing taxonomic measurements of 150 iris flowers, fifty from each of three species. The task is to classify the species given the measurements.

As of October 2020, the second most popular dataset in the UCI repository is the Adult dataset. Extracted from the 1994 Census database, it features nearly 50,000 instances describing individuals in the United States, each having fourteen attributes. The task is to classify whether an individual earns more than or less than $50,000. The Adult dataset remains popular in the algorithmic fairness community, largely because it is one of the few publicly available datasets that features demographic information including *gender* (coded in binary as male/female), as well as *race* (coded as Amer-Indian-Eskimo, Asian-Pac-Islander, Black, Other, and White).

Unfortunately, the data has some idiosyncrasies that make it less than ideal for understanding biases in machine learning models. Due to the age of the data, and the income cutoff at $50,000, almost all instances labeled *Black* are below the cut-off, as are almost all instances labeled *female*. Indeed, a standard logistic regression model trained on the data achieves about 85 percent accuracy overall, while the same model achieves 91 percent accuracy on Black instances, and nearly 93 percent accuracy on female instances. Likewise, the ROC curves for the latter two groups enclose actually more area than the ROC curve for male instances. This

Figure 9.1
A sample of MNIST digits.

is an atypical situation: more often, machine learning models perform worse on historically disadvantaged groups.

MNIST

The MNIST dataset contains images of handwritten digits. Its most common version has 60,000 training images and 10,000 test images, each having 28×28 black and white pixels (figure 9.1).

MNIST was created by researchers Burges, Cortes, and LeCun from an earlier dataset released by the National Institute of Standards and Technology (NIST). The dataset was introduced in a research paper in 1998 to showcase the use of gradient-based deep learning methods for document recognition tasks [478]. Since then cited over 30,000 times, MNIST became a highly influential benchmark in the computer vision community. Two decades later, researchers continue to use the data actively.

The original NIST data had the property that training and test data came from two different populations. The former featured the handwriting of two thousand American Census Bureau employees, whereas the latter came from 500 American high school students [479]. The creators of MNIST reshuffled these two data sources and split them into training and test set. Moreover, they scaled and centered the digits. The exact procedure to derive MNIST from NIST was lost, but recently reconstructed by matching images from both data sources [480].

The original MNIST test set was of the same size as the training set, but the smaller test set became standard in research use. The 50,000 digits in the original

Table 9.2
A snapshot of the original MNIST leaderboard from February 2, 1999
Source: Internet Archive (retrieved December 4, 2020).

Method	Test error (%)
linear classifier (1-layer NN)	12.0
linear classifier (1-layer NN) [deskewing]	8.4
pairwise linear classifier	7.6
K-nearest-neighbors, Euclidean	5.0
K-nearest-neighbors, Euclidean, deskewed	2.4
40 PCA + quadratic classifier	3.3
1,000 RBF + linear classifier	3.6
K-NN, Tangent Distance, 16×16	1.1
SVM deg 4 polynomial	1.1
Reduced Set SVM deg 5 polynomial	1.0
Virtual SVM deg 9 poly [distortions]	0.8
2-layer NN, 300 hidden units	4.7
2-layer NN, 300 HU [distortions]	3.6
2-layer NN, 300 HU [deskewing]	1.6
2-layer NN, 1000 hidden units	4.5
2-layer NN, 1000 HU [distortions]	3.8
3-layer NN, 300 + 100 hidden units	3.05
3-layer NN, 300 + 100 HU [distortions]	2.5
3-layer NN, 500 + 150 hidden units	2.95
3-layer NN, 500 + 150 HU [distortions]	2.45
LeNet-1 [with 16×16 input]	1.7
LeNet-4	1.1
LeNet-4 with K-NN instead of last layer	1.1
LeNet-4 with local learning instead of ll	1.1
LeNet-5 [no distortions]	0.95
LeNet-5 [huge distortions]	0.85
LeNet-5 [distortions]	0.8
Boosted LeNet-4 [distortions]	0.7

test set that didn't make it into the smaller test set were later identified and dubbed *the lost digits*. [480]

From the beginning, MNIST was intended to be a benchmark used to compare the strengths of different methods. For several years, LeCun maintained an informal leaderboard on a personal website that listed the best accuracy numbers that different learning algorithms achieved on MNIST.

In its capacity as a benchmark, it became a showcase for the emerging kernel methods of the early 2000s that temporarily achieved top performance on MNIST [481]. Today, it is not difficult to achieve less than 0.5 percent classification error with a wide range of convolutional neural network architectures. The best models classify all but a few pathological test instances correctly. As a result, MNIST is widely considered too easy for today's research tasks.

MNIST wasn't the first dataset of handwritten digits in use for machine learning research. Earlier, the US Postal Service (USPS) had released a dataset of 9,298 images (7,291 for training, and 2,007 for testing). The USPS data were actually a fair bit harder to classify than MNIST. A nonnegligible fraction of the USPS digits look unrecognizable to humans [482], whereas humans recognize essentially all digits in MNIST.

ImageNet

ImageNet is a large repository of labeled images that has been highly influential in computer vision research over the last decade. The image labels correspond to nouns from the WordNet lexical database of the English language [483]. WordNet groups nouns into cognitive synonyms, called *synsets*. The words *car* and *automobile*, for example, would fall into the same synset. On top of these categories, WordNet provides a hierarchical tree structure according to a supersubordinate relationship between synsets. The synset for *chair*, for example, is a child of the synset for *furniture* in the WordNet hierarchy. WordNet existed before ImageNet and in part inspired the creation of ImageNet.

The initial release of ImageNet included about 5,000 image categories, each corresponding to a synset in WordNet. These ImageNet categories averaged about 600 images per category [484]. ImageNet grew over time and its fall 2011 release had reached about 32,000 categories.

The construction of ImageNet required two essential steps: retrieving candidate images for each synset and labeling the retrieved images. This first step utilized online search engines and photo sharing platforms with a search interface, specifically Flickr. Candidate images were taken from the image search results associated with the synset nouns for each category.

For the second labeling step, the creators of ImageNet turned to Amazon's Mechanical Turk platform (MTurk). MTurk is an online labor market that allows individuals and corporations to hire on-demand workers to perform simple tasks. In this case, MTurk workers were presented with candidate images and had to decide whether or not the candidate image was indeed an image corresponding to the category that it was putatively associated with.

It is important to distinguish between this ImageNet database and a popular machine learning benchmark and competition, called ImageNet Large Scale Visual Recognition Challenge (ILSVRC), that was derived from it [485]. The competition was organized yearly from 2010 until 2017, reaching significant notoriety in both industry and academia, especially as a benchmark for emerging deep learning models.

When machine learning practitioners say "ImageNet" they typically refer to the data used for the image classification task in the 2012 ILSVRC benchmark. The competition included other tasks, such as object recognition, but image classification has become the most popular task for the dataset. Expressions such as "a model trained on ImageNet" typically refer to training an image classification model on the benchmark data set from 2012.

Another common practice involving the ILSVRC data is *pretraining*. Often a practitioner has a specific classification problem in mind whose label set differs from the 1,000 classes present in the data. It's possible nonetheless to use the data to create useful features that can then be used in the target classification problem. Where ILSVRC enters real-world applications it's often to support pretraining.

This colloquial use of the word ImageNet can lead to some confusion, not least because the ILSVRC-2012 dataset differs significantly from the broader database. It includes only a subset of 1,000 categories. Moreover, these categories are a rather skewed subset of the broader ImageNet hierarchy. For example, of these 1,000 categories only three are in the *person* branch of the WordNet hierarchy, specifically, *groom*, *baseball player*, and *scuba diver*. Yet, more than 100 of the 1,000 categories correspond to different dog breeds. The number is 118, to be exact, not counting wolves, foxes, and wild dogs, which are also present among the categories.

What motivated the exact choice of these 1,000 categories is not entirely clear. The apparent canine inclination, however, isn't just a quirk either. At the time, there was an interest in the computer vision community in making progress on prediction with many classes, some of which are very similar. This reflects a broader pattern in the machine learning community. The creation of datasets is often driven by an intuitive sense of what the technical challenges are for the field. In the case of ImageNet, another important consideration was scale, both in terms of the number of images and the number of classes.

The large-scale annotation and labeling that went into ImageNet falls into a category of labor that Mary Gray and Siddharth Suri call *ghost work* in their book of the same name [486]. They point out: "MTurk workers are the AI revolution's unsung heroes."

Indeed, ImageNet was labeled by about 49,000 MTurk workers from 167 countries over the course of multiple years.

The Netflix Prize

The Netflix Prize was one of the most famous machine learning competitions. Starting on October 2, 2006, the competition ran for nearly three years, ending with a grand prize of $1 million, announced on September 18, 2009. Over the years, the competition saw 44,014 submissions from 5,169 teams.

The Netflix training data contained roughly 100 million movie ratings from nearly 500 thousand Netflix subscribers on a set of 17,770 movies. Each data point corresponds to a tuple `<user, movie, date of rating, rating>`. At about 650 megabytes in size, the dataset was just small enough to fit on a CD-ROM, but large enough to be pose a challenge at the time.

The Netflix data can be thought of as a matrix with $n = 480{,}189$ rows and $m = 17{,}770$ columns. Each row corresponds to a Netflix subscriber and each column to a movie. The only entries present in the matrix are those for which a given subscriber rated a given movie with rating in $\{1, 2, 3, 4, 5\}$. All other entries—that is, the vast majority—are missing. The objective of the participants was to predict the missing entries of the matrix, a problem known as "matrix completion," or somewhat more broadly, "collaborative filtering." In fact, the Netflix challenge did so much to popularize this problem that it is sometimes called the Netflix problem. The idea is that if we could predict missing entries, we would be able to recommend unseen movies to users accordingly.

The holdout data that Netflix kept secret consisted of about three million ratings. Half of them were used to compute a running leaderboard throughout the competition. The other half determined the final winner.

The Netflix competition was hugely influential. Not only did it attract significant participation, it also fueled much academic interest in collaborative filtering for years to come. Moreover, it popularized the competition format as an appealing way for companies to engage with the machine learning community. A startup called Kaggle, founded in April 2010, organized hundreds of machine learning competitions for various companies and organizations before its acquisition by Google in 2017.

But the Netflix competition became infamous for another reason. Although Netflix had replaced usernames by pseudonymous numbers, researchers Arvind Narayanan and Vitaly Shmatikov were able to reidentify some of the Netflix subscribers whose movie ratings were in the dataset [487] by linking those ratings with publicly available movie ratings on IMDb, an online movie database. Some Netflix subscribers had also publicly rated an overlapping set of movies on IMDb under their real identities. In the privacy literature, this is called a *linkage attack* and it's one of the ways that seemingly anonymized data can be deanonymized [488].

What followed were multiple class action lawsuits against Netflix, as well as an inquiry by the Federal Trade Commission (FTC) over privacy concerns. As a consequence, Netflix canceled plans for a second competition, which it had announced on August 6, 2009.

To this day, privacy concerns are a legitimate obstacle to public data release and dataset creation. Deanonymization techniques are mature and efficient. There provably is no algorithm that could take a dataset and provide a rigorous privacy guarantee to all participants while being useful for all analyses and machine learning purposes. Cynthia Dwork and Aaron Roth call this the Fundamental Law of Information Recovery: "overly accurate answers to too many questions will destroy privacy in a spectacular way" [489].

Roles Datasets Play

In machine learning research and engineering, datasets play a different and more prominent set of roles than they do in most other fields. We have mentioned several of these above but let us now examine them in more detail. Understanding these is critical to figuring out which technical and cultural aspects of benchmarks are essential, how harms arise, and how to mitigate them.

A Source of Real Data

Edgar Anderson was a botanist and horticulturist who spent much of the 1920s and 1930s collecting and analyzing data on irises to study biological and taxonomic questions. The iris dataset in the UCI machine learning repository mentioned above is the result of Anderson's labors—or a tiny sliver of them, as most of the observations in the dataset came from a single day of fieldwork. The dataset contains fifty observations each of three iris plants; the task is to distinguish the species based on four physical attributes (sepal length and width; petal length and width). Most of the tens of thousands of researchers who have used this dataset are not interested in taxonomy, let alone irises. What, then, are they using the dataset for?

Although the data were collected by Anderson, they were actually published in the paper "The Use of Multiple Measurements in Taxonomic Problems" by Ronald Fisher, who was a founder of modern statistics as well as a eugenicist [490]. The eugenics connection is not accidental: other central figures in the development of modern statistics, such as Francis Galton and Karl Pearson, were algo eugenicists [491, 492]. Fisher was Anderson's collaborator. Although Fisher had some interest in taxonomy, he was primarily interested in using the data to develop statistical techniques (with an eye toward applications for eugenics). In the 1936 paper, Fisher introduces linear discriminant analysis (LDA) and shows that it performs well on this task.

The reason the iris dataset proved to be a good application of LDA is that there exists a linear projection of the four features that seems to result in a mixture of Gaussians (one for each of the three species), and the means of the three distributions are relatively far apart; one of the species is in fact perfectly separable from the other two. Every learning algorithm implicitly makes assumptions about the data-generating process: without assumptions, there is no basis for making predictions on unseen points [493]. If we could perfectly mathematically describe the data-generating process behind the physical characteristics of irises (or any other population), we wouldn't need a dataset—we could mathematically work out how well an algorithm would perform. In practice, for complex phenomena, such perfect mathematical descriptions rarely exist. Different communities place different values on attempting to discover the true data-generating process. Machine learning places relatively little emphasis on this goal [494]. Ultimately, the usefulness of a learning algorithm is established by testing it on real datasets.

The reliance on benchmark datasets as a source of real data was a gradual development in machine learning research. For example, Rosenblatt's perceptron experiments in the 1950s used two artificial stimuli (the characters E and X), with numerous variants of each created by rotation and other transformations [495]. The controlled input was considered useful to understand the behavior of the system. Writing in 1988, Pat Langley advocates for a hybrid approach, pointing out that "successful runs on a number of different natural domains provide evidence of generality," but he also highlights the use of artificial data for better understanding [496]. Especially after the establishment of the UCI repository around this time, it has become common to evaluate new algorithms on widely used benchmark datasets as a way of establishing that the researcher is not "cheating" by picking contrived inputs.

To summarize, when a researcher seeks to present evidence that an algorithmic innovation is useful, the use of a real dataset as opposed to artificial data ensures that the researcher didn't make up data to suit the algorithm. Further, the use of prominent benchmark datasets wards off skepticism that the researcher may have cherry-picked a dataset with specific properties that make the algorithm effective. Finally, the use of multiple benchmark datasets from different domains suggests that the algorithm is highly general.

Perversely, domain ignorance is treated almost as a virtue rather than a drawback. For example, researchers who achieve state-of-the-art performance on, say, Chinese-to-English translation may point out that none of them speaks Chinese. The subtext is that they couldn't have knowingly or unknowingly picked a model that works well only when the source language is linguistically similar to Chinese.

A Catalyst and Measure of Domain-Specific Progress

Algorithmic innovations that are highly portable across domains, while important, are rare. Much of the progress in machine learning is instead tailored to specific domains and problems. The commonest way to demonstrate such progress is to show that the innovation in question can be used to achieve "state of the art" performance on a benchmark dataset for that task.

The idea that datasets spur algorithmic innovation requires some explanation. For example, the Netflix Prize is commonly credited as being responsible for the discovery of the effectiveness of matrix factorization in recommender systems (often attributed to Simon Funk, a pseudonymous contestant [497]). Yet, the technique had been proposed in the context of movie recommendation as early as 1998 [498] and for search as early as 1990 [499]. However, it was not previously apparent that it outperformed neighborhood-based methods and that it could discover meaningful latent factors. The clarity of the Netflix leaderboard and the credibility of the dataset helped establish the significance of matrix factorization [500].

Somewhat separately from the role of spurring algorithmic innovation, benchmark datasets also offer a convenient way to measure its results (hence the term benchmark). The progression of state-of-the-art accuracy on a benchmark dataset and task can be a useful indicator. A relatively flat curve of accuracy over time may indicate that progress has stalled, while a discontinuous jump may indicate a breakthrough. Reaching an error rate that is close to zero or at least lower than the "human error" for perception tasks is often considered a sign that the task is "solved" and that it is time for the community to move on to a harder challenge.

While these are appealing heuristics, there are also pitfalls. In particular, a statement such as "the state-of-the-art accuracy for image classification is 95 percent" is not a scientifically meaningful claim that can be assigned a truth value because the number is highly sensitive to the data distribution.

A notable illustration of this phenomenon comes from a paper by Benjamin Recht, Rebecca Roelofs, Ludwig Schmidt et al. They carefully recreated new test sets for the CIFAR-10 and ImageNet classification benchmarks according to the very same procedure as the original test sets [501]. They then took a large collection of representative models proposed over the years and evaluated all of them on the new test sets. All models suffered a significant drop in performance on the new test set, corresponding to about five years of progress in image classification. They found that this was because the new test set represents a slightly different distribution. This is despite the researchers' careful efforts to replicate the data collection procedure; we should expect that test sets created by different procedures should result in much greater performance differences (figure 9.2).

The same graphs also provide a striking illustration of why benchmark datasets are a practical necessity for performance comparison in machine learning. Consider a hypothetical alternative approach analogous to the norm in many other branches of science: a researcher evaluating a claim (algorithm) describes in detail their procedure for sampling the data; other researchers working on the same problem sample their own datasets based on the published procedure. Some reuse of datasets occurs, but there is no standardization. The graphs show that even extremely careful efforts to sample a new dataset from the same distribution would shift the distribution sufficiently to make performance comparison hopeless.

In other words, reported accuracy figures from benchmark datasets do not constitute generalizable scientific knowledge because they don't have external validity beyond the specific dataset. While the paper by Recht et al. is limited to image classification, it seems scientifically prudent to assume a lack of external validity for other machine learning tasks as well, unless there is evidence to the contrary.

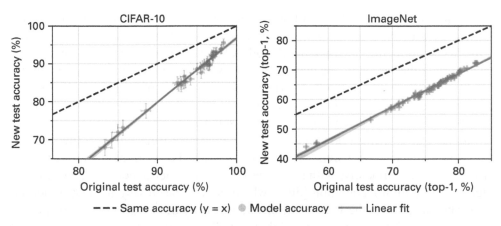

Figure 9.2
Model accuracy on the original test sets vs. new test sets for CIFAR-10 and ImageNet. Each data point corresponds to one model in a test bed of representative models (shown with 95 percent Clopper-Pearson confidence intervals). The plots reveal two main phenomena: (1) there is generally a significant drop in accuracy from the original to the new test sets; (2) the model accuracies closely follow a linear function, meaning that models that perform well on the old test set also tend to perform well on the new test set. The narrow shaded region is a 95 percent confidence region for the linear fit.

Yet the two graphs above hint at a different type of knowledge that seems to transfer almost perfectly to the new test set: the *relative* performance of models. Indeed, another paper showed evidence that relative performance is stable on many datasets across a much wider range of distribution shifts, with strong correlations between in-domain and out-of-domain performance [502].

The relative performance of models for a given task is a very useful type of practitioner-oriented knowledge that can be gained from benchmark leaderboards. A question that practitioners often face is, "Which class of models should I use for [given task] and how should I optimize it"? A benchmark dataset (together with the associated task definition) can be seen as a proxy for answering this question in a constrained setting, analogous to laboratory studies in other branches of science. The hope is that algorithms (and model classes or architectures) identified as state of the art based on benchmark evaluation are also the ones that will be effective on the practitioner's test set. In other words, practitioners can outsource the laborious task of model selection to the benchmark leaderboard.

To be clear, this is an oversimplification. Practitioners have many concerns in addition to that of accuracy, such as the computational cost (of both training and prediction), interpretability, and, increasingly, fairness and environmental cost. Thus, benchmark performance is useful to practitioners but far from the only consideration for model selection.

We can imagine a spectrum of how similar the new test set is to the benchmark set. At the one extreme, if the new test set is truly a new sample from the exact same distribution, then the ranking of model classes should be the same for the two sets. At the other extreme, the distributions may be so different that they constitute essentially different tasks, so that performance on one is not a useful guide to performance on the other. In between these extremes is a big gray area that is not well understood, and it is currently more art than science.

The lack of clarity on how much we can generalize from one or a few benchmarks is associated with well-known controversies. For example, support vector machines were competitive with neural networks on earlier-generation benchmarks such as NIST digit recognition [503], which is one reason why interest in neural networks dwindled in the 1990s. The clear superiority of neural networks on newer benchmarks such as ImageNet was only belatedly recognized.

A Source of (Pre)Training Data

Above, we have envisaged that practitioners use the benchmark leaderboard as a guide to model selection but then train the selected models from scratch on their own (often proprietary) data sources. But practitioners often can and do go further.

In some cases, it may be possible to train on a benchmark dataset and directly use the resulting model in one's application. This depends on the domain and the task, and is more suitable when the distribution shift is minimal and the set of class labels is stable. For example, it is reasonable to deploy a digit recognizer pretrained on MNIST, but not so much an image classifier pretrained on ILSVRC (without some type of adaptation to the target domain). Indeed, ILSVRC consists of a rather arbitrary subset of 1,000 classes of ImageNet, and a pretrained model is correspondingly limited in the set of labels it is able to output. The ImageNet Roulette project was a telling demonstration of what happens when a model trained on the (full) ImageNet dataset is applied to a different test distribution, one consisting primarily of images of people. The results were grotesque. The demonstration has been discontinued, but many archived results may be found in articles about the project [504]. Finally, consider a recommendation system benchmark dataset. There is no way to even attempt to use it directly as training data because the users about whom one wants to make predictions are highly unlikely to be present in the training set.

In most cases, the creators of benchmark datasets do not intend them to be used as a source of training data, although benchmark datasets are often misused for this purpose. A rare exception is The Pile, a large (800GB) English text corpus that is explicitly targeted at training language models. To improve the generalization capabilities of models trained on this corpus, the authors included diverse text from twenty-two different sources [505].

Even when benchmark datasets are not useful as training data for the above-mentioned reasons, they can be useful as pretraining data for transfer learning. Transfer learning refers to using an existing model as a starting point for building a new model. A new model may be needed because the data distribution has shifted compared to what the existing model was optimized for, or because it aims to solve a different task altogether. For example, a model pretrained on ImageNet (or ILSVRC) may be adapted via further training for recognizing different species (distribution shift) or as part of an image captioning model (a different task).

There are different intuitions to explain why transfer learning is often effective. One is that the final layers of a neural network correspond to semantically high-level representations of the input. Pretraining is a way of learning these representations that tend to be useful for many tasks. Another intuition is that pretraining is a way of initializing weights that offers an improvement over random initialization in that it requires fewer samples from the target domain for convergence.

Pretraining offers the practical benefit of being able to share the knowledge contained in a dataset without releasing the raw data. Many datasets, especially those created by companies using customer data, cannot be published due to privacy or confidentiality concerns. The release of pretrained models is thus an important avenue of knowledge sharing from industry to academia. Sharing pretrained models is also helpful to users for whom training from scratch is cost prohibitive. However, privacy and data protection concerns surface in the context of sharing pretrained models due to the possibility that personal data used for training can be reconstructed from the pretrained model [506].

Let's wrap up our analysis of the roles of benchmark datasets. We identified six distinct roles: (1) providing data sampled from real-world occurring distributions that enables largely domain-agnostic investigations of learning algorithms; (2) enabling domain-specific progress by providing datasets that are representative of real-world tasks in that domain yet that abstract away unnecessary detail; (3) providing a convenient albeit crude numerical way to track scientific progress on a problem; (4) enabling model comparison and allowing practitioners to outsource model selection to public leaderboards; (5) providing a source of pretraining data for representation learning, weight initialization, and so on; (6) providing a source of training data. The progression of these six roles is generally toward increasing domain and task specificity, and from science-oriented to practice-oriented.

The Scientific Basis of Machine Learning Benchmarks

Now we examine a seeming mystery: whether and why the benchmark approach works despite the practice of repeated testing on the same data.

Methodologically, much of modern machine learning practice rests on a variant of trial and error, which we call the train-test paradigm. Practitioners repeatedly build models using any number of heuristics and test their performance to see what works. Anything goes as far as training is concerned, subject only to computational constraints, so long as the performance looks good in testing. Trial and error is sound so long as the testing protocol is robust enough to absorb the pressure placed on it. We will examine to what extent this is the case in machine learning.

From a theoretical perspective, the best way to test the performance of a classifier is to collect a sufficiently large fresh dataset and to compute the average error on that test set. Data collection, however, is a difficult and costly task. In most applications, practitioners cannot sample fresh data for each model they would like to try out. A different practice has therefore become the defacto standard. Practitioners split their dataset into typically two parts, a *training set*, used for training a model, and a *test set*, used for evaluating its performance. Often the split is determined when the dataset is created. Datasets used for benchmarks in particular have one fixed split persistent throughout time. A number of variations on this theme go under the name *holdout method*.

Machine learning competitions have adopted the same format. The company Kaggle, for example, has organized hundreds of competitions since it was founded. In a competition, a holdout set is kept secret and is used to rank participants on a public leaderboard as the competition unfolds. In the end, the final winner is whoever scores highest on a separate secret test set not used to that point.

In all applications of the holdout method the hope is that the test set will serve as a fresh sample that provides good performance estimates for all the models. The central problem is that practitioners don't use the test data just once, to retire it immediately thereafter. The test data are used incrementally for building one model at a time while incorporating feedback received previously from the test data. This leads to the fear that eventually models begin to *overfit* to the test data. This type of overfitting is sometimes called *adaptive overfitting* or *human-in-the-loop overfitting*.

Richard Duda, Peter Hart, and David Stork summarize the problem aptly in their 1973 textbook [507]:

> In the early work on pattern recognition, when experiments were often done with very small numbers of samples, the same data were often used for designing and testing the classifier. This mistake is frequently referred to as "testing on the training data." A related but less obvious problem arises when a classifier undergoes a long series of refinements guided by the results of repeated testing on the same data. This form of "training on the testing data" often escapes attention until new test samples are obtained.

Nearly half a century later, Trevor Hastie, Robert Tibshirani, and Jerome Friedman still caution in the 2017 edition of their influential textbook [508]:

Ideally, the test set should be kept in a "vault," and be brought out only at the end of the data analysis. Suppose instead that we use the test-set repeatedly, choosing the model with smallest test-set error. Then the test set error of the final chosen model will underestimate the true test error, sometimes substantially.

While the suggestion to keep the test data in a "vault" is safe, it couldn't be further from the reality of modern practice. Popular test datasets often see tens of thousands of evaluations.

Yet adaptive overfitting doesn't seem to be happening. Recall the scatter plots by Recht et al. above: the plots admit a clean linear fit with positive slope. In other words, the better a model is on the old test set, the better it is on the new test set. But notice that newer models, that is, those with higher performance on the original test set, had *more* time to adapt to the test set and to incorporate more information about it. Nonetheless, the better a model performed on the old test set the better it performs on the new set. Moreover, on CIFAR-10 we even see clearly that the absolute performance drop diminishes with increasing accuracy on the old test set. In particular, if our goal was to do well on the new test set, seemingly our best strategy is to continue to inch forward on the old test set.

The theoretical understanding of why machine learning practice has not resulted in overfitting is still catching up. Here, we highlight one of many potential explanations, called the *leaderboard principle*. It is a subtle effect in which publication biases force researchers to chase state-of-the-art results, and they publish models only if they see significant improvements over prior models. This cultural practice can be formalized by Avrim Blum and Moritz Hardt's *Ladder algorithm*. For each given classifier, it compares the classifier's holdout error to the previously smallest holdout error achieved by any classifier encountered so far. If the error is below the previous best by some margin, it announces the holdout error of the current classifier and notes it as the best seen so far. Importantly, if the error is not smaller by a margin, the algorithm releases the previous best (rather than the new error). It can be proven that the Ladder algorithm avoids overfitting in the sense that it accurately measures the error of the best performing classifier among those encountered [509].

Benchmark Praxis and Culture

The above discussion hints at the importance of cultural practices for a full understanding of benchmark datasets. Let us now discuss these in more detail, highlighting both dataset creators and users. These practices have helped make the benchmark-oriented approach successful but they also impact the harms associated with data. Let's start with creators.

Benchmark creators define the task. This involves, among other things, selecting the high-level problem and defining the target variable, the procedure for sampling the data, and the scoring function. If manually annotating of the data is necessary, the dataset creator must develop a codebook or rubric for doing so and orchestrate crowd-work if needed. Data cleaning to ensure high-quality labels is usually required.

In defining the task, benchmark developers navigate a tricky balance: a task that is seen as too easy using existing techniques will not spur innovation while a task that is seen as too hard may be demotivating. Finding the sweet spot requires expertise, judgment, and some luck. If the right balance is achieved, the benchmark drives progress on the problem. In this way, benchmark creators play an outsized role in defining the vision and agenda for machine learning communities. The selection of tasks in benchmarks is known to affect the ranking of models, which influences and biases the direction of progress in the community [510]. This effect may be getting more pronounced over time due to increasing concentration on fewer datasets [511].

As an example of the kinds of decisions benchmark developers must make, and how they influence the direction of research, consider MNIST. As discussed above, it was derived from a previous dataset released by NIST in which the training and test set were drawn from different sources, but MNIST eliminated this distribution shift. The MNIST creators argued that this was necessary because

> drawing sensible conclusions from learning experiments requires that the result be independent of the choice of training set and test among the complete set of samples. (http://yann.lecun.com/exdb/mnist/)

In other words, if an algorithm performs well on NIST it is unclear how much of this is due to its ability to learn the training distribution and how much of it is due to its ability to ignore the differences between the train and test distributions. MNIST allows researchers to focus selectively on the former question. This was a fruitful approach in 1995. Decades later, when problems like MNIST classification are effectively solved, the attention of benchmark dataset creators has turned toward methods for handling distribution shift that LeCun et al. justifiably chose to ignore [512].

Another tricky balance is between abstracting away domain details so that the task is approachable for a broad swath of machine learning experts, and preserving enough details so that the methods that work in the benchmark setting will translate to production settings. One reason the Netflix Prize was so popular is because the data just from a matrix, and it is possible to achieve good performance (in the sense of beating Netflix's baseline) without really thinking about what the data means. No understanding of film or user psychology was necessary—or helpful, as it turned out. It is possible that domain expertise would have proved essential if the problem had been formulated differently—say, if it required explainability or emphasized good performance even for users with very few previous ratings.

Another challenge for dataset creators is to avoid leakage. In an apocryphal story from the early days of computer vision, a classifier was trained to discriminate between images of Russian and American tanks with seemingly high accuracy, but it turned out that this was only because the Russian tanks had been photographed on a cloudy day and the American ones on a sunny day [513]. Data leakage refers to a spurious relationship between the feature vector and the target variable that is an artifact of the data collection or sampling strategy. Since the spurious relationship won't be present when the model is deployed, leakage usually leads to inflated estimates of model performance. Shachar Kaufman, Saharon Rosset, Claudia Perlich et al. present an overview of leakage in machine learning [514].

Yet another critical responsibility of benchmark dataset creators is to implement a train-test framework. Most contests have various restrictions in place in an attempt to prevent both accidental overfitting to the leaderboard test set and intentional reverse engineering. Although, as we described above, benchmark praxis differs from the textbook version of the holdout method, practitioners have arrived at a set of techniques that have worked in practice, even if our theoretical understanding of why they work is still catching up.

Taking a step back, in any scientific endeavor there are the difficult tasks of framing the problem, ensuring that the methods have internal and external validity, and interpreting the results. Benchmark dataset creators handle as many of these hard tasks as possible, simplifying the goal of dataset users to the point where if a researcher beats the state-of-the-art performance, there is a good chance that there is a scientific insight somewhere in the methods, although extracting what this insight is may still require work. Further simplifying things for dataset users is the fact that there are no restrictions other than computational constraints on how the researcher uses the training data, as long as performance on the test set looks good.

To be clear, this approach has many pitfalls. Researchers rarely perform the statistical hypothesis tests needed to have confidence in the claim that one model performs better than another [515]. Our understanding of how to account for the numerous sources of variance in these performance measurements is still evolving; a 2021 paper that aims to do so argues that many of the claims of state-of-the-art performance in natural language performance and computer vision don't hold up when subjected to such tests [516].

There have long been articles noting the limitations of what researchers and practitioners can learn from benchmark performance evaluation [517, 518]. David Aha, co-creator of the UCI repository, recalls that these limitations were well understood as early as 1995, just a few years after the repository was established [519].

While it is important to acknowledge the limitations, it is also worth highlighting that this approach works at all. One reason for this success is that the scientific questions are primarily about algorithms and not the populations that the datasets are sampled from.

Indeed, there is a case to be made that other scientific communities should adopt the machine learning community's approach, sometimes called the Common Task Method [472]. Diverse scientific fields including economics, political science, psychology, genetics, and many others have seen an infusion of machine learning methods alongside a new focus on maximizing predictive accuracy as a research objective. These shifts have been accompanied by a rash of reproducibility failures, with large fractions of published papers falling prey to pitfalls such as data leakage [520]. Use of the benchmark dataset approach could have avoided most of these pitfalls.

Now let us transition to dataset users. Benchmark users have embraced the freedom afforded by the approach. As a result, the community of users is large—for example, the data science platform Kaggle has over five million registered users, of whom over 130,000 have participated in a competition. There is less gatekeeping in machine learning research than in other disciplines. Many prominent findings bypass peer review. If a technique performs well on the leaderboard, that is considered to speak for itself. Many people who contribute these findings are not formally affiliated with research institutions.

Overall, the culture of progress in machine learning combines the culture of academic scholarship, engineering, and even gaming with a community of hobbyists and practitioners sharing tips and tricks on forums and engaging in friendly competition. This freewheeling culture may seem jarring to some observers, especially given the sensitivity of some of the datasets involved. The lack of gatekeeping means fewer opportunities for ethical training.

There is another aspect of benchmark culture that amplifies the harms associated with data: collecting data without informed consent and distributing it widely without adequate context. Many modern datasets, especially in computer vision and natural language processing, are scraped from the web. In such cases, it is infeasible to obtain informed consent from the individual authors of the content. What about a dataset such as the Netflix Prize, where a company releases data from its own platform? Even if companies disclose in their terms of service that data might be used for research, it is doubtful that informed consent has been obtained since few users read and understand "terms of Service" documents and because of the complexity of the issues involved.

When an individual's data become part of a benchmark dataset, they gets distributed widely. Popular benchmark datasets are downloaded by thousands of researchers, students, developers, and hobbyists. Scientific norms also call for the data to be preserved indefinitely in the interest of transparency and reproducibility. Thus, not only might individual pieces of data in these datasets be distributed and viewed widely, they may be viewed in a form that strips them of their original context. A joke in bad taste written on social media and later deleted may be captured alongside documents from the Library of Congress.

Harms Associated with Data

Now we will discuss a few important types of harms associated with benchmark datasets and how to mitigate them. We don't mean to imply that all of these harms are the "fault" of dataset creators, but understanding how data play into these harms will bring clarity on how to intervene.

Downstream and Representational Harms

A dataset's downstream harms are those that arise from the models trained on it. This is a type of harm that readily comes to mind: bad data may lead to bad models, which can cause harm to the people they purportedly serve. For instance, biased criminal risk prediction systems disproportionately harm Black, minority, and overpoliced populations among others.

Properties of datasets that sometimes (but not always and not in easily predictable ways) propagate downstream include imbalance, biases, stereotypes, and categorization. By imbalance we mean unequal representation of different groups. For example, Joy Buolamwini and Timnit Gebru pointed out that two facial analysis benchmarks, IJB-A and Adience, overwhelmingly featured lighter-skinned subjects [332]. By dataset biases we mean incorrect associations, especially those corresponding to social and historical prejudices. For example, a dataset that measures arrests as a proxy for crime may reflect the biases of policing and discriminatory laws. By stereotypes we mean associations that accurately reflect a property of the world (or a specific culture at a specific point in time) that is thought to be the result of social and historical prejudice. For example, gender-occupation associations can be called stereotypes. By categorization we mean assigning discrete (often binary) labels to complex aspects of identity such as gender and race.

Representational harms occur when systems reinforce the subordination of some groups along the lines of identity. Representational harms could be downstream harms—such as when models apply offensive labels to people from some groups—but they could be inherent in the dataset. For example, ImageNet contains numerous slurs and offensive labels inherited from WordNet and pornographic images of people who did not consent to their inclusion in the dataset [521, 522].

While downstream and representational harms are two categories that have drawn a lot of attention and criticism, there are many other harms that often arise including the environmental cost of training models on unnecessarily large datasets [523] and the erasure of the labor of subjects who contributed the data [519] or the annotators who labeled it [486]. For an overview of ethical concerns associated with datasets, see the survey by Amandalynne Paullada, Inioluwa Deborah Raji, Emily Bender et al. [524]

Mitigating Harms: An Overview

Approaches for mitigating the harms associated with data are quickly developing. Here we review a few selected ideas.

One approach targets the fact that many machine learning datasets are poorly documented, and details about their creation are often missing. This leads to a range of issues from lack of reproducibility and concerns of scientific validity to misuse and ethical concerns. In response, *datasheets for datasets* is a template and initiative by Timnit Gebru, Jamie Morgenstern, Briana Vecchione et al. to promote more detailed and systematic annotation for datasets [525]. A datasheet requires the creator of a dataset to answer questions relating to several areas of interest: Motivation, composition, collection process, preprocessing/cleaning/labeling, uses, distribution, maintenance. One goal is that the process of creating a datasheet will help anticipate ethical issues with the dataset. But datasheets also aim to make data practices more reproducible, and help practitioners select more adequate data sources.

Going a step beyond datasheets, Eun Seo Jo and Gebru Timnit [526] draw lessons from archival and library sciences for the construction and documentation of machine learning datasets. These lessons draw attention to issues of consent, inclusivity, power, transparency, ethics and privacy.

Other approaches stay within the paradigm of minimally curated data collection but aim to modify or sanitize content deemed problematic in datasets. The ImageNet creators have made efforts to remove slurs and harmful terms as well as categories considered nonimageable, or unable to be characterized using images. "Vegetarian" and "philanthropist" are two such categories that were removed [522]. The REVISE tool aims to partially automate the process of identifying various kinds of biases in visual datasets [527].

Mitigating Harms by Separating the Roles of Datasets

Our analysis of the different roles datasets play allows greater clarity in mitigating harms while preserving benefits. This analysis is not intended as an alternative to the many approaches that have already been proposed for mitigating harms. Rather, it can sharpen our thinking and strengthen other harm-mitigation strategies.

Our main observation is that the reuse of scientific benchmark datasets in engineering pipelines complicates efforts to address biases and harms. Attempts to address harms in such dual-use datasets leave creators with a conundrum. On the one hand, benchmark datasets need to be long-lived: many benchmark datasets created decades ago continue to be useful and widely used today. Thus, modifying a dataset down the line when new harms become known will compromise its scientific utility, as performance on the modified dataset may not be meaningfully comparable to performance on the older dataset.

On the other hand, attempting to anticipate all possible harms during dataset creation is infeasible if the dataset is going to be used as training or pretraining data. Experience shows that datasets turn out to be useful for an ever-expanding set of downstream tasks, some of which were not even conceived of at the time of dataset creation.

Better trade-offs are possible if there is a clear separation between scientific benchmarks and production-oriented datasets. In cases where the same dataset can be potentially useful for both purposes, creators should consider making two versions or forks of the data, because many of the harm-mitigation strategies that apply to one don't apply to the other, and vice versa.

To enforce this separation, benchmark dataset creators should consider avoiding the use of the dataset in production pipelines by explicitly prohibiting it in the terms of use. Currently the licenses of many benchmark datasets prohibit commercial uses. This restriction has a similar effect, but it is not the best way to make this distinction. After all, production models may be noncommercial: they may be built by researchers or governments, with the latter category having an especially high potential for harm. At the same time, prohibiting commercial uses is arguably too strict, as it prohibits the use of the dataset as a guide to model selection, a use that does not raise the same risks of downstream harm.

One reason why there are fairness interventions applicable to scientific benchmark datasets but not production datasets is that, as we've argued, most of the scientific utility of benchmarks is captured by the *relative* performance of models. The fact that interventions that hurt absolute performance may be acceptable gives greater leeway for harm-mitigation efforts. Consider image classification benchmarks. We hypothesize that the relative ranking of models will be only minimally affected if the dataset is modified to remove all images containing people (keeping high-level properties including the number of classes and images the same). Such an intervention would avoid a wide swath of the harms associated with datasets while preserving much of its scientific utility.

Conversely, one reason why there are fairness interventions applicable to production datasets but not scientific benchmarks is that interventions for production datasets can be strongly guided by an understanding of their downstream impacts in specific applications. Language and images, in particular, capture such a variety of cultural stereotypes that sanitizing all of them has proved infeasible [528]. It is much easier to design interventions once we fix an application and the cultural context(s) in which it will be deployed. Different interventions may be applicable to the same dataset used in different applications. Unlike scientific benchmarks, dataset standardization is not necessary in engineering settings.

In fact, the best locus of intervention even for dataset biases may be downstream of the data. For example, it has been observed for many years that online translation systems perpetuate gender stereotypes when translating gender-neutral pronouns. The text "O bir doctor. O bir hemşire" may be translated from Turkish to English as "He is a doctor; she is a nurse." Google Translate mitigated this by showing multiple translations in such cases [529, 530]. Compared to data interventions, this has the benefit of making the potential bias (or, in some cases, erroneous translation) more visible to the user.

Our analysis points to many areas where further research could help clarify ethical implications. In particular, the pretraining role of benchmark datasets occupies a gray area where it is not clear when and to what extent data biases propagate to the target task/domain. Research on this area is nascent [531]; this research is vital because the (mis)use of scientific benchmarks for pretraining in production pipelines is common today and unlikely to cease in the near future.

Datasets should not be seen as static, neutral technical artifacts. The harms that could arise from a dataset depend not just on its contents but also the rules, norms, and culture surrounding its usage. Thus, modifying these cultural practices is one potential way to mitigate harms. As we discussed above, lack of domain knowledge by dataset users has come to be seen almost as a virtue in machine learning. This attitude should be reconsidered as it has a tendency to accentuate ethical blind spots.

Datasets require stewardship, whether by the dataset creator or by another designated entity or set of entities. Consider the problem of derivatives: popular benchmark datasets are often extended by other researchers with additional features, and these derived datasets can introduce the possibility of harms not present in the original (to the same extent). For example, the Labeled Faces in the Wild (LFW) dataset of faces was annotated by other researchers with characteristics as race, gender, and attractiveness [532, 533]. Regardless of the ethics of LFW itself, the derived dataset enables new applications that classify people by appearance in harmful ways. Of course, not all derivatives are ethically problematic. Adjudicating and enforcing such ethical distinctions is possible only if there is a governance mechanism in place.

Beyond Datasets

In this final section, we discuss important scientific and ethical questions that are relevant to datasets but also go beyond datasets, pervading machine learning: validity, problem framing, and limits to prediction.

Lessons from Measurement

Measurement theory is an established science with ancient roots. In short, measurement is about assigning numbers to objects in the real world in a way that reflects relationships between these objects. Measurement draws an important distinction between a *construct* that we wish to measure and the measurement procedure that we used to create a numerical representation of the construct.

For example, we can think of a well-designed math exam as measuring the mathematical abilities of a student. A student with greater mathematical ability than another is expected to score higher on the exam. Viewed this way, an exam is a *measurement procedure* that assigns numbers to students. The *mathematical ability* of a student is the construct we hope to measure. We desire that the ordering of these numbers reflects the sorting of students by their mathematical abilities. A measurement procedure operationalizes a construct.

Every prediction problem has a target variable, the thing we're trying to predict. By viewing the target variable as a construct, we can apply measurement theory to understand what makes a good target variable.

The choice of a poor target variable cannot be ironed out with additional data. In fact, the more data we feed into our model, the better it gets at capturing the flawed target variable. Improved data quality or diversity are no cure either.

All formal fairness criteria that involve the target variable—separation and sufficiency being two prominent examples—are either meaningless or downright misleading when the target variable itself is the locus of discrimination.

But what makes a target variable good or bad? Let's get a better grasp on this question by considering a few examples.

1. Predicting the value of the Standard and Poor 500 Index (S&P 500) at the close of the New York Stock Exchange tomorrow.
2. Predicting whether an individual is going to default on a loan.
3. Predicting whether an individual is going to commit a crime.

The first example is rather innocuous. It references a fairly robust target variable, even though it relies on a number of social facts.

The second example is a common application of statistical modeling that underlies much of modern credit scoring in the United States. At first sight a default event seems like a clean-cut target variable. But the reality is different. In a dataset published by the Federal Reserve [116], default events are coded by a so-called *performance* variable that measures a *serious delinquency in at least one credit line of a certain time period*. More specifically, the Federal Reserve report states that the

measure is based on the performance of new or existing accounts and measures whether individuals have been late 90 days or more on one or more of their accounts or had a public record item or a new collection agency account during the performance period.

Our third example runs into the most concerning measurement problem. How do we determine if an individual committed a crime? What we can determine with certainty is whether or not an individual was arrested and found guilty of a crime. But this depends crucially on who is likely to be policed in the first place and who is able to maneuver the criminal justice system successfully following an arrest.

Sorting out what a good target variable is, in full generality, can involve the whole apparatus of measurement theory. The scope of measurement theory, however, goes beyond defining reliable and valid target variables for prediction. Measurement comes in whenever we create features for a machine learning problem and should therefore be an essential part of the data creation process [534].

Judging the quality of a measurement procedure is a difficult task. Measurement theory has two important conceptual frameworks for arguing about what makes measurement *good*: one is *reliability*, the other is *validity*.

Reliability describes the differences observed in multiple measurements of the same object under identical conditions. Thinking of the measurement variable as a random variable, reliability is about the variance between independent identically distributed measurements. As such, reliability can be analogized with the statistical notion of variance.

Validity is concerned with how well the measurement procedure in principle captures the concept that we try to measure. If reliability is analogous to variance, it is tempting to see validity as analogous to bias. But the situation is a bit more complicated. There is no simple formal criterion that we could use to establish validity. In practice, validity is based to a large extent on human expertise and subjective judgments.

One approach to formalize validity is to ask how well a score predicts some external criterion. This is called *external validity*. For example, we could judge a measure of creditworthiness by how well it predicts default in a lending scenario. While external validity leads to concrete technical criteria, it essentially identifies good measurement with predictive accuracy. However, that's certainly not all there is to validity.

Construct validity is a framework for discussing validity that includes numerous different types of evidence. Samuel Messick highlights six aspects of construct validity [535]:

- Content: How well does the content of the measurement instrument, such as the items on a questionnaire, measure the construct of interest?
- Substantive: Is the construct supported by a sound theoretical foundation?

- Structural: Does the score express relationships in the construct domain?
- Generalizability: Does the score generalize across different populations, settings, and tasks?
- External: Does the score successfully predict external criteria?
- Consequential: What are the potential risks of using the score with regards to bias, fairness, and distributive justice?

Of these different criteria, external validity is the one most familiar to the machine learning practitioner. But machine learning practice would do well to embrace the other, more qualitative, criteria as well. The consequential criterion has been controversial, but Messick forcefully defends its inclusion as an aspect of validity. Ultimately, measurement forces us to grapple with the often surprisingly uncomfortable question: What are we even trying to do when we predict something?

Problem Framing: Comparisons with Humans

A long-standing ambition of artificial intelligence research is to match or exceed human cognitive abilities by an algorithm. This desire often leads to comparisons between humans and machines on various tasks. Judgments about human accuracy often also enter the debate around when to use statistical models in high-stakes decision-making settings.

The comparison between human decision makers and statistical models is by no means new. For decades, researchers have compared the accuracy of human judgments with that of statistical models [536].

Even within machine learning, the debate dates back a long way. A 1991 paper by Bromley and Sackinger explicitly compared the performance of artificial neural networks to a measure of human accuracy on the USPS digits dataset that predates the famous MNIST data [482]. A first experiment put the human accuracy at 2.5 percent, a second experiment found the number 1.51 percent, while a third reported the number 2.37 percent [537].

Comparison with so-called human baselines has since become widely accepted in the machine learning community. The Electronic Frontier Foundation (EFF), for example, hosts a major repository of AI progress measures that compares the performance of machine learning models to reported human accuracies on numerous benchmarks.

For the ILSVRC 2012 data, the reported human accuracy is 5.1 percent. This often-quoted number corresponds to the performance of a single human annotator who was "trained on 500 images and annotated 1500 test images" [485]. A second annotator, who was "trained on 100 images and then annotated 258 test images," achieved an accuracy of 12 percent. Based on this number of 5.1 percent, researchers announced in 2015 that their model was "the first to surpass human-level performance" [538]. Not surprisingly, this claim received significant attention throughout the media.

However, a later more careful investigation into "human accuracy" on ImageNet revealed a very different picture [539]. The researchers found that only models from 2020 are actually on par with the strongest human labeler. Moreover, when restricting the data to 590 object classes out of 1,000 classes in total, the best human labeler performed much better, at less than 1 percent error, than even the best predictive models. Recall that the ILSVRC 2012 data featured 118 different dog breeds alone, some of which are extremely hard to distinguish for anyone who is not a trained dog expert. In fact, the researchers had to consult with experts from the American Kennel Club to disambiguate challenging cases of different dog breeds. Simply removing dog classes alone increases the performance of the best human labeler to less than 1.3 percent error.

There is another troubling fact. Small variations in the data collection protocol turn out to have a significant effect on the performance of machine classifiers: "the accuracy scores of even the best image classifiers are still highly sensitive to minutiae of the data cleaning process" [501].

These results cast doubt not only on how me measure human accuracy but also on the validity of the presumed theoretical construct of "human accuracy" itself. However, the machine learning community has adopted a rather casual approach to measuring human accuracy. Many researchers assume that the construct of *human accuracy* exists unambiguously and it is whatever number comes out of some ad hoc testing protocol for some set of human beings. These ad hoc protocols often result in anecdotal comparisons of questionable scientific value.

Invalid judgments about human performance relative to machines are not just a scientific error, they also have the potential to create narratives that support poor policy choices in high-stakes policy questions around the use of predictive models in consequential decisions. For example, criminal justice policy is being driven by claims that statistical methods are superior to judges at predicting risk of recidivism or failure to appear in court. However, these comparisons are dubious because judges are not solving pure prediction problems but rather incorporate other factors such as leniency toward younger defendants [2].

Problem Framing: Focusing on a Single Optimization Objective
Real-life problems rarely involve optimizing a single objective and more commonly involve some kind of trade-off between multiple objectives. How best to formulate this as a statistical optimization problem is both an art and a science. However, benchmark tasks, especially those with leaderboards, tend to pick a single objective. For high-profile benchmarks, the resulting "overfitting to the problem formulation"

may result in scientific blind spots and limit the applicability of published findings to practical settings.

For example, it was well known at the time Netflix launched its prize that recommendation is not just a matter of maximizing predictive accuracy and, even to the extent that it is, there isn't one single measure that's always appropriate [540]. Yet the contest focused purely on prediction accuracy evaluated by a single metric. A few years after the contest ended, Netflix revealed that most of the work that went into the leaderboard had not translated into production models. Part of the reason was that the contest did not capture the range of Netflix's objectives and constraints: the tight dependence of recommendations on the user interface, the fact that "users" are typically households made of members with differing tastes, explainability, freshness, and many more [541].

If many of the insights from the leaderboard did not even generalize to Netflix's own production setting, the gap between Netflix and other recommendation-oriented platforms is far greater. Notably, as a movie platform, Netflix is unusual in that it has a relatively static inventory compared to those with user-generated content such as YouTube or Facebook. When the content pool is dynamic, a different class of algorithms is needed. The pull that the Netflix Prize exerted on recommender systems research may have diverted attention away from the latter type of algorithm for many years, although it is hard to know for sure because the counterfactual is unobservable.

Formal machine learning competitions, even if they cause blind spots due to the need to pick a single optimization objective, are at least carefully structured to promote scientific progress in some narrow sense. Arguably more damaging are the informal competitions that seem to inevitably emerge in the presence of a prominent benchmark dataset, resulting in unfortunate outcomes such as insightful papers being rejected because they failed to beat the state of the art, or unoriginal papers being published because they did beat the state of the art by (scientifically insignificant) application of greater computing power.

Another downside to a field oriented on one-dimensional, competitive pursuit is that it becomes structurally difficult to address biases in models and classifiers. If a contestant takes steps to prevent dataset bias from propagating to their models, there will be an accuracy drop (because accuracy is judged on a biased dataset) and fewer people will pay attention to the work.

As fairness issues in machine learning have gained prominence, fairness-focused benchmarks datasets have proliferated, such as the Pilot Parliamentarians Benchmark for facial analysis [332] and the Equity Evaluation Corpus for sentiment analysis [329]. An advantage of this approach is that the scientific and cultural machinery of benchmark-oriented innovation can be repurposed for fairness research. A potential danger is Goodhart's law, which states, in its broad form,

"When a measure becomes a target, it ceases to be a good measure." As we've emphasized in this book, fairness is multifaceted, and benchmarks can capture only narrow notions of fairness. While these can be useful diagnostics, if they are misconstrued as targets in their own right, then research that is focused on optimizing for these benchmarks may not result in fairness in a more substantive sense. In addition, the construction of these datasets has often been haphazard, without adequate attention to issues of validity [542].

In addition to creating fairness-focused benchmarks, the algorithmic fairness community has also repurposed earlier benchmarks toward the study of fairness questions. Consider the Census dataset from the UCI repository discussed earlier. It originally gained popularity as a source of real-world data. Its use is acceptable for studying algorithmic questions such as, say, the relative strengths of decision trees and logistic regression. We expect the answers to be insensitive to issues like the cultural context of the data. But now it is being used for studying fairness questions such as how classification accuracy tends to vary by race or gender. For such questions, the answers are sensitive to the details of the subpopulations. Further, the classification task associated with the benchmark (prediction of income treated as a binary variable) is artificial and does not correspond to any real-life application. Thus, accuracy disparities (and other fairness-related measurements) may look different for a different task, or if the data had been sampled differently, or if it came from a different time or place. Using benchmark datasets to make generalizable claims about fairness requires careful attention to issues of context, sampling, and validity. Michelle Bao, Angela Zhou, Samantha Zottola et al. question whether benchmark datasets for sociotechnical systems like criminal justice are useful. They point out that benchmark culture—where the focus is on methods, with the dataset being secondary and the context ignored—is at odds with the actual needs of fairness and justice, where attention to context is paramount [543].

Limits of Data and Prediction
Machine learning fails in many scenarios and it's important to understand the failure cases as much as the success stories.

The Fragile Families Challenge was a machine learning competition based on the Fragile Families and Child Wellbeing Study (FFCWS) [544]. Starting from a random sample of hospital births between 1998 and 2000, the FFCWS followed thousand of American families over the course of fifteen years, collecting detailed information, about the families' children, their parents, educational outcomes, and the larger social environment. Once a family agreed to participate in the study, data were collected when the child was born, and then at ages one, three, five, nine, and fifteen.

The Fragile Families Challenge concluded in 2017. The underlying dataset for the competition contains 4,242 rows, one for each family, and 12,943 columns, one for each variable plus an ID number for each family. Of the 12,942 variables, 2,358 are constant (i.e., they had the same value for all rows), mostly due to redactions for privacy and ethics concerns. Of the approximately fifty-five million (4,242 $\times$ 12,942) entries in the dataset, about 73 percent do not have a value. Missing values have many possible reasons, including nonresponse from surveyed families, drop out of study participants, as well as logical relationships between features that imply certain fields are missing depending on how others are set. There are six outcome variables, measured at age fifteen: (1) *child grade point average (GPA)*, (2) *child grit*, (3) *household eviction*, (4) *household material hardship*, (5) *caregiver layoff*, and (6) *caregiver participation in job training*.

The goal of the competition was to predict the value of the outcome variables at age fifteen given the data from ages one through nine. As is common for competitions, the challenge featured a three-way data split: training, leaderboard, and test sets. The training set is publicly available to all participants, the leaderboard data support a leaderboard throughout the competition, and the test set is used to determine a final winner.

The outcome of the prediction challenge was disappointing. Even the winning models performed hardly better than a simple baseline; their predictions didn't differ much compared to predicting the mean of each outcome.

What caused the poor performance of machine learning on the fragile families data? One obvious possibility is that none of the contestants hit on the right machine learning techniques for this task. But the fact that 160 teams of motivated experts submitted thousands of models over the course of five months makes this highly unlikely. Besides, models from disparate model classes all made very similar (and equally erroneous) predictions, suggesting that learning algorithms weren't the limitation [90]. There are a few other technical possibilities that could explain the disappointing performance, including the sample size, the study design, and the missing values.

But there is also a more fundamental reason that remains plausible. Perhaps the dynamics of life trajectories are inherently unpredictable over the six-year time delay between measurement of the covariates and measurement of the outcome. This six-year gap, for example, included the Great Recession, a period of economic shocks and decline between 2007 and 2009, which might have changed trajectories in unforeseeable ways.

In fact, there's an important reason why even the performance of models in the challenge, dismal as they were, may overestimate what we can expect in a real-world setting. That's because the models were allowed to peek into the future, so to speak. The training and test sets were drawn from the same distribution and, in particular, the same time period, as is the standard practice in machine learning research. Thus, the data already incorporate information about the effect of the Great Recession and other global shocks during this period. In a real application, models must be trained on data from the past whereas predictions are about the future. Thus, there is always some drift—a change in the relationship between the covariates and the outcome. This puts a further limit on model performance.

Machine learning works best in a static and stable world where the past looks like the future. Prediction alone can be a poor choice when we're anticipating dynamic changes, or when we are trying to reason about the effect that hypothetical actions would have in the real world.

Summary

Benchmark datasets are central to machine learning. They play many roles, including enabling algorithmic innovation, measuring progress, and providing training data. Since its systematization in the late 1980s, performance evaluation on benchmarks has gradually become a ubiquitous practice because it makes it harder for researchers to cheat intentionally or unintentionally.

But an excessive focus on benchmarks brings many drawbacks. Researchers spend prodigious amounts of effort optimizing models to achieve state-of-the-art performance. The results are often both scientifically uninteresting and of little relevance to practitioners because benchmarks omit many real-world details. The approach also amplifies the harms associated with data, including downstream harms, representational harms, and privacy violations.

As we write this book, the benchmark approach is coming under scrutiny because of these ethical concerns. While the benefits and drawbacks of benchmarks are both well known, our overarching goal in this chapter has been to provide a single framework that can help analyze both. Our position is that the core of the benchmark approach is worth preserving, but we envision a future where benchmarks play a more modest role as one of many ways to advance knowledge. To mitigate the harms associated with data, we believe that substantial changes to the practices of dataset creation, use, and governance are necessary. We have outlined a few ways to do this, adding to the emerging literature on this topic.

Chapter Notes

This chapter was developed and first published by Moritz Hardt and Benjamin Recht in the textbook *Patterns, Predictions, and Actions: Foundations of Machine Learning* [119]. With permission from the authors, we include a large part of the original text here with only slight modifications. We removed a significant amount of material on adaptive data analysis and the problem of overfitting in machine learning benchmarks. We added new material on the roles that datasets play, as well as discussion about fairness and ethical concerns relating to datasets.

Adaptivity in holdout reuse was studied by Cynthia Dwork, Vitaly Feldman, Moritz Hardt et al. [545] and there has been subsequent work in the area of adaptive data analysis. Similar concerns go under the name of *inference after selection* in the statistics community.

The collection and use of large ad hoc datasets (once referred to as "big data") have been scrutinized in several important works; see, for example, Danah Boyd and Kate Crawford [546], as well as Zeynep Tufekci [547, 548]. More recently, Nick Couldry and Ulises Mejias [549] use the term *data colonialism* to emphasize the processes by which data are appropriated and marginalized communities are exploited through data collection. Alexandra Olteanu, Carlos Castillo, Fernando Diaz et al. [550] discuss biases, methodological pitfalls, and ethical questions in the context of social data analysis. In particular, the article provides taxonomies of biases and issues that can arise in the sourcing, collection, processing, and analysis of social data. Geoffrey Bowker and Susan Leigh Star's classic text explains why categorization is a morally laden activity [191]. For a discussion of the harms of category systems embedded in machine learning datasets, see *The Atlas of AI* [551].

The benefits of the benchmark dataset approach are discussed in a talk by Mark Liberman, who calls it the common task method [552]. Amandalynne Paullada, Inioluwa Deborah Raji, Emily M. Bender et al. survey dataset development and use cases in machine learning research [553]. A survey by Alessandro Fabris, Stefano Messina, Gianmaria Silvello et al. lists and discusses numerous datasets uses throughout the fairness literature [554]. Emily Denton, Alex Hanna, Razvan Amironesei et al. provide a genealogy of ImageNet through a critical lens [555]. Inioluwa Deborah Raji, Emily M, Bender, Amandalynne Paullada et al. give an overview of concerns arising from basing our understanding of progress on a small collection of influential benchmarks [556]. The EFF AI metrics project is available at https://www.eff.org/ai/metrics.

For an introduction to measurement theory, not specific to the social sciences, see the books by David Hand [557, 558]. The textbook by Deborah Bandalos [559] focuses on applications to the social science, including a chapter on fairness. Thomas Liao, Rohan Taori, Inioluwa Deborah Raji et al. provide a taxonomy of evaluation failures across many subfields of machine learning, encompassing both internal and external validity issues [560].

References

[1] Gates, Susan Wharton, Perry, Vanessa Gail, and Zorn, Peter M. 2002. "Automated underwriting in mortgage lending: Good news for the underserved?" *Housing Policy Debate*, 13(2):369–391.

[2] Stevenson, Megan T., and Doleac, Jennifer L. 2022. "Algorithmic risk assessment in the hands of humans." *Available at SSRN*: https://ssrn.com/abstract=3489440 or http://dx.doi.org/10.2139/ssrn.3489440.

[3] Ingold, David, and Soper, Spencer. 2016. "Amazon doesn't consider the race of its customers. Should it?" *Bloomberg*, April 21, 2016. https://www.bloomberg.com/graphics/2016-amazon-same-day/.

[4] Crawford, Kate. 2013. "The hidden biases in big data." *Harvard Business Review*, April 1, 2013, 1. https://hbr.org/2013/04/the-hidden-biases-in-big-data.

[5] Kaggle. 2012. "The Hewlett Foundation: Automated essay scoring." https://www.kaggle.com/c/asap-aes.

[6] Hanna, Rema N., and Linden, Leigh L. 2012. "Discrimination in grading." *American Economic Journal: Economic Policy*, 4(4):146–168.

[7] Sprietsma, Maresa. 2013. "Discrimination in grading: Experimental evidence from primary school teachers." *Empirical Economics*, 45(1):523–538.

[8] Ashkenas, Jeremy, Park, Haeyoun, and Pearce, Adam. 2017. "Even with affirmative action, Blacks and Hispanics are more underrepresented at top colleges than 35 years ago." *New York Times*, August 24, 2017. https://www.nytimes.com/interactive/2017/08/24/us/affirmative-action.html.

[9] Barocas, Solon and Selbst, Andrew D. 2016. "Big data's disparate impact." *California Law Review*, 104:671–732.

[10] Plaugic, Lizzie. 2017. "Faceapp's creator apologizes for the app's skin-lightening 'hot' filter." *The Verge*, April 25, 2017. https://www.theverge.com/2017/4/25/15419522/faceapp-hot-filter-racist-apology.

[11] Manthorpe, Rowland. 2017. "The beauty.ai robot beauty contest is back." *Wired UK*. https://www.wired.co.uk/article/robot-beauty-contest-beauty-ai.

[12] Lo, Puck, and Corey, Ethan. 2019. "The 'failure to appear' fallacy." *The Appeal*, January 9, 2019. https://theappeal.org/the-failure-to-appear-fallacy/.

[13] Deng, J., Dong, W., Socher, R., Li, L.-J., Li, K., and Fei-Fei, L. 2009. "ImageNet: A large-scale hierarchical image database." In *Proceedings of the IEEE Conference on Computer Vision and Pattern Recognition*.

[14] Miller, George A. 1995. "Wordnet: A lexical database for english." *Communications of the ACM*, 38(11):39–41.

[15] Birhane, Abeba, and Prabhu, Vinay Uday. 2021. "Large image datasets: A pyrrhic win for computer vision?" In *2021 IEEE Winter Conference on Applications of Computer Vision*, 1536–1546.

[16] Roth, Lorna. 2009. "Looking at Shirley, the ultimate norm: Colour balance, image technologies, and cognitive equity." *Canadian Journal of Communication*, 34(1):111–136.

[17] Torralba, Antonio, and Efros, Alexei A. 2011. "Unbiased look at dataset bias." In *Proceedings of the IEEE Conference on Computer Vision and Pattern Recognition*, 1521–1528.

[18] Liu, Zicheng, Zhang, Cha, and Zhang, Zhengyou. 2007. "Learning-based perceptual image quality improvement for video conferencing." In *2007 IEEE International Conference on Multimedia and Expo*, 1035–1038.

[19] Kaufman, Liad, Lischinski, Dani, and Werman, Michael. 2012. "Content-aware automatic photo enhancement." In *Computer Graphics Forum*, 31(8):2528–2540.

[20] Kalantari, Nima Khademi, and Ramamoorthi, Ravi. 2017. "Deep high dynamic range imaging of dynamic scenes." *ACM Transactions on Graphics*, 36(4):1–12.

[21] Bonham, Vence L., Callier, Shawneequa L., and Royal, Charmaine D. 2016. "Will precision medicine move us beyond race?" *New England Journal of Medicine*, 374(21):2003–2005.

[22] Wilson, James F., Weale, Michael E., Smith, Alice C., Gratrix, Fiona, Fletcher, Benjamin, Thomas, Mark G., Bradman, Neil, and Goldstein, David B. 2001. "Population genetic structure of variable drug response." *Nature Genetics*, 29(3):265–269.

[23] Caliskan, Aylin, Bryson, Joanna J., and Narayanan, Arvind. 2017. "Semantics derived automatically from language corpora contain human-like biases." *Science*, 356(6334):183–186.

[24] Danesi, Marcel. 2014. *Dictionary of media and communications*. Routledge.

[25] Hardt, Moritz. 2014. "How big data is unfair." *Medium*, September 26, 2014. https://medium.com/@mrtz/how-big-data-is-unfair-9aa544d739de.

[26] Caruana, Rich, Lou, Yin, Gehrke, Johannes, Koch, Paul, Sturm, Marc, and Elhadad, Noemie. 2015. "Intelligible models for healthcare: Predicting pneumonia risk and hospital 30-day readmission." In *Proceedings of the 21st ACM SIGKDD International Conference on Knowledge Discovery and Data Mining*, 1721–1730.

[27] Joachims, Thorsten, Swaminathan, Adith, and Schnabel, Tobias. 2017. "Unbiased learning-to-rank with biased feedback." In *Proceedings of the 10 International Conference on Web Search and Data Mining*, 781–789. ACM.

[28] Sweeney, Latanya. 2013. "Discrimination in online ad delivery." *Queue*, 11(3):10:10–10:29.

[29] Dobbie, Will, Goldin, Jacob, and Yang, Crystal. 2016. "The effects of pre-trial detention on conviction, future crime, and employment: Evidence from randomly assigned judges." Working Paper 22511, National Bureau of Economic Research, Washington, DC.

[30] Lum, Kristian, and Isaac, William. 2016. "To predict and serve?" *Significance*, 13(5):14–19.

[31] Ensign, Danielle, Friedler, Sorelle A., Neville, Scott, Scheidegger, Carlos, and Venkatasubramanian, Suresh. 2018. "Runaway feedback loops in predictive policing." *Proceedings of Machine Learning Research*, 81:1–12.

[32] Zhang, Junzhe, and Bareinboim, Elias. 2018. "Fairness in decision-making: The causal explanation formula." In *Proceedings of the 32nd AAAI Conference on Artificial Intelligence*, 32(1).

[33] Rock, David and Grant, Heidi. 2016. "Why diverse teams are smarter." *Harvard Business Review*, November 4, 2016. https://hbr.org/2016/11/why-diverse-teams-are-smarter.

[34] Freeman, Richard B., and Huang, Wei. 2015. "Collaborating with people like me: Ethnic coauthorship within the united states." *Journal of Labor Economics*, 33(S1):S289–S318.

[35] Barocas, Solon. 2014. "Putting data to work." In *Data and discrimination: Collected essays*, edited by Seeta Peña Gangadharan, Virginia Eubanks, and Solon Barocas, 59–62. New America Foundation.

[36] Jackson, John W., and VanderWeele, Tyler J. 2018. "Decomposition analysis to identify intervention targets for reducing disparities." *Epidemiology*, 29(6): 825–835.

[37] Kay, Matthew, Matuszek, Cynthia, and Munson, Sean A. 2015. "Unequal representation and gender stereotypes in image search results for occupations." In *Proceedings of the 33rd Conference on Human Factors in Computing Systems*, 3819–3828. ACM.

[38] Crawford, Kate. 2017. "The trouble with bias." NeurIPS Keynote, December 10, 2017. https://www.youtube.com/watch?v=fMym_BKWQzk.

[39] Huszár, Ferenc, Ktena, Sofia Ira, O'Brien, Conor, Belli, Luca, Schlaikjer, Andrew, and Hardt, Moritz. 2022. "Algorithmic amplification of politics on Twitter." *Proceedings of the National Academy of Sciences*, 119(1):e2025334119.

[40] Reisman, Dillon, Schultz, Jason, Crawford, Kate, and Whittaker, Meredith. 2018. "Algorithmic impact assessments: A practical framework for public agency accountability." AINow. https://ainowinstitute.org/aiareport2018.pdf.

[41] Munoz, Cecilia, Smith, Megan, and Patil, D. 2016. "Big data: A report on algorithmic systems, opportunity, and civil rights." Executive Office of the President. The White House. https://obamawhitehouse.archives .gov/sites/default/files/microsites/ostp/2016_0504_data_discrimination.pdf.

[42] Campolo, Alex, Sanfilippo, Madelyn, Whittaker, Meredith, and Crawford, Kate. 2017. "AI now 2017 report." AI Now Institute at New York University. https://ainowinstitute.org/AI_Now_2017_Report.pdf.

[43] Friedman, Batya, and Nissenbaum, Helen. 1996. "Bias in computer systems." *ACM Transactions on Information Systems (TOIS)*, 14(3):330–347.

[44] Pedreschi, Dino, Ruggieri, Salvatore, and Turini, Franco. 2008. "Discrimination-aware data mining." In *Proceedings of the 14th SIGKDD*. ACM.

[45] Pasquale, Frank. 2015. *The black box society: The secret algorithms that control money and information.* Harvard University Press.

[46] O'Neil, Cathy. 2016. *Weapons of math destruction: How big data increases inequality and threatens democracy.* Broadway.

[47] Eubanks, Virginia. 2018. *Automating inequality: How high-tech tools profile, police, and punish the poor.* St. Martin's.

[48] Noble, Safiya Umoja. 2018. *Algorithms of oppression: How search engines reinforce racism.* NYU Press.

[49] Konger, Kate and Chen, Brian X. 2022. "A change by Apple is tormenting Internet companies, especially Meta." *New York Times*, February 3, 2022. https://www.nytimes.com/2022/02/03/technology/apple-privacy -changes-meta.html.

[50] Richardson, Willliam Jamal. 2017. "Against black inclusion in facial recognition." *Digital Talking Drum*, August 15, 2017. https://digitaltalkingdrum.com/2017/08/15/against-black-inclusion-in-facial -recognition/.

[51] Powles, Julia, and Nissenbaum, Helen. 2018. "The seductive diversion of 'solving' bias in artificial intelligence." Medium, December 7, 2018. https://onezero.medium.com/the-seductive-diversion-of-solving-bias -in-artificial-intelligence-890df5e5ef53.

[52] Weber, Max. 2019. *Economy and society.* Harvard University Press.

[53] Strandburg, Katherine J. 2019. "Rulemaking and inscrutable automated decision tools." *Columbia Law Review*, 119(7):1851–1886.

[54] Creel, Kathleen, and Hellman, Deborah. 2021. "The algorithmic Leviathan: Arbitrariness, fairness, and opportunity in algorithmic decision making systems." *Canadian Journal of Philosophy*, 52(1):26–43.

[55] Kroll, Joshua A., Huey, Joanna, Barocas, Solon, Felten, Edward W., Reidenberg, Joel R., Robinson, David G., and Yu, Harlan. 2017. "Accountable algorithms." *University of Pennsylvania Law Review*, 165(3):633–705.

[56] Citron, Danielle Keats. 2008. "Technological due process." *Washington University Law Review*, 85(6):1249–1313.

[57] Christie, James. 2020. "The Post Office Horizon IT scandal and the presumption of the dependability of computer evidence." *Digital Evidence & Electronic Signature Law Review*, 17:49.

[58] Kaplow, Louis. 1992. "Rules versus standards: An economic analysis." *Duke Law Journal*, 42(3):557–629.

[59] Alkhatib, Ali, and Bernstein, Michael. 2019. "Street-level algorithms: A theory at the gaps between policy and decisions." In *Conference on Human Factors in Computing Systems*, Glasgow (May 4–9):1–13.

[60] Lecher, Colin. 2018. "What happens when an algorithm cuts your health care." *The Verge*, March 21, 2018. https://www.theverge.com/2018/3/21/17144260/healthcare-medicaid-algorithm-arkansas-cerebral-palsy.

[61] Clarke, Roger. 1988. "Information technology and dataveillance." *Communications of the ACM*, 31(5): 498–512.

[62] Kaminski, Margot E., and Urban, Jennifer M. 2021. "The right to contest AI." *Columbia Law Review*, 121(7):1957–2048.

[63] Gilman, Michele. 2020. "Poverty lawgorithms." Technical report, Data & Society.

[64] Nissenbaum, Helen. 1996. "Accountability in a computerized society." *Science and Engineering Ethics*, 2(1):25–42.

[65] Binns, Reuben, Kleek, Max Van, Veale, Michael, Lyngs, Ulrik, Zhao, Jun, and Shadbolt, Nigel. 2018, " 'It's reducing a human being to a percentage': Perceptions of justice in algorithmic decisions." In *Proceedings of the 2018 CHI Conference on Human Factors in Computing Systems*, 1–14. ACM.

[66] Collins, Harry. 1991. *Artificial experts: Social knowledge and intelligent machines*. MIT Press.

[67] Forsythe, Diana E. 2000. *Studying those who study us: An anthropologist in the world of artificial intelligence*. Stanford University Press.

[68] Hand, David J. 2006. "Classifier technology and the illusion of progress." *Statistical Science*, 21(1): 1–14.

[69] Burrell, Jenna. 2016. "How the machine 'thinks': Understanding opacity in machine learning algorithms." *Big Data & Society*, 3(1).

[70] Ribeiro, Marco Tulio, Singh, Sameer, and Guestrin, Carlos. 2016. "'Why should I trust you?': Explaining the predictions of any classifier." In *Proceedings of the 22nd ACM SIGKDD International Conference on Knowledge Discovery and Data Mining*, 1135–1144. ACM.

[71] Perelman, Les. 2012. "Construct validity, length, score, and time in holistically graded writing assessments: The case against automated essay scoring (AES)." In *International advances in writing research: Cultures, places, measures*, edited by Charles Bazerman, Chris Dean, Jessica Early, Karen Lunsford, Suzie Null, Paul Rogers, and Amanda Stansell, 121–131. WAC Clearinghouse.

[72] Johnson, Rebecca Ann, and Zhang, Simone. 2022. "What is the bureaucratic counterfactual? Categorical versus algorithmic prioritization in U.S. social policy." In *ACM Conference on Fairness, Accountability, and Transparency*, 1671–1682. ACM.

[73] Abebe, Rediet, Barocas, Solon, Kleinberg, Jon, Levy, Karen, Raghavan, Manish, and Robinson, David G. 2020. "Roles for computing in social change." In *ACM Conference on Fairness, Accountability, and Transparency*, 252–260. ACM.

[74] Tyler, Tom R. 1988. "What is procedural justice?: Criteria used by citizens to assess the fairness of legal procedures." *Law & Society Review*, 22(1):103–135.

[75] Passi, Samir, and Barocas, Solon. 2019. "Problem formulation and fairness." In *Proceedings of the Conference on Fairness, Accountability, and Transparency*, 39–48. ACM.

[76] Hand, David J. 1994. "Deconstructing statistical questions." *Journal of the Royal Statistical Society: Series A (Statistics in Society)*, 157(3):317–338.

[77] Richardson, Rashida. 2022. "Racial segregation and the data-driven society: How our failure to reckon with root causes perpetuates separate and unequal realities." *Berkeley Technology Law Journal*, 36(3):1051–1090.

[78] Lum, Kristian, and Isaac, William. 2016. "To predict and serve?" *Significance*, 13(5):14–19.

[79] Harcourt, Bernard E. 2008. *Against prediction: Profiling, policing, and punishing in an actuarial age*. University of Chicago Press.

[80] Gandy, Oscar H. 2010. "Engaging rational discrimination: Exploring reasons for placing regulatory constraints on decision support systems." *Ethics and Information Technology*, 12(1):29–42.

[81] Obermeyer, Ziad, Powers, Brian, Vogeli, Christine, and Mullainathan, Sendhil. 2019. "Dissecting racial bias in an algorithm used to manage the health of populations." *Science*, 366(6464):447–453.

[82] Schauer, Frederick. 2006. *Profiles, probabilities, and stereotypes*. Harvard University Press.

[83] Lippert-Rasmussen, Kasper. 2011. "'We are all different': Statistical discrimination and the right to be treated as an individual." *Journal of Ethics*, 15(1–2):47–59.

[84] Mitchell, Tom M. 1980. "The need for biases in learning generalizations." Technical report, Department of Computer Science, Laboratory for Computer Science Research, Rutgers University, New Brunswick.

[85] Oreskes, Naomi, Shrader-Frechette, Kristin, and Belitz, Kenneth. 1994. "Verification, validation, and confirmation of numerical models in the earth sciences." *Science*, 263(5147):641–646.

[86] Malik, Momin M. 2020. "A hierarchy of limitations in machine learning." *arXiv preprint* arXiv:2002.05193.

[87] Mayer-Schönberger, Viktor, and Cukier, Kenneth. 2013. *Big data: A revolution that will transform how we live, work, and think*. Harper Business.

[88] Pasquale, Frank. 2018. "When machine learning is facially invalid." *Communications of the ACM*, 61(9): 25–27.

[89] Kim, Pauline T., and Hanson, Erika. 2016. "People analytics and the regulation of information under the Fair Credit Reporting Act." *Saint Louis University Law Journal*, 61(1):17–34.

[90] Salganik, Matthew J., Lundberg, Ian, Kindel, Alexander T., Ahearn, Caitlin E., Al-Ghoneim, Khaled, Almaatouq, Abdullah, Altschul, Drew M., Brand, Jennie E., Carnegie, Nicole Bohme, Compton, Ryan James, et al. 2020. "Measuring the predictability of life outcomes with a scientific mass collaboration." *Proceedings of the National Academy of Sciences*, 117(15):8398–8403.

[91] Chouldechova, Alexandra, Putnam-Hornstein, Emily, Benavides-Prado, Diana, Fialko, Oleksandr, and Vaithianathan, Rhema. 2017. "A case study of algorithm-assisted decision making in child maltreatment hotline screening decisions." *Proceedings of Machine Learning Research*, 81:1–15.

[92] Huq, Aziz Z. 2020. "A right to a human decision." *Virginia Law Review*, 106(3):611–688.

[93] Ustun, Berk, Spangher, Alexander, and Liu, Yang. 2019. "Actionable recourse in linear classification." In *Proceedings of the Conference on Fairness, Accountability, and Transparency*, 10–19. ACM.

[94] Milli, Smitha, Miller, John, Dragan, Anca D., and Hardt, Moritz. 2019. "The social cost of strategic classification." In *Proceedings of the Conference on Fairness, Accountability, and Transparency*, 230–239. ACM.

[95] Hu, Lily, Immorlica, Nicole, and Vaughan, Jennifer Wortman. 2019. "The disparate effects of strategic manipulation." In *Proceedings of the Conference on Fairness, Accountability, and Transparency*, 259–268. ACM.

[96] Karimi, Amir-Hossein, Barthe, Gilles, Schölkopf, Bernhard, and Valera, Isabel. 2022. "A survey of algorithmic recourse: Contrastive explanations and consequential recommendations." *ACM Computing Surveys*, 55(5):1–29.

[97] Kiviat, Barbara. 2019. "The moral limits of predictive practices: The case of credit-based insurance scores." *American Sociological Review*, 84(6):1134–1158.

[98] Miller, John, Milli, Smitha, and Hardt, Moritz. 2020. "Strategic classification is causal modeling in disguise." In *Proceedings of the 37th International Conference on Machine Learning*, 6917–6926. ACM.

[99] O'Neil, Cathy. 2017. *Weapons of math destruction*. Penguin Random House.

[100] Bambauer, Jane, and Zarsky, Tal. 2018. "The algorithm game." *Notre Dame Law Review*, 94(1):1–48.

[101] Chen, Irene Y., Daumé, Hal, III, and Barocas, Solon. 2021. "The many roles that causal reasoning plays in reasoning about fairness in machine learning." In *NeurIPS Workshop on Algorithmic Fairness through the lens of Causality and Robustness*.

[102] Desrosières, Alain. 1998. *The politics of large numbers: A history of statistical reasoning*. Harvard University Press.

[103] Porter, Theodore M. 2020. *The rise of statistical thinking, 1820–1900*. Princeton University Press.

[104] Bouk, Dan. 2015. *How our days became numbered: Risk and the rise of the statistical individual*. University of Chicago Press.

[105] Crenshaw, Kimberlé W. 2017. *On intersectionality: Essential writings*. New Press.

[106] Poplin, Ryan, Varadarajan, Avinash V., Blumer, Katy, Liu, Yun, McConnell, Michael V., Corrado, Greg S., Peng, Lily, and Webster, Dale R. 2018. "Prediction of cardiovascular risk factors from retinal fundus photographs via deep learning." *Nature Biomedical Engineering*, 2(3):158–164.

[107] Feldman, Michael, Friedler, Sorelle A., Moeller, John, Scheidegger, Carlos, and Venkatasubramanian, Suresh. 2015. "Certifying and removing disparate impact." In *Proceedings of the 21st ACM SIGKDD International Conference on Knowledge Discovery and Data Mining*, 259–268. ACM.

[108] Ryan, Michelle K., and Haslam, S., Alexander. 2005. "The glass cliff: Evidence that women are over-represented in precarious leadership positions." *British Journal of Management*, 16(2):81–90.

[109] Cook, Alison, and Glass, Christy. 2014. "Above the glass ceiling: When are women and racial/ethnic minorities promoted to CEO?" *Strategic Management Journal*, 35(7):1080–1089.

[110] Platt, John. 1999. "Probabilistic outputs for support vector machines and comparisons to regularized likelihood methods." *Advances in Large Margin Classifiers*, 10(3):61–74.

[111] Ding, Frances, Hardt, Moritz, Miller, John, and Schmidt, Ludwig. 2021. "Retiring adult: New datasets for fair machine learning." In *Proceedings of the 34th NeurIPS conference*, 1–13.

[112] Liu, Lydia T., Simchowitz, Max, and Hardt, Moritz. 2019. "The implicit fairness criterion of unconstrained learning." In *Proceedings of the 36th International Conference on Machine Learning*, 4051–4060. PMLR.

[113] Cover, Thomas M. 1999. *Elements of information theory*. John Wiley.

[114] Hardt, Moritz, Price, Eric, and Srebro, Nati. 2016. "Equality of opportunity in supervised learning." In *Advances in Neural Information Processing Systems*, 29:3315–3323.

[115] Wasserman, Larry. 2010. *All of statistics: A concise course in statistical inference*. Springer.

[116] Federal Reserve Board. 2007. "Report to the Congress on credit scoring and its effects on the availability and affordability of credit." https://www.federalreserve.gov/boarddocs/rptcongress/creditscore/. Accessed: 2018-05-29.

[117] Hacking, Ian. 1990. *The taming of chance*. Cambridge University Press.

[118] Hacking, Ian. 2006. *The emergence of probability: A philosophical study of early ideas about probability, induction and statistical inference*. Cambridge University Press.

[119] Hardt, Moritz, and Recht, Benjamin. 2022. *Patterns, predictions, and actions: Foundations of machine learning*. Princeton University Press.

[120] Hutchinson, Ben, and Mitchell, Margaret. 2019. "50 years of test (un)fairness: Lessons for machine learning." In *Proceedings of the Conference on Fairness, Accountability, and Transparency*, 49–58. ACM.

[121] Cleary, T. Anne. 1966. "Test bias: Validity of the scholastic aptitude test for Negro and White students in integrated colleges." *ETS Research Bulletin Series*, 1966(2):i–23.

[122] Cleary, T. Anne. 1968. "Test bias: Prediction of grades of Negro and White students in integrated colleges." *Journal of Educational Measurement*, 5(2):115–124.

[123] Darlington, Richard B. 1971. "Another look at 'cultural fairness.'" *Journal of Educational Measurement*, 8(2):71–82.

[124] Einhorn, Hillel J., and Bass, Alan R. 1971. "Methodological considerations relevant to discrimination in employment testing." *Psychological Bulletin*, 75(4):261–269.

[125] Thorndike, Robert L. 1971. "Concepts of culture-fairness." *Journal of Educational Measurement*, 8(2): 63–70.

[126] Lewis, Mary A. 1978. "A comparison of three models for determining test fairness." Technical report, Office of Aviation Medicine, Federal Aviation Administration, Washington, DC.

[127] Calders, Toon, Kamiran, Faisal, and Pechenizkiy, Mykola. 2009. "Building classifiers with independency constraints." In *Proceedings of the IEEE International Conference on Data Mining Workshops*, 13–18.

[128] Kamiran, Faisal and Calders, Toon. 2009. "Classifying without discriminating." In *Proceedings of the 2nd International Conference on Computer, Control and Communication*.

[129] Zemel, Richard S., Wu, Yu, Swersky, Kevin, Pitassi, Toniann, and Dwork, Cynthia. 2013. "Learning fair representations." *Proceedings of Machine Learning Research*, 28(3):325–333.

[130] Dwork, Cynthia, Hardt, Moritz, Pitassi, Toniann, Reingold, Omer, and Zemel, Richard. 2012. "Fairness through awareness." In *Proceedings of the 3rd Innovations in Theoretical Computer Science*, 14–226. ACM.

[131] Zafar, Muhammad Bilal, Valera, Isabel, Gómez Rodriguez, Manuel, and Gummadi, Krishna P. 2017. "Fairness beyond disparate treatment & disparate impact: Learning classification without disparate mistreatment." In *Proceedings of the 26th International World Wide Web Conference*, 1171–1180.

[132] Woodworth, Blake E., Gunasekar, Suriya, Ohannessian, Mesrob I., and Srebro, Nathan. 2017. "Learning non-discriminatory predictors." In *Proceedings of the 30th Conference on Learning Theory*, 1920–1953.

[133] Angwin, Julia, Larson, Jeff, Mattu, Surya, and Kirchner, Lauren. 2016. "Machine bias." *ProPublica*, May 23, 2016. https://www.propublica.org/article/machine-bias-risk-assessments-in-criminal-sentencing.

[134] Dieterich, William, Mendoza, Christina, and Brennan, Tim. 2016. "Compas OMPAS risk scales: Demonstrating accuracy equity and predictive parity." Research paper, Northpointe.

[135] Berk, Richard, Heidari, Hoda, Jabbari, Shahin, Kearns, Michael, and Roth, Aaron. 2017. "Fairness in criminal justice risk assessments: The state of the art." *arXiv e-print* arXiv: 1703.09207.

[136] Chouldechova, Alexandra. 2017. "Fair prediction with disparate impact: A study of bias in recidivism prediction instruments." *Big Data*, 5(2):153–163.

[137] Kleinberg, Jon M., Mullainathan, Sendhil, and Raghavan, Manish. 2017. "Inherent trade-offs in the fair determination of risk scores." In *Proceedings of the 8th Innovations in Theoretical Computer Science Conference.*

[138] Pleiss, Geoff, Raghavan, Manish, Wu, Felix, Kleinberg, Jon, and Weinberger, Kilian Q. 2017. "On fairness and calibration." In *Advances in Neural Information Processing Systems*, 30.

[139] Hellman, Deborah. 2007. *When is discrimination wrong?* Harvard University Press.

[140] Lippert-Rasmussen, Kasper. 2013. *Born free and equal? A philosophical inquiry into the nature of discrimination.* Oxford University Press.

[141] Sunstein, Cass R. 1994. "The anticaste principle." *Michigan Law Review*, 92(8):2410.

[142] Singer, Peter. 1978. "Is racial discrimination arbitrary?" *Philosophia*, 8(2–3):185–203.

[143] Alexander, Larry. 1992. "What makes wrongful discrimination wrong? Biases, preferences, stereotypes, and proxies." *University of Pennsylvania Law Review*, 141(1):149.

[144] Arneson, Richard J. 2006. "What is wrongful discrimination ." *San Diego Law Review*, 43(4):775–808.

[145] Eidelson, Benjamin. 2015. *Discrimination and disrespect.* Oxford University Press.

[146] Balkin, J. M. 1997. "The constitution of status." *The Yale Law Journal*, 106(8):2313–2374.

[147] Hoffman, Sharona. 2011. "The importance of immutability in employment discrimination law." *William & Mary Law Review*, 52(5):1483–1546.

[148] Clarke, Jessica A. 2015. "Against immutability." *Yale Law Journal*, 125(1):1–102.

[149] Hellman, Deborah. 2020. "Indirect discrimination and the duty to avoid compounding injustice." In *Foundations of Indirect Discrimination Law*, edited by Hugh Collins and Tarunabh Khaitan, 105–122. Bloomsbury.

[150] Schauer, Frederick. 2017. "Statistical (and non-statistical) discrimination." In *The Routledge Handbook of the Ethics of Discrimination*, edited by Kasper Lippert-Rasmussen, 42–53. Routledge.

[151] Hellman, Deborah. 2017. "Discrimination and social meaning." In *The Routledge Handbook of the Ethics of Discrimination*, edited by Kasper Lippert-Rasmussen, 97–107. Routledge.

[152] Prince, Anya E. R., and Schwarcz, Daniel. 2020. "Proxy discrimination in the age of artificial intelligence and big data." *Iowa Law Review*, 105(3):1257–1318.

[153] Anderson, Elizabeth S. 1999. "What is the point of equality?" *Ethics*, 109(2):287–337.

[154] Rawls, John. 1998. *A theory of justice.* Harvard University Press, Cambridge, MA.

[155] Arneson, Richard. 2018. "Four conceptions of equal opportunity." *Economic Journal*, 128(612):F152–F173.

[156] Fishkin, Joseph. 2013. *Bottlenecks.* Oxford University Press.

[157] Roemer, John. 2000. *Equality of opportunity.* Harvard University Press.

[158] Selbst, Andrew D., Boyd, Danah, Friedler, Sorelle A., Venkatasubramanian, Suresh, and Vertesi, Janet. 2019. "Fairness and abstraction in sociotechnical systems." In *Proceedings of the Conference on Fairness, Accountability, and Transparency*, 59–68. ACM.

[159] Liu, Lydia T., Dean, Sarah, Rolf, Esther, Simchowitz, Max, and Hardt, Moritz. 2017. "Delayed impact of fair machine learning." In *Proceedings of the 35th International Conference on Machine Learning*, 80: 3150–3158.

[160] Phillips, Anne. 2004. "Defending equality of outcome." *Journal of Political Philosophy*, 12(1):1–19.

[161] Rieke, Aaron, and Koepke, Logan. 2015. "Led astray: Online lead generation and payday loans." Technical report, Upturn, Washington, DC.

[162] Joseph, Matthew, Kearns, Michael, Morgenstern, Jamie H, and Roth, Aaron. 2016. "Fairness in learning: Classic and contextual bandits." In *Advances in Neural Information Processing Systems*, 29:325–333.

[163] Perry, Ronen, and Zarsky, Tal. 2015. "'May the odds be ever in your favor': Lotteries in law." *Alabama Law Review*, 66(5):1035–1098.

[164] Creel, Kathleen, and Hellman, Deborah. 2022. "The algorithmic Leviathan: Arbitrariness, fairness, and opportunity in algorithmic decision-making systems." *Canadian Journal of Philosophy*, 52(1): 26–43.

[165] Matwin, Stan, Yu, Shipeng, Farooq, Faisal, Corbett-Davies, Sam, Pierson, Emma, Feller, Avi, Goel, Sharad, and Huq, Aziz. 2017. "Algorithmic decision making and the cost of fairness." In *Proceedings of the 23rd ACM SIGFDD International Conference on Knowledge Discovery and Data Mining*, 797–806. ACM.

[166] Hellman, Deborah. 2022. "Measuring algorithmic fairness." *Virginia Law Review*, 106(4):811–866.

[167] Krishnapriya, K. S., Vangara, Kushal, King, Michael C., Albiero, Vitor, and Bowyer, Kevin. 2019. "Characterizing the variability in face recognition accuracy relative to race." In *2019 IEEE/CVF Conference on Computer Vision and Pattern Recognition Workshops*, 2278–2285.

[168] Blodgett, Su Lin, Green, Lisa, and O'Connor, Brendan. 2016. "Demographic dialectal variation in social media: A case study of African-American English." In *Proceedings of the 2016 Conference on Empirical Methods in Natural Language Processing*, 1119–1130. Assocation for Computational Linguistics.

[169] Huq, Aziz Z. 2018. "Racial equity in algorithmic criminal justice." *Duke Law Journal*, 68:1043–1134.

[170] Passi, Samir, and Barocas, Solon. 2019. "Problem formulation and fairness." In *Proceedings of the Conference on Fairness, Accountability, and Transparency*, 39–48. ACM.

[171] Black, Emily, Raghavan, Manish, and Barocas, Solon. 2022. "Model multiplicity: Opportunities, concerns, and solutions." In *Proceedings of the Conference on Fairness, Accountability, and Transparency*, 850–863. ACM.

[172] Rodolfa, Kit T., Lamba, Hemank, and Ghani, Rayid. 2021. "Empirical observation of negligible fairness–accuracy trade-offs in machine learning for public policy." *Nature Machine Intelligence*, 3(10): 896–904.

[173] Friedler, Sorelle A., Scheidegger, Carlos, and Venkatasubramanian, Suresh. 2021. "The (im)possibility of fairness." *Communications of the ACM*, 64(4):136–143.

[174] Hannah-Jones, Nikole. 2020. "What is owed." *New York Times Magazine*, June 25, 2020.

[175] Bickel, Peter J., Hammel, Eugene A., and O'Connell, J. William. 1975. "Sex bias in graduate admissions: Data from Berkeley." *Science*, 187(4175):398–404.

[176] Humphrey, Linda L., Chan, Benjamin K. S., and Sox, Harold C.. 2002. "Postmenopausal hormone replacement therapy and the primary prevention of cardiovascular disease." *Annals of Internal Medicine*, 137(4):273–284.

[177] Berkson, Joseph. 2014. "Limitations of the application of fourfold table analysis to hospital data." *International Journal of Epidemiology*, 43(2):511–515.

[178] Moneta-Koehler, Liane, Brown, Abigail M., Petrie, Kimberly A., Evans, Brent J., and Chalkley, Roger. 2017. "The limitations of the GRE in predicting success in biomedical graduate school." *PLOS ONE*, 12(1):1–17.

[179] Hall, Joshua D., O'Connell, Anna B., and Cook, Jeanette G. 2017. "Predictors of student productivity in biomedical graduate school applications." *PLOS ONE*, 12(1):1–14.

[180] Pearl, Judea. 2009. *Causality*. Cambridge University Press.

[181] Deaton, Angus, and Cartwright, Nancy. 2018. "Understanding and misunderstanding randomized controlled trials." *Social Science & Medicine*, 210:2–21.

[182] Pearl, Judea, and Mackenzie, Dana. 2018. *The book of why: The new science of cause and effect*. Basic Books.

[183] Glymour, M. Maria. 2006. "Using causal diagrams to understand common problems in social epidemiology." In *Methods in Social Epidemiology*, edited by J. Michael Oakes and Jay S. Kaufman, 393–428. Jossey-Bass/Wiley.

[184] Krieger, Nancy. 2011. *Epidemiology and the people's health: Theory and context*. Oxford University Press.

[185] Peters, Jonas, Janzing, Dominik, and Schölkopf, Bernhard. 2017. *Elements of causal inference*. MIT Press.

[186] Baron, Reuben M., and Kenny, David A. 1986. "The moderator–mediator variable distinction in social psychological research: Conceptual, strategic, and statistical considerations." *Journal of Personality and Social Psychology*, 51(6):1173.

[187] Kusner, Matt J., Loftus, Joshua R., Russell, Chris, and Silva, Ricardo. 2017. "Counterfactual fairness." In *Advances in Neural Information Processing Systems*, 30:4069–4079.

[188] Pearl, Judea, Glymour, Madelyn, and Jewell, Nicholas P. 2016. *Causal inference in statistics: A primer*. Wiley.

[189] Egan, Patrick J. 2020. "Identity as dependent variable: How Americans shift their identities to align with their politics." *American Journal of Political Science*, 64(3):699–716.

[190] Glasgow, Joshua, Haslanger, Sally, Jeffers, Chike, and Spencer, Quayshawn. 2019. *What is race? Four philosophical views*. Oxford University Press.

[191] Bowker, Geoffrey C., and Star, Susan Leigh. 2000. *Sorting things out: Classification and its consequences*. MIT Press.

[192] Fields, Karen E., and Fields, Barbara J.. 2014. *Racecraft: The soul of inequality in American life*. Verso.

[193] Benjamin, Ruha. 2019. *Race after technology*. Polity.

[194] Hacking, Ian. 2000. *The social construction of what?* Harvard University Press.

[195] Haslanger, Sally. 2012. *Resisting reality: Social construction and social critique*. Oxford University Press.

[196] Mallon, Ron. 2018. *The construction of human kinds*. Oxford University Press.

[197] Cartwright, Nancy. 2006. *Hunting causes and using them, too*. Cambridge University Press.

[198] Hacking, Ian. 2006. "Making up people." *London Review of Books*, 28(16).

[199] Holland, Paul W. 1986. "Statistics and causal inference." *Journal of the American Statistical Association*, 81(396):945–970.

[200] VanderWeele, Tyler J., and Robinson, Whitney R. 2014. "On causal interpretation of race in regressions adjusting for confounding and mediating variables." *Epidemiology*, 25(4):473–483.

[201] Greiner, D. James, and Rubin, Donald B. 2011. "Causal effects of perceived immutable characteristics." *Review of Economics and Statistics*, 93(3):775–785.

[202] Kohler-Hausmann, Issa. 2019. "Eddie Murphy and the dangers of counterfactual causal thinking about detecting racial discrimination." *Northwestern University Law Review*, 113(5):1163–1228.

[203] Spirtes, Peter, Glymour, Clark N., Scheines, Richard, Heckerman, David, Meek, Christopher, Cooper, Gregory, and Richardson, Thomas. 2000. *Causation, prediction, and search*. MIT Press.

[204] Morgan, Stephen L., and Winship, Christopher. 2014. *Counterfactuals and causal inference*. Cambridge University Press.

[205] Imbens, Guido W., and Rubin, Donald B. 2015. *Causal inference for statistics, social, and biomedical sciences*. Cambridge University Press.

[206] Angrist, Joshua D., and Jörn-Steffen, Pischke. 2009. *Mostly harmless econometrics: An empiricist's companion*. Princeton University Press.

[207] Hernán, Miguel, and Robins, James. 2019. *Causal inference*. Chapman & Hall/CRC, Boca Raton. Forthcoming.

[208] Simpson, Edward H. 1951. "The interpretation of interaction in contingency tables." *Journal of the Royal Statistical Society: Series B (Methodological)*, 13(2):238–241.

[209] Hernán, Miguel A., Clayton, David, and Keiding, Niels. 2011. "The Simpson's paradox unraveled." *International Journal of Epidemiology*, 40(3):780–785.

[210] Zhang, Lu, Wu, Yongkai, and Wu, Xintao. 2017. "A causal framework for discovering and removing direct and indirect discrimination." In *Proceedings of the 26th International Joint Conference on Artificial Intelligence*, 3929–3935.

[211] Russell, Chris, Kusner, Matt J., Loftus, Joshua R., and Silva, Ricardo. 2017. "When worlds collide: Integrating different counterfactual assumptions in fairness." In *Advances in Neural Information Processing Systems*, 30:6417–6426.

[212] Chiappa, Silvia. 2019. "Path-specific counterfactual fairness." In *Proceedings of the 33rd AAAI Conference on Artificial Intelligence*, 33:7801–7808.

[213] Kilbertus, Niki, Rojas-Carulla, Mateo, Parascandolo, Giambattista, Hardt, Moritz, Janzing, Dominik, and Schölkopf, Bernhard. 2017. "Avoiding discrimination through causal reasoning." In *Advances in Neural Information Processing Systems*, 30:656–666.

[214] Nabi, Razieh, and Shpitser, Ilya. 2018. "Fair inference on outcomes." In *Proceedings of the 32nd AAAI Conference on Artificial Intelligence*, 32:1931–1940.

[215] Chiappa, Silvia, and Isaac, William S. 2019. "A causal Bayesian networks viewpoint on fairness." *arXiv preprint*, arXiv:1907.06430.

[216] Kasirzadeh, Atoosa, and Smart, Andrew. 2021. "The use and misuse of counterfactuals in ethical machine learning." In *Proceedings of the Conference on Fairness, Accountability, and Transparency*, 228–236. ACM.

[217] Krieger, Nancy. 2014. "On the causal interpretation of race." *Epidemiology*, 25(6):937.

[218] Krieger, Nancy. 2014. "Discrimination and health inequities." *International Journal of Health Services*, 44(4):643–710.

[219] 1896. Plessy v. Ferguson. 163 US 537.

[220] Wilkerson, Isabel. 2011. *The warmth of other suns: The epic story of America's great migration*. Penguin Random House.

[221] Keele, Luke, Cubbison, William, and White, Ismail. 2021. "Suppressing Black votes: A historical case study of voting restrictions in Louisiana." *American Political Science Review*, 115(2):694–700.

[222] Rothstein, Richard. 2018. *The color of law: A forgotten history of how our government segregated America*. W. W. Norton.

[223] Walker, Juliet E. K.. 1998. *History of black business in America: capitalism, race, entrepreneurship*. Twayne.

[224] Vaas, Francis J. 1965. "Title VII: Legislative history." *Boston College Industrial & Commercial Law Review*, 7(3):431–458.

[225] Friedan, Betty. 2001. *The feminine mystique*. Norton.

[226] 1973. Roe v. Wade. 410 US 113.

[227] 1976. Senate Report no. 589, 94th Congress.

[228] Eskridge, William N.. 2008. *Dishonorable passions: Sodomy laws in America, 1861–2003*. Viking, New York, NY.

[229] 2003. Lawrence v. Texas. 539 US 558.

[230] 2015. Obergefell v. Hodges. 576 US 644.

[231] 2020. Bostock v. Clayton County, Georgia. 590 US —.

[232] Okoro, Catherine A., Hollis, NaTasha D., Cyrus, Alissa C., and Griffin-Blake, Shannon. 2018. "Prevalence of disabilities and health care access by disability status and type among adults—United States, 2016." *Morbidity and Mortality Weekly Report*, 67(32):882–887.

[233] Fleischer, Doris Zames, and Zames, Frieda. 2013. *The disability rights movement: From charity to confrontation*. Temple University Press.

[234] Settlement Agreement, United States v. Meta Platforms, Inc., No. 1:22-CV-05187 (S.D.N.Y. June 21, 2022), ECF No. 5-1.

[235] Rosenblat, Alex, Levy, Karen EC, Barocas, Solon, and Hwang, Tim. 2016. "Discriminating tastes: Customer ratings as vehicles for bias." *Data & Society*, October 19, 2016, 1–21.

[236] Dau-Schmidt, Kenneth Glenn, and Sherman, Ryland. 2013. "The employment and economic advancement of African-Americans in the twentieth century." *Jindal Journal of Public Policy*, 1(2):95–116.

[237] 1941. Executive Order no. 8802. General Records of the United States Government, record group 11, National Archives, Washington, DC.

[238] 2015. "Questions and answers about the EEOC's enforcement guidance on pregnancy discrimination and related issues." Washington, DC. https://www.eeoc.gov/laws/guidance/questions-and-answers-about-eeocs-enforcement-guidance-pregnancy-discrimination.

[239] Valentin, Iram. 1997. "Title IX: A brief history." Technical report, Equity Resource Center, Newton, MA.

[240] Graham, Hugh Davis. 1998. "The storm over Grove City College: Civil rights regulation, higher education, and the Reagan administration." *History of Education Quarterly*, 38(4):407–429.

[241] Melnick, R. Shep. 2020. "Analyzing the Department of Education's final Title IX rules on sexual misconduct." Technical report, Brookings Institution, Washington, DC.

[242] Office for Civil Rights, US Department of Education. 2021. "Enforcement of Title IX of the education amendments of 1972 with respect to discrimination based on sexual orientation and gender identity in light of Bostock v. Clayton County." 86 FR 32637.

[243] Price, Robert N. 2016. "Griggs v. Duke Power Co.: The first landmark under Title VII of the Civil Rights Act of 1964." *Southwestern Law Journal*, 25(3):484–493.

[244] Eyer, Katie. 2021. "The but-for theory of anti-discrimination law." *Virginia Law Review*, 107(8):1621–1710.

[245] Civil Rights Division, US Department of Justice. 2021. "Title VI legal manual." Technical report, Washington, DC.

[246] 1963. House of Representatives Report no. 914, pt. 2, 88th Congress, 1st session.

[247] 2015. Texas Department of Housing and Community Affairs v. Inclusive Communities Project, Inc. 135 S. Ct. 2507.

[248] Porter, Nicole Buonocore. 2019. "A new look at the ADA's undue hardship defense." *Missouri Law Review*, 84(1):121–176.

[249] Menand, Louis. 2020. "The changing meaning of affirmative action." *New Yorker*. January 13, 2020.

[250] Bagenstos, Samuel R. 2014. "Formalism and employer liability under Title VII." *University of Chicago Legal Forum*, 2014(1):145–176.

[251] Cairns, John W. 1976. "Credit equality comes to women: An analysis of the Equal Credit Opportunity Act." *San Diego Law Review*, 13(4):960–977.

[252] Chay, Kenneth Y. 1998. "The impact of federal civil rights policy on black economic progress: Evidence from the Equal Employment Opportunity Act of 1972." *Industrial & Labor Relations Review*, 51(4): 608–632.

[253] Moss, Scott A. 2007. "Fighting discrimination while fighting litigation: A tale of two supreme courts." *Fordham Law Review*, 76(2):981–1013.

[254] Sperino, Sandra F., and Thomas, Suja A. 2017. *Unequal: How America's courts undermine discrimination law*. Oxford University Press.

[255] Ajunwa, Ifeoma. 2020. "The paradox of automation as anti-bias intervention." *Cardozo Law Review*, 41(5):1671–1742.

[256] Selmi, Michael. 2001. "Why are employment discrimination cases so hard to win?" *Louisiana Law Review*, 61(3):555–575.

[257] Eyer, Katie R. 2012. "That's not discrimination: American beliefs and the limits of anti-discrimination law." *Minnesota Law Review*, 96(4):1275–1362.

[258] Halpern, Stephen C. 1995. *On the limits of the law: The ironic legacy of Title VI of the 1964 Civil Rights Act*. Johns Hopkins University Press.

[259] Edelman, Lauren B., Krieger, Linda H., Eliason, Scott R., Albiston, Catherine R., and Mellema, Virginia. 2011. "When organizations rule: Judicial deference to institutionalized employment structures." *American Journal of Sociology*, 117(3):888–954.

[260] 1979. Steelworkers v. Weber. 443 US 193.

[261] Kim, Pauline. 2022. "Race-aware algorithms: Fairness, nondiscrimination and affirmative action." *California Law Review*, 10(5):1539–1596.

[262] Bent, Jason R. 2020. "Is algorithmic affirmative action legal?" *Georgetown Law Journal*, 108(4):803–853.

[263] 1971. Griggs v. Duke Power Co. 401 US 424.

[264] Kim, Pauline T. 2017. "Data-driven discrimination at work." *William & Mary Law Review*, 58(3):857–936.

[265] Raghavan, Manish, Barocas, Solon, Kleinberg, Jon, and Levy, Karen. 2020. "Mitigating bias in algorithmic hiring: Evaluating claims and practices." In *Proceedings of the Conference on Fairness, Accountability, and Transparency*, 469–481. ACM.

[266] Duhigg, Charles. 2014. *The power of habit: Why we do what we do in life and business paperback*. Random House.

[267] Ali, Muhammad, Sapiezynski, Piotr, Bogen, Miranda, Korolova, Aleksandra, Mislove, Alan, and Rieke, Aaron. 2019. "Discrimination through optimization: How Facebook's ad delivery can lead to biased outcomes." *Proceedings of the ACM on Human-Computer Interaction*, 3(CSCW):199. ACM.

[268] Agan, Amanda, and Starr, Sonja. 2017. "Ban the Box, criminal records, and racial discrimination: A field experiment." *Quarterly Journal of Economics*, 133(1):191–235.

[269] Barocas, Solon and Nissenbaum, Helen. 2014. "Big data's end run around anonymity and consent." In *Privacy, Big Data, and the Public Good: Frameworks for Engagement*, edited by Julia Lane, Victoria Stodden, Stefan Bender, and Helen Nissenbaum, 44–75. Cambridge University Press.

[270] Areheart, Bradley A., and Roberts, Jessica L. 2019. "GINA, big data, and the future of employee privacy." *Yale Law Journal*, 128(3):710–790.

[271] Strandburg, Katherine J. 1999. "Rulemaking and inscrutable automated decision tools." *Columbia Law Review*, 119(7):1851–1886.

[272] Strandburg, Katherine J. 2021. "Adjudicating with inscrutable decision tools." In *Machines We Trust: Perspectives on Dependable AI*, edited by Marcello Pelillo and Teresa Scantamburlo, 61–86 MIT Press.

[273] Citron, Danielle Keats. 2008. "Technological Due Process." *Washington University Law Review*, 85(6): 1249–1313.

[274] Selbst, Andrew D., and Barocas, Solon. 2018. "The intuitive appeal of explainable machines." *Fordham Law Review*, 87(3):1085–1139.

[275] Andreou, Athanasios, Venkatadri, Giridhari, Goga, Oana, Gummadi, Krishna, Loiseau, Patrick, and Mislove, Alan. 2018. "Investigating ad transparency mechanisms in social media: A case study of Facebook's explanations." In *NDSS 2018—Network and Distributed System Security Symposium*, 1–15.

[276] Kaminski, Margot E., and Malgieri, Gianclaudio. 2020. "Algorithmic impact assessments under the GDPR: Producing multi-layered explanations." *International Data Privacy Law*, 11(2):125–144.

[277] Selbst, Andrew D. 2021. "An institutional view of algorithmic impact assessments." *Harvard Journal of Law & Technology*, 35(1):117–191.

[278] Lovelace, Ada, and DataKind, UK. 2020. "Examining the black box: Tools for assessing algorithmic systems." Technical report, Ada Lovelace Institute. https://www.adalovelaceinstitute.org/wp-content/uploads /2020/04/Ada-Lovelace-Institute-DataKind-UK-Examining-the-Black-Box-Report-2020.pdf.

[279] Costanza-Chock, Sasha, Raji, Inioluwa Deborah, and Buolamwini, Joy. 2022. "Who audits the auditors? Recommendations from a field scan of the algorithmic auditing ecosystem." In *Proceedings of the Conference on Fairness, Accountability, and Transparency*, 1571–1583. ACM.

[280] Ajunwa, Ifeoma. 2021. "An auditing imperative for automated hiring." *Harvard Journal of Law & Technology*, 34(2):621–299.

[281] Malgieri, Gianclaudio, and Pasquale, Frank A. 2022. "From transparency to justification: Toward ex ante accountability for AI." *Brooklyn Law School, Legal Studies Paper*, (712).

[282] Sinclair, Upton. 2006. *The jungle*. Penguin.

[283] Hoofnagle, Chris Jay. 2016. *Federal trade commission privacy law and policy*. Cambridge University Press.

[284] Slaughter, Rebecca Kelly. 2020. "Statement of commissioner Rebecca Kelly Slaughter in the matter of Liberty Chevrolet, Inc. d/b/a Bronx Honda." Federal Trade Commission, matter number 162 3238. https://www.ftc.gov/system/files/documents/public_statements/1576006/bronx_honda_2020-5-27_bx_ho nda_rks_concurrence_for_publication.pdf.

[285] Selbst, Andrew D., and Barocas, Solon. 2023. "Unfair artificial intelligence: How FTC intervention can overcome the limitations of discrimination law." *University of Pennsylvania Law Review*, 171.

[286] Federal Trade Commission. 2021. "Decision and Order, In re Everalbum, Inc." Federal Trade Commission, matter number 192 3172. https://www.ftc.gov/system/files/documents/cases/1923172_-_everalbum _decision_final.pdf.

[287] Pitofsky, Robert. 2005. "Past, present, and future of antitrust enforcement at the Federal Trade Commission." *University of Chicago Law Review*, 72(1):209–227.

[288] Jillson, Elisa. 2021. "Aiming for truth, fairness, and equity in your company's use of AI." Federal Trade Commission, Business blog, April 19, 2021.

[289] Slaughter, Rebecca Kelly, Kopec, Janice, and Batal, Mohamad. 2020. "Algorithms and economic justice: A taxonomy of harms and a path forward for the Federal Trade Commission." *Special issue, Yale Journal of Law & Technology*, 23:1.

[290] Richards, Neil, and Hartzog, Woodrow. 2021. "A duty of loyalty for privacy law." *Washington University Law Review*, 99(3):961–1021.

[291] Bogen, Miranda, and Rieke, Aaron. 2018. "Help wanted: An examination of hiring algorithms, equity, and bias." Technical report, Upturn, Washington, DC.

[292] Wienk, Ronald E., Reid, Clifford E., Simonson, John C., and Eggers, Frederick J. 1979. "Measuring racial discrimination in American housing markets: The housing market practices survey." Technical report, Housing Market Practices Survey, US Department of Housing and Development.

[293] Ayres, Ian, and Siegelman, Peter. 1995. "Race and gender discrimination in bargaining for a new car." *American Economic Review*, 85(3):304–321.

[294] Freeman, Jonathan B., Penner, Andrew M., Saperstein, Aliya, Scheutz, Matthias, and Ambady, Nalini. 2011. "Looking the part: Social status cues shape race perception." *PLOS ONE*, 6(9):e25107.

[295] Bertrand, Marianne, and Mullainathan, Sendhil. 2004. "Are Emily and Greg more employable than Lakisha and Jamal? A field experiment on labor market discrimination." *American Economic Review*, 94(4): 991–1013.

[296] Pager, Devah. 2007. "The use of field experiments for studies of employment discrimination: Contributions, critiques, and directions for the future." *Annals of the American Academy of Political and Social Science*, 609(1):104–133.

[297] Kohler-Hausmann, Issa. 2019. "Eddie Murphy and the dangers of counterfactual causal thinking about detecting racial discrimination." *Northwestern University Law Review*, 113(5):1163–1228.

[298] Bertrand, Marianne., and Duflo, Esther. 2017. "Field experiments on discrimination." In *Handbook of Economic Field Experiments*, vol. 1, edited by Esther Duflo and Abhijit Banerjee, 309–393. Elsevier.

[299] Quillian, Lincoln, Pager, Devah, Hexel, Ole, and Midtbøen, Arnfinn H. 2017. "Meta-analysis of field experiments shows no change in racial discrimination in hiring over time." *Proceedings of the National Academy of Sciences*, 114(41):10870–10875.

[300] Blank, Rebecca M. 1991. "The effects of double-blind versus single-blind reviewing: Experimental evidence from the *American Economic Review*." *American Economic Review*, 81(5):1041–1067.

[301] Pischke, Jorn-Steffen. 2005. "Empirical methods in applied economics: Lecture notes." https://econ.lse.ac.uk/staff/spischke/ec524/evaluation3.pdf.

[302] Bertrand, Marianne, Duflo, Esther, and Mullainathan, Sendhil. 2004. "How much should we trust differences-in-differences estimates?" *The Quarterly Journal of Economics*, 119(1):249–275.

[303] Kang, Sonia K., DeCelles, Katherine A., Tilcsik, András, and Jun, Sora. 2016. "Whitened resumes: Race and self-presentation in the labor market." *Administrative Science Quarterly*, 61(3):469–502.

[304] Eren, Ozkan, and Mocan, Naci. 2018. "Emotional judges and unlucky juveniles." *American Economic Journal: Applied Economics*, 10(3):171–205.

[305] Danziger, Shai, Levav, Jonathan, and Avnaim-Pesso, Liora. 2011. "Extraneous factors in judicial decisions." *Proceedings of the National Academy of Sciences*, 108(17):6889–6892.

[306] Lakens, Daniel. 2017. "Impossibly hungry judges." https://daniellakens.blogspot.com/2017/07/impossibly-hungry-judges.html.

[307] Weinshall-Margel, Keren, and Shapard, John. 2011. "Overlooked factors in the analysis of parole decisions." *Proceedings of the National Academy of Sciences*, 108(42):E833–E833.

[308] Norton, Helen. 2010. "The Supreme Court's post-racial turn towards a zero-sum understanding of equality." *William & Mary Law Review*, 52(1):197–259.

[309] Ayres, Ian. 2005. "Three tests for measuring unjustified disparate impacts in organ transplantation: The problem of 'included variable' bias." *Perspectives in Biology and Medicine*, 48(1):S68–87.

[310] Simoiu, Camelia, Corbett-Davies, Sam, and Goel, Sharad. 2017. "The problem of infra-marginality in outcome tests for discrimination." *Annals of Applied Statistics*, 11(3):1193–1216.

[311] Lakkaraju, Himabindu, Kleinberg, Jon, Leskovec, Jure, Ludwig, Jens, and Mullainathan, Sendhil. 2017. "The selective labels problem: Evaluating algorithmic predictions in the presence of unobservables." In *Proceedings of the 23rd ACM International Conference on Knowledge Discovery and Data Mining*, 275–284. ACM.

[312] Bird, Sarah, Barocas, Solon, Crawford, Kate, Diaz, Fernando, and Wallach, Hanna. 2016. "Exploring or exploiting? Social and ethical implications of autonomous experimentation in AI." In *Workshop on Fairness, Accountability, and Transparency in Machine Learning*. https://www.fatml.org/schedule/2016/page/papers%202016%20Papers%20::%20FAT%20ML.

[313] Becker, Gary S. 1957. *The economics of discrimination*. University of Chicago Press.

[314] Phelps, Edmund S. 1972. "The statistical theory of racism and sexism." *American Economic Review*, 62(4):659–661.

[315] Arrow, Kenneth J. 1973. "The theory of discrimination." In *Discrimination in Labor Markets*, edited by Orley Ashenfelter and Albert Rees, 3–33. Princeton University Press.

[316] Agan, Amanda, and Starr, Sonja. 2017. "Ban the box, criminal records, and racial discrimination: A field experiment." *Quarterly Journal of Economics*, 133(1):191–235.

[317] Williams, Wendy M., and Ceci, Stephen J. 2015. "National hiring experiments reveal 2:1 faculty preference for women on STEM tenure track." *Proceedings of the National Academy of Sciences*, 112(17):5360–5365.

[318] Neckerman, Kathryn M., and Kirschenman, Joleen. 1991. "Hiring strategies, racial bias, and inner-city workers." *Social Problems*, 38(4):433–447.

[319] Pager, Devah, and Shepherd, Hana. 2008. "The sociology of discrimination: Racial discrimination in employment, housing, credit, and consumer markets." *Annual Review of Sociology*, 34:181–209.

[320] Rivera, Lauren A. 2016. *Pedigree: How elite students get elite jobs*. Princeton University Press.

[321] Posselt, Julie R. 2016. *Inside graduate admissions*. Harvard University Press.

[322] Roth, Alvin E. 2003. "The origins, history, and design of the resident match." *Journal of the American Medical Association*, 289(7):909–912.

[323] Dastin, Jeffrey. 2018. "Amazon scraps secret AI recruiting tool that showed bias against women." *Reuters*, October 10, 2018.

[324] Buranyi, Stephen. 2018. "How to persuade a robot that you should get the job." *The Guardian*, March 4, 2018.

[325] De-Arteaga, Maria, Romanov, Alexey, Wallach, Hanna, Chayes, Jennifer, Borgs, Christian, Chouldechova, Alexandra, Geyik, Sahin, Kenthapadi, Krishnaram, and Kalai, Adam Tauman. 2019. "Bias in bios: A case study of semantic representation bias in a high-stakes setting." In *Proceedings of the Conference on Fairness, Accountability, and Transparency*, 120–128. ACM.

[326] Ramineni, Chaitanya and Williamson, David. 2018. "Understanding mean score differences between the e-rater automated scoring engine and humans for demographically based groups in the GRE general test." *ETS Research Report Series*, 2018(1):1–31.

[327] Amorim, Evelin, Cançado, Marcia, and Veloso, Adriano. 2018. "Automated essay scoring in the presence of biased ratings." In *Proceedings of the 2018 Conference of the North American Chapter of the Association for Computational Linguistics*: Human Language Technologies, vol. 1, 229–237.

[328] Sap, Maarten, Card, Dallas, Gabriel, Saadia, Choi, Yejin, and Smith, Noah A. 2019. "The risk of racial bias in hate speech detection." In *Proceedings of the 57th Annual Meeting of the Association for Computational Linguistics*, 1668–1678. Association for Computational Linguistics.

[329] Kiritchenko, Svetlana, and Mohammad, Saif. 2018. "Examining gender and race bias in two hundred sentiment analysis systems." In *Conference on Lexical and Computational Semantics*, 43–53. Association for Computational Linguistics.

[330] Tatman, Rachael. 2017. "Gender and dialect bias in YouTube's automatic captions." In *Proceedings of the First ACL Workshop on Ethics in Natural Language Processing*, 53–59. Association for Computational Linguistics.

[331] Solaiman, Irene, Brundage, Miles, Clark, Jack, Askell, Amanda, Herbert-Voss, Ariel, Wu, Jeff, Radford, Alec, and Wang, Jasmine. 2019. "Release strategies and the social impacts of language models." *arXiv preprint* arXiv:1908.09203.

[332] Buolamwini, Joy, and Gebru, Timnit. 2018. "Gender shades: Intersectional accuracy disparities in commercial gender classification." In *Proceedings of the Conference on Fairness, Accountability, and Transparency*, 77–91. ACM.

[333] De Vries, Terrance, Misra, Ishan, Wang, Changhan, and van der Maaten, Laurens. 2019. "Does object recognition work for everyone?" In *Conference on Computer Vision and Pattern Recognition Workshops*, 52–59.

[334] Shankar, Shreya, Halpern, Yoni, Breck, Eric, Atwood, James, Wilson, Jimbo, and Sculley, D. 2017. "No classification without representation: Assessing geodiversity issues in open data sets for the developing world." In *NeurIPS 2017 Workshop: Machine Learning for the Developing World*, 1–5.

[335] Simonite, Tom. 2018. "When it comes to gorillas, Google photos remains blind." *Wired*, January 11, 2018, 13. https://www.wired.com/story/when-it-comes-to-gorillas-google-photos-remains-blind/.

[336] Hern, Alex. 2015. "Flickr faces complaints over 'offensive' auto-tagging for photos." *The Guardian*, May 20, 2015, 20.

[337] Martineau, Paris. 2019. "Cities examine proper—and improper—uses of facial recognition." *Wired*, October 11, 2019. https://www.wired.com/story/cities-examine-proper-improper-facial-recognition/.

[338] O'Toole, Alice J., Deffenbacher, Kenneth, Abdi, Hervé, and Bartlett, James C. 1991. "Simulating the 'other-race effect' as a problem in perceptual learning." *Connection Science*, 3(2):163–178.

[339] Frucci, Adam. 2009. "HP face-tracking webcams don't recognize black people." *Gizmodo*, December 21, 2009. https://gizmodo.com/hp-face-tracking-webcams-dont-recognize-black-people-5431190.

[340] McEntegart, Jane. 2010. "Kinect may have issues with dark-skinned users." *Tom's Guide*, November 5, 2010. https://www.tomsguide.com/us/Microsoft-Kinect-Dark-Skin-Facial-Recognition,news-8638.html.

[341] Wilson, Benjamin, Hoffman, Judy, and Morgenstern, Jamie. 2019. "Predictive inequity in object detection." *arXiv preprint* arXiv:1902.11097.

[342] Turow, Joseph, King, Jennifer, Hoofnagle, Chris Jay, Bleakley, Amy, and Hennessy, Michael. 2009. "Americans reject tailored advertising and three activities that enable it." Available at SSRN 1478214.

[343] Raghavan, Manish, Barocas, Solon, Kleinberg, Jon, and Levy, Karen. 2019. "Mitigating bias in algorithmic employment screening: Evaluating claims and practices." *arXiv preprint* arXiv:1906.09208.

[344] Yao, Sirui, and Huang, Bert. 2017. "Beyond parity: Fairness objectives for collaborative filtering." In *Advances in Neural Information Processing Systems*, 30:2921–2930.

[345] Mehrotra, Rishabh, Anderson, Ashton, Diaz, Fernando, Sharma, Amit, Wallach, Hanna, and Yilmaz, Emine. 2017. "Auditing search engines for differential satisfaction across demographics." In *Proceedings of the 26th International Conference on World Wide Web*, 626–633. ACM.

[346] Susser, Daniel, Roessler, Beate, and Nissenbaum, Helen. 2018. "Online manipulation: Hidden influences in a digital world." *Georgetown Law Technology Review*, 4(1).

[347] Coltrane, Scott, and Messineo, Melinda. 2000. "The perpetuation of subtle prejudice: Race and gender imagery in 1990s television advertising." *Sex Roles*, 42(5–6):363–389.

[348] Angwin, Julia, Varner, Madeleine, and Tobin, Ariana. 2017. "Facebook enabled advertisers to reach 'jew haters'." *ProPublica*, September 14, 2017. https://www.propublica.org/article/facebook-enabled-advertisers-to-reach-jew-haters.

[349] Lambrecht, Anja, and Tucker, Catherine. 2019. "Algorithmic bias? An empirical study of apparent gender-based discrimination in the display of stem career ads." *Management Science*, 65(7):2966–2981.

[350] Datta, Amit, Tschantz, Michael Carl, and Datta, Anupam. 2015. "Automated experiments on ad privacy settings." *Proceedings on Privacy Enhancing Technologies*, 2015(1):92–112.

[351] Andreou, Athanasios, Goga, Oana, Gummadi, Krishna, Loiseau, Patrick, and Mislove, Alan. 2017. "Adanalyst." https://adanalyst.mpi-sws.org/.

[352] Hutson, Jevan A., Taft, Jessie G., Barocas, Solon, and Levy, Karen. 2018. "Debiasing desire: Addressing bias & discrimination on intimate platforms." *Proceedings of the ACM on Human-Computer Interaction*, 2(CSCW):73. ACM.

[353] Ayres, Ian, Banaji, Mahzarin, and Jolls, Christine. 2015. "Race effects on ebay." *RAND Journal of Economics*, 46(4):891–917.

[354] Lee, Min Kyung, Kusbit, Daniel, Metsky, Evan, and Dabbish, Laura. 2015. "Working with machines: The impact of algorithmic and data-driven management on human workers." In *Proceedings of the 33rd Annual ACM Conference on Human Factors in Computing Systems*, 1603–1612. ACM.

[355] Edelman, Benjamin, Luca, Michael, and Svirsky, Dan. 2017. "Racial discrimination in the sharing economy: Evidence from a field experiment." *American Economic Journal: Applied Economics*, 9(2): 1–22.

[356] Thebault-Spieker, Jacob, Terveen, Loren, and Hecht, Brent. 2017. "Toward a geographic understanding of the sharing economy: Systemic biases in Uberx and Taskrabbit." *ACM Transactions on Computer-Human Interaction*, 24(3):1–40.

[357] Ge, Yanbo, Knittel, Christopher R, MacKenzie, Don, and Zoepf, Stephen. 2016. "Racial and gender discrimination in transportation network companies." Working paper no. 22776, National Bureau of Economic Research.

[358] Levy, Karen and Barocas, Solon. 2017. "Designing against discrimination in online markets." *Berkeley Technology Law Journal*, 32:1183–1238.

[359] Tjaden, Jasper Dag, Schwemmer, Carsten, and Khadjavi, Menusch. 2018. "Ride with me: Ethnic discrimination, social markets, and the sharing economy." *European Sociological Review*, 34(4):418–432.

[360] Muthukumar, Vidya, Pedapati, Tejaswini, Ratha, Nalini, Sattigeri, Prasanna, Wu, Chai-Wah, Kingsbury, Brian, Kumar, Abhishek, Thomas, Samuel, Mojsilovic, Aleksandra, and Varshney, Kush R. 2018. "Understanding unequal gender classification accuracy from face images." *arXiv preprint* arXiv:1812.00099.

[361] Robertson, Ronald E, Jiang, Shan, Joseph, Kenneth, Friedland, Lisa, Lazer, David, and Wilson, Christo. 2018. "Auditing partisan audience bias within google search." *Proceedings of the ACM on Human-Computer Interaction*, 2(CSCW):148. ACM.

[362] D'Onfro, Jillian. 2019. "We sat in on an internal Google meeting where they talked about changing the search algorithm: Here's what we learned." https://www.cnbc.com/2018/09/17/google-tests-changes-to-its-search-algorithm-how-search-works.html.

[363] Hannák, Anikó, Sapieżyński, Piotr, Molavi Kakhki, Arash, Krishnamurthy, Balachander, Lazer, David, Mislove, Alan, and Wilson, Christo. 2013. "Measuring personalization of web search." In *Proceedings of the 22nd International Conference on World Wide Web*, 527–538. ACM.

[364] Tripodi, Francesca. 2018. "Searching for alternative facts: Analyzing scriptural inference in conservative news practices." Technical report, Data & Society.

[365] Valentino-DeVries, Jennifer, Singer-Vine, Jeremy, and Soltani, Ashkan. 2012. "Websites vary prices, deals based on users' information." *Wall Street Journal*, December 24, 2012, 60–68.

[366] Ojala, Markus and Garriga, Gemma C. 2010. "Permutation tests for studying classifier performance." *Journal of Machine Learning Research*, 11:1833–1863.

[367] Venkatadri, Giridhari, Lucherini, Elena, Sapiezynski, Piotr, and Mislove, Alan. 2019. "Investigating sources of PII used in Facebook's targeted advertising." *Proceedings on Privacy Enhancing Technologies*, 2019(1):227–244.

[368] Bashir, Muhammad Ahmad, Arshad, Sajjad, Robertson, William, and Wilson, Christo. 2016. "Tracing information flows between ad exchanges using retargeted ads." In *Proceedings of the 25th USENIX Security Symposium*, 481–496. USENIX.

[369] Singer-Vine, Jeremy, Valentino-DeVries, Jennifer, and Soltani, Ashkan. 2012. "How the Journal tested prices and deals online." *Wall Street Journal*, December 23, 2012. http://blogs. wsj. com/digits/2012/12/23/how-the-journal-tested-prices-and-deals-online.

[370] Chen, Le, Mislove, Alan, and Wilson, Christo. 2015. "Peeking beneath the hood of Uber." In *Proceedings of the 2015 Internet Measurement Conference*, 495–508. ACM.

[371] Salganik, Matthew. 2019. *Bit by bit: Social research in the digital age*. Princeton University Press.

[372] Bennett, James, and Lanning, Stan. 2007. "The Netflix Prize." In *Proceedings of KDD Cup and Workshop*, 35. ACM.

[373] Chaney, Allison J.B., Stewart, Brandon M., and Engelhardt, Barbara E. 2018. "How algorithmic confounding in recommendation systems increases homogeneity and decreases utility." In *Proceedings of the 12th ACM Conference on Recommender Systems*, 224–232. ACM.

[374] Obermeyer, Ziad, Powers, Brian, Vogeli, Christine, and Mullainathan, Sendhil. 2019. "Dissecting racial bias in an algorithm used to manage the health of populations." *Science*, 366(6464):447–453.

[375] Chouldechova, Alexandra, Benavides-Prado, Diana, Fialko, Oleksandr, and Vaithianathan, Rhema. 2018. "A case study of algorithm-assisted decision making in child maltreatment hotline screening decisions." In *Proceedings of the Conference on Fairness, Accountability and Transparency*, 134–148. ACM.

[376] Narayanan, Arvind. 2022. "The limits of the quantitative approach to discrimination." James Baldwin Lecture (transcript), Princeton University.

[377] Gaddis, S. Michael. 2018. "An introduction to audit studies in the social sciences." In *Audit Studies: Behind the Scenes with Theory, Method, and Nuance*, edited by S. Michael Gaddes, 3–44. Springer.

[378] Sandvig, Christian, et al. 2014. "Auditing algorithms: Research methods for detecting discrimination on internet platforms." *Data and discrimination: converting critical concerns into productive inquiry* 22: 4349–4357.

[379] Vecchione, Briana, Levy, Karen, and Barocas, Solon. 2021. "Algorithmic auditing and social justice: Lessons from the history of audit studies." In *Equity and Access in Algorithms, Mechanisms, and Optimization*, 1–9. ACM.

[380] Brundage, Miles, Avin, Shahar, Wang, Jasmine, Belfield, Haydn, Krueger, Gretchen, Hadfield, Gillian, Khlaaf, Heidy, Yang, Jingying, Toner, Helen, Fong, Ruth, et al. 2020. "Toward trustworthy AI development: Mechanisms for supporting verifiable claims." *arXiv preprint* arXiv:2004.07213.

[381] Wu, Tim. 2010. *The master switch: The rise and fall of information empires*. Vintage.

[382] Gillespie, Tarleton. 2010. "The politics of 'platforms'." *New Media & Society*, 12(3):347–364.

[383] Gillespie, Tarleton. 2018. *Custodians of the Internet: Platforms, content moderation, and the hidden decisions that shape social media*. Yale University Press.

[384] Klonick, Kate. 2017. "The new governors: The people, rules, and processes governing online speech." *Harvard Law Review*, 131:1598.

[385] Cook, Cody, Diamond, Rebecca, Hall, Jonathan, List, John A, and Oyer, Paul. 2018. "The gender earnings gap in the gig economy: Evidence from over a million rideshare drivers." Working paper no. 24732, National Bureau of Economic Research.

[386] Winner, Langdon. 2017. *Do artifacts have politics?* Routledge.

[387] Bullard, Robert Doyle, Johnson, Glenn Steve, and Torres, Angel O. 2004. *Highway robbery: Transportation racism and new routes to equity*. South End Press.

[388] Small, Mario L., and Pager, Devah. 2020. "Sociological perspectives on racial discrimination." *Journal of Economic Perspectives*, 34(2):49–67.

[389] Wikipedia. 2021. "Same-sex marriage." https://en.wikipedia.org/wiki/Same-sex_marriage.

[390] Rothstein, Richard. 2017. *The color of law: A forgotten history of how our government segregated America*. Liveright.

[391] Fellner, Jamie. 2009. "Race, drugs, and law enforcement in the United States." *Stanford Law & Policy Review*, 20(2):257–292.

[392] Mouw, Ted. 2002. "Are Black workers missing the connection? The effect of spatial distance and employee referrals on interfirm racial segregation". *Demography*, 39(3):507–528.

[393] Leslie, Sarah-Jane, Cimpian, Andrei, Meyer, Meredith, and Freeland, Edward. 2015. "Expectations of brilliance underlie gender distributions across academic disciplines." *Science*, 347(6219):262–265.

[394] Bian, Lin, Leslie, Sarah-Jane, and Cimpian, Andrei. 2017. "Gender stereotypes about intellectual ability emerge early and influence children's interests." *Science*, 355(6323):389–391.

[395] West, Candace and Zimmerman, Don H. 1987. "Doing gender." *Gender & Society*, 1(2):125–151.

[396] Valencia Caicedo, Felipe. 2019. "The mission: Human capital transmission, economic persistence, and culture in south america." *Quarterly Journal of Economics*, 134(1):507–556.

[397] Dell, Melissa. 2010. "The persistent effects of Peru's mining *mita*." *Econometrica*, 78(6):1863–1903.

[398] Barone, Guglielmo, and Mocetti, Sauro. 2016. "Intergenerational mobility in the very long run: Florence 1427–2011." Bank of Italy Temi di Discussione (Working Paper) no. 1060.

[399] Davidai, Shai, and Gilovich, Thomas. 2015. "Building a more mobile America—One income quintile at a time." *Perspectives on Psychological Science*, 10(1):60–71.

[400] Chetty, Raj, Hendren, Nathaniel, Jones, Maggie R., and Porter, Sonya R. 2020. "Race and economic opportunity in the United States: An intergenerational perspective." *Quarterly Journal of Economics*, 135(2): 711–783.

[401] Kochhar, Rakesh, and Cilluffo, Anthony. 2018. "Key findings on the rise in income inequality within america's racial and ethnic groups." Pew Research Center, July 12, 2018.

[402] Derenoncourt, Ellora, Kim, Chi Hyun, Kuhn, Moritz, and Schularick, Moritz. 2022. "Wealth of two nations: The US racial wealth gap, 1860–2020." Working paper no. 30101, National Bureau of Economic Research.

[403] Kraus, Michael W., Onyeador, Ivuoma N., Daumeyer, Natalie M., Rucker, Julian M., and Richeson, Jennifer A. 2019. "The misperception of racial economic inequality." *Perspectives on Psychological Science*, 14(6):899–921.

[404] Semega, Jessica L., Fontenot, Kayla R., and Kollar, Melissa A. 2017. "Income and poverty in the United States: 2016." Current Population Reports, P60–259, Economics and Statistics Administration, US Department of Commerce.

[405] Pendall, Rolf, and Hedman, Carl. 2015. "Worlds apart: Inequality between America's most and least affluent neighborhoods." Urban Institute.

[406] Merill, Jeremy. 2020. "Does Facebook still sell discriminatory ads?" *The Markup*, August 25, 2020.

[407] Gandy, Oscar H. 2016. *Coming to terms with chance: Engaging rational discrimination and cumulative disadvantage*. Routledge.

[408] Hellman, Deborah. 2020. "Sex, causation, and algorithms: How equal protection prohibits compounding prior injustice." *Washington University Law Review*, 98(2):481–523.

[409] Myrdal, Gunnar. 2017. *An American dilemma: The negro problem and modern democracy*, Vol. 2. Routledge.

[410] Boyd, Danah. 2012. "White flight in networked publics? How race and class shaped American teen engagement with MySpace and Facebook." In *Race after the Internet*, edited by Lisa Nakamura and Peter A. Chow-White, 203–222. Routledge.

[411] Kleinberg, Jon, and Raghavan, Manish. 2021. "Algorithmic monoculture and social welfare." *Proceedings of the National Academy of Sciences*, 118(22):e2018340118.

[412] O'Neil, Cathy. 2016. "How algorithms rule our working lives." *The Guardian*, September 1, 2016, 16.

[413] Gillespie, Tarleton. 2020. "Content moderation, AI, and the question of scale." *Big Data & Society*, 7(2): 2053951720943234.

[414] Kalluri, Pratyusha. 2019. "The values of machine learning." Neural Information Processing Systems (NeurIPS) Queer in AI workshop.

[415] Lokugamage, Amali U., Taylor, Sharon, and Rayner, Clare. 2020. "Patients' experiences of 'longcovid' are missing from the NHS narrative." *BMJ Opinion*, July 10, 2020.

[416] Callard, Felicity, and Perego, Elisa. 2021. "How and why patients made Long Covid." *Social Science & Medicine*, 268:113426.

[417] Sambasivan, Nithya, and Veeraraghavan, Rajesh. 2022. "The deskilling of domain expertise in AI development." In *Proceedings of the 2022 CHI Conference on Human Factors in Computing Systems*, 1–14. ACM.

[418] Lipsky, Michael. 2010. *Street-level bureaucracy: Dilemmas of the individual in public service*. Russell Sage Foundation.

[419] Jenkins, Jimmy 2021. "Whistleblowers: Software bug keeping hundreds of inmates in Arizona prisons beyond release dates." *KJZZ*, February 22, 2021.

[420] 2015. "US prisoners released early by software bug." *BBC News*, December 23, 2015.

[421] Mulligan, Deirdre K., and Bamberger, Kenneth A. 2019. "Procurement as policy: Administrative process for machine learning." *Berkeley Technology Law Journal*, 34:773–852.

[422] Barabas, Chelsea, Virza, Madars, Dinakar, Karthik, Ito, Joichi, and Zittrain, Jonathan. 2018. "Interventions over predictions: Reframing the ethical debate for actuarial risk assessment." In *Proceedings of the Conference on Fairness, Accountability and Transparency*, 62–76. PMLR.

[423] Akbar, Amna A. 2020. "An abolitionist horizon for police (reform)." *California Law Review*, 108(6).

[424] Roberts, Dorothy. 2022. *Torn apart: How the child welfare system destroys black families–and how abolition can build a safer world*. Basic Books.

[425] Karabel, Jerome. 2005. *The chosen: The hidden history of admission and exclusion at Harvard, Yale, and Princeton*. Houghton Mifflin Harcourt.

[426] Harwell, Drew. 2019. "Doorbell-camera firm ring has partnered with 400 police forces, extending surveillance concerns." *Washington Post*, August 29, 2019.

[427] Whittaker, Meredith, Crawford, Kate, Dobbe, Roel, Fried, Genevieve, Kaziunas, Elizabeth, Mathur, Varoon, West, Sarah Mysers, Richardson, Rashida, Schultz, Jason, and Schwartz, Oscar. 2018. *AI Now report 2018*. AI Now Institute at New York University.

[428] Barocas, Solon, and Levy, Karen. 2020. "Privacy dependencies." *Washington Law Review*, 95(2):555–616.

[429] Sloane, Mona, Moss, Emanuel, Awomolo, Olaitan, and Forlano, Laura. 2020. "Participation is not a design fix for machine learning." *arXiv preprint* arXiv:2007.02423.

[430] Hoffmann, Anna Lauren. 2019. "Where fairness fails: Data, algorithms, and the limits of antidiscrimination discourse." *Information, Communication & Society*, 22(7):900–915.

[431] Jobin, Anna, Ienca, Marcello, and Vayena, Effy. 2019. "The global landscape of AI ethics guidelines." *Nature Machine Intelligence*, 1(9):389–399.

[432] Judd, Sarah. 2020. "Activities for building understanding: How AI4ALL teaches AI to diverse high school students." In *Proceedings of the 51st ACM Technical Symposium on Computer Science Education*, 633–634. ACM.

[433] Tech Equity Collaborative. 2021. "Announcing the contract worker disparity project." Tech Equity Collaborative blog, April 20, 2021.

[434] Cowgill, Bo, Dell'Acqua, Fabrizio, Deng, Samuel, Hsu, Daniel, Verma, Nakul, and Chaintreau, Augustin. 2020. "Biased programmers? or Biased data? A field experiment in operationalizing AI ethics." In *Proceedings of the 21st ACM Conference on Economics and Computation*, 679–681.

[435] Fiesler, Casey, Garrett, Natalie, and Beard, Nathan. 2020. "What do we teach when we teach tech ethics? A syllabi analysis." In *Proceedings of the 51st ACM Technical Symposium on Computer Science Education*, 289–295. ACM.

[436] Martin, C. Dianne, Huff, Chuck, Gotterbarn, Donald, and Miller, Keith. 1996. "Implementing a tenth strand in the CS curriculum." *Communications of the ACM*, 39(12):75–84.

[437] Wikipedia. 2021. "Certified software development professional." https://en.wikipedia.org/wiki/Certified _Software_Development_Professional.

[438] Birhane, Abeba, Kalluri, Pratyusha, Card, Dallas, Agnew, William, Dotan, Ravit, and Bao, Michelle. 2022. "The values encoded in machine learning research." In *Proceedings of the Conference on Fairness, Accountability, and Transparency*, 173–184. ACM.

[439] Nanayakkara, Priyanka, Hullman, Jessica, and Diakopoulos, Nicholas. 2021. "Unpacking the expressed consequences of AI research in broader impact statements." *arXiv preprint* arXiv:2105.04760.

[440] Kasy, Maximilian, and Abebe, Rediet. 2021. "Fairness, equality, and power in algorithmic decision-making." In *Proceedings of the Conference on Fairness, Accountability, and Transparency*, 576–586. ACM.

[441] Paluck, Elizabeth Levy, and Green, Donald P. 2009. "Prejudice reduction: What works? A review and assessment of research and practice." *Annual Review of Psychology*, 60:339–367.

[442] Paluck, Elizabeth Levy, Porat, Roni, Clark, Chelsey S, and Green, Donald P. 2020. "Prejudice reduction: Progress and challenges." *Annual Review of Psychology*, 72:533–560.

[443] Chohlas-Wood, Alex, Nudell, Joe, Lin, Zhiyuan Jerry, Nyarko, Julian, and Goel, Sharad. 2021. "Blind justice: Algorithmically masking race in charging decisions." In *Proceedings of the 2021 AAAI/ACM Conference on AI, ethics and society*, 35–45.

[444] Squires, Gregory D. 1994. *Capital and communities in black and white: the intersections of race, class, and uneven development.* SUNY Press.

[445] Stevenson, Megan. 2018. "Assessing risk assessment in action." *Minn. L. Rev.* 103: 303.

[446] Albright, Alex. 2019. "If you give a judge a risk score: Evidence from Kentucky bail decisions." *Law, Economics, and Business Fellows' Discussion Paper Series* 85.

[447] Raso, Jennifer. 2017. "Displacement as regulation: New regulatory technologies and front-line decision-making in Ontario works." *Canadian Journal of Law and Society*, 32(1):75–95.

[448] Brayne, Sarah. 2020. *Predict and surveil: Data, discretion, and the future of policing.* Oxford University Press.

[449] Frankel, Marvin E. 1973. "Criminal sentences: Law without order."

[450] Palamar, Joseph J., Davies, Shelby, Ompad, Danielle C., Cleland, Charles M., and Weitzman, Michael. 2015. "Powder cocaine and crack use in the United States: An examination of risk for arrest and socioeconomic disparities in use." *Drug and Alcohol Dependence*, 149:108–116.

[451] Google re:work team. 2021. "Guide: Hire by committee." https://rework.withgoogle.com/print/guides /6053596147744768/.

[452] Natasha Tiku. 2021. "Google's approach to historically Black schools helps explain why there are few Black engineers in Big Tech." *Washington Post*, March 4, 2021.

[453] Edelman, Lauren B. 2005. "Law at work: The endogenous construction of civil rights." In *Handbook of employment discrimination research*, edited by Laura Beth Nielsen and Robert L. Nelson, 337–352. Springer.

[454] Soper, Spencer 2021. "Fired by bot at Amazon: 'It's you against the machine'." *Bloomberg*, June 28, 2021.

[455] Dynarski, Susan, Libassi, C.J., Michelmore, Katherine, and Owen, Stephanie. 2018. "Closing the gap: The effect of a targeted, tuition-free promise on college choices of high-achieving, low-income students." Working paper no. 25349, National Bureau of Economic Research.

[456] Antecol, Heather, Bedard, Kelly, and Stearns, Jenna. 2018. "Equal but inequitable: Who benefits from gender-neutral tenure clock stopping policies?" *American Economic Review*, 108(9):2420–41.

[457] Fishbane, Alissa, Ouss, Aurelie, and Shah, Anuj K. 2020. "Behavioral nudges reduce failure to appear for court." *Science*, 370(6517).

[458] Hellman, Deborah. 2016. "Two concepts of discrimination." *Virginia Law Review*, 102(4):895–952.

[459] Stevenson, Megan T., and Mayson, Sandra G. 2021. "Pretrial detention and the value of liberty." *Virginia Law Review*, 108:709–782.

[460] Lang, Kevin, and Kahn-Lang Spitzer, Ariella. 2020. "Race discrimination: An economic perspective." *Journal of Economic Perspectives* 34(2):68–89.

[461] Eaglin, Jessica M. 2017. "Constructing recidivism risk." *Emory Law Journal*, 67(1):59–122.

[462] Aragon, Cecilia, Guha, Shion, Kogan, Marina, Mulle, Michael, and Neff, Gina. 2022. *Human-centered data science: An introduction.* MIT Press.

[463] Green, Ben. 2021. "Data science as political action: Grounding data science in a politics of justice." *Journal of Social Computing*, 2(3):249–265.

[464] Kohler-Hausmann, Issa. 2018. *Misdemeanorland: Criminal courts and social control in an age of broken windows policing.* Princeton University Press.

[465] Paluck, Elizabeth Levy, Green, Seth A., and Green, Donald P. 2019. "The contact hypothesis re-evaluated." *Behavioural Public Policy*, 3(2):129–158.

[466] Lundberg, Shelly, and Startz, Richard. 1998. "On the persistence of racial inequality." *Journal of Labor Economics*, 16(2):292–323.

[467] Rosenblat, Alex. 2018. *Uberland: How algorithms are rewriting the rules of work.* University of California Press.

[468] Bowles, Samuel, and Sethi, Rajiv. 2006. "Social segregation and the dynamics of group inequality." Economics Department Working paper no. 2006–02, UMass Amherst.

[469] Loury, Glenn C. 1976. "A dynamic theory of racial income differences." Discussion paper no. 225, Center for Mathematical Studies in Economics and Management Science, Northwestern University.

[470] Massey, Douglas S., Rothwell, Jonathan, and Domina, Thurston. 2009. "The changing bases of segregation in the United States." *Annals of the American Academy of Political and Social Science*, 626(1):74–90.

[471] Li, Xiaochang, and Mills, Mara. 2019. "Vocal features: From voice identification to speech recognition by machine." *Technology and Culture*, 60(2):S129–S160.

[472] Liberman, Mark. 2010. "Fred Jelinek." *Computational Linguistics*, 36(4):595–599.

[473] Church, Kenneth Ward. 2018. "Emerging trends: A tribute to Charles Wayne." *Natural Language Engineering*, 24(1):155–160.

[474] Liberman, Mark, and Wayne, Charles. 2020. "Human language technology." *AI Magazine*, 41(2).

[475] Garofolo, John S., Lamel, Lori F., Fisher, William M., Fiscus, Jonathan G., and Pallett, David S. 1993. "DARPA TIMIT acoustic-phonetic continuous speech corpus CD-ROM." NIST Speech Disc 1–1.1. Technical report no. 93:27403. National Institute of Standards and Technology.

[476] Koenecke, Allison, Nam, Andrew, Lake, Emily, Nudell, Joe, Quartey, Minnie, Mengesha, Zion, Toups, Connor, Rickford, John R, Jurafsky, Dan, and Goel, Sharad. 2020. "Racial disparities in automated speech recognition." *Proceedings of the National Academy of Sciences*, 117(14):7684–7689.

[477] Langley, Pat. 2011. "The changing science of machine learning." *Machine Learning*, 82:275–279.

[478] LeCun, Yann, Bottou, Léon, Bengio, Yoshua, and Haffner, Patrick. 1998. "Gradient-based learning applied to document recognition." *Proceedings of the IEEE*, 86(11):2278–2324.

[479] Grother, Patrick J. 1995. "NIST Special Database 19." Handprinted forms and characters database, National Institute of Standards and Technology, 10.

[480] Yadav, Chhavi, and Bottou, Léon. 2019. "Cold case: The lost MNIST digits." *arXiv preprint* arXiv: 1905.10498.

[481] DeCoste, Dennis, and Schölkopf, Bernhard. 2002. "Training invariant support vector machines." *Machine Learning*, 46(1):161–190.

[482] Bromley, J., and Sackinger, E. 1991. "Neural-network and k-nearest-neighbor classifiers." *Rapport Technique*, 11359–910819.

[483] Miller, George A. 1998. *WordNet: An electronic lexical database*. MIT Press.

[484] Deng, Jia, Dong, Wei, Socher, Richard, Li, Li-Jia, Li, Kai, and Fei-Fei, Li. 2009. "ImageNet: A large-scale hierarchical image database." In *2009 IEEE Conference on Computer Vision and Pattern Recognition*, 248–255.

[485] Russakovsky, Olga, Deng, Jia, Su, Hao, Krause, Jonathan, Satheesh, Sanjeev, Ma, Sean, Huang, Zhiheng, Karpathy, Andrej, Khosla, Aditya, Bernstein, Michael, et al. 2015. "ImageNet large scale visual recognition challenge." *International Journal of Computer Vision*, 115(3):211–252.

[486] Gray, Mary L., and Suri, Siddharth. 2019. *Ghost work: How to stop Silicon Valley from building a new global underclass*. Eamon Dolan.

[487] Narayanan, Arvind, and Shmatikov, Vitaly. 2008. "Robust de-anonymization of large sparse datasets." In *2008 IEEE Symposium on Security and Privacy*, 111–125.

[488] Dwork, Cynthia, Smith, Adam, Steinke, Thomas, and Ullman, Jonathan. 2017. "Exposed! A survey of attacks on private data." *Annual Review of Statistics and Its Application*, 4:61–84.

[489] Dwork, Cynthia, and Roth, Aaron. 2014. "The algorithmic foundations of differential privacy." *Foundations and Trends in Theoretical Computer Science*, 9(3–4):211–407.

[490] Fisher, Ronald A. 1936. "The use of multiple measurements in taxonomic problems." *Annals of Eugenics*, 7(2):179–188.

[491] Evans, Richard J. 2020. "RA Fisher and the science of hatred." *New Statesman*, July 28, 2020.

[492] Louçã, Francisco. 2009. "Emancipation through interaction: How eugenics and statistics converged and diverged." *Journal of the History of Biology*, 42(4):649–684.

[493] Mitchell, Tom M. 1980. "The need for biases in learning generalizations." Rutgers CS technical report CBM-TR-117, Department of Computer Science, Laboratory for Computer Science Research, Rutgers University.

[494] Breiman, Leo. 2001. "Statistical modeling: The two cultures (with comments and a rejoinder by the author)." *Statistical Science*, 16(3):199–231.

[495] Rosenblatt, Frank. 1960. "Perceptron simulation experiments." *Proceedings of the IRE*, 48(3):301–309.

[496] Langley, Pat. 1988. "Machine learning as an experimental science." *Machine Learning*, 3:5–8.

[497] Funk, Simon. 2006. "Try this at home." http://sifter.org/~simon/journal/2006.

[498] Billsus, Daniel and Pazzani, Michael J. 1998. "Learning collaborative information filters." In *Proceedings of the Fifteenth International Conference on Machine Learning*, 46–54.

[499] Deerwester, Scott, Dumais, Susan T., Furnas, George W., Landauer, Thomas K., and Harshman, Richard. 1990. "Indexing by latent semantic analysis." *Journal of the American Society for Information Science*, 41(6):391–407.

[500] Koren, Yehuda, Bell, Robert, and Volinsky, Chris. 2009. "Matrix factorization techniques for recommender systems." *Computer*, 42(8):30–37.

[501] Recht, Benjamin, Roelofs, Rebecca, Schmidt, Ludwig, and Shankar, Vaishaal. 2019. "Do ImageNet classifiers generalize to ImageNet?" In *Proceedings of the 36th International Conference on Machine Learning*, 5389–5400. PMLR.

[502] Miller, John P., Taori, Rohan, Raghunathan, Aditi, Sagawa, Shiori, Koh, Pang Wei, Shankar, Vaishaal, Liang, Percy, Carmon, Yair, and Schmidt, Ludwig. 2021. "Accuracy on the line: On the strong correlation between out-of-distribution and in-distribution generalization." In *Proceedings of the 38th International Conference on Machine Learning*, 7721–7735. PMLR.

[503] Cortes, Corinna, and Vapnik, Vladimir. 1995. "Support-vector networks." *Machine Learning*, 20(3): 273–297.

[504] Crawford, Kate, and Paglen, Trevor. 2019. "Excavating AI: The politics of training sets for machine learning." *Excavating AI*, September 19, 2019. https://excavating.ai/.

[505] Gao, Leo, Biderman, Stella, Black, Sid, Golding, Laurence, Hoppe, Travis, Foster, Charles, Phang, Jason, He, Horace, Thite, Anish, Nabeshima, Noa, et al. 2020. "The Pile: An 800GB dataset of diverse text for language modeling." *arXiv preprint* arXiv:2101.00027.

[506] Veale, Michael, Binns, Reuben, and Edwards, Lilian. 2018. "Algorithms that remember: Model inversion attacks and data protection law." *Philosophical Transactions of the Royal Society A: Mathematical, Physical and Engineering Sciences*, 376(2133):20180083.

[507] Duda, Richard O., Hart, Peter E., and Stork, David G. 1973. *Pattern classification and scene analysis*. Vol. 3. Wiley.

[508] Hastie, Trevor, Tibshirani, Robert, and Friedman, Jerome. 2017. *The elements of statistical learning: Data mining, inference, and prediction*. Springer.

[509] Blum, Avrim, and Hardt, Moritz. 2015. "The Ladder: A reliable leaderboard for machine learning competitions." In *Proceedings of the 32nd International Conference on Machine Learning*, 1006–1014. PMLR.

[510] Dehghani, Mostafa, Tay, Yi, Gritsenko, Alexey A, Zhao, Zhe, Houlsby, Neil, Diaz, Fernando, Metzler, Donald, and Vinyals, Oriol. 2021. "The benchmark lottery." *arXiv preprint* arXiv:2107.07002.

[511] Koch, Bernard, Denton, Emily, Hanna, Alex, and Foster, Jacob G. 2021. "Reduced, reused and recycled: The life of a dataset in machine learning research." *arXiv preprint* arXiv:2112.01716.

[512] Koh, Pang Wei, Sagawa, Shiori, Marklund, Henrik, Xie, Sang Michael, Zhang, Marvin, Balsubramani, Akshay, Hu, Weihua, Yasunaga, Michihiro, Phillips, Richard Lanas, Gao, Irena, et al. 2020. "Wilds: A benchmark of in-the-wild distribution shifts." *arXiv preprint* arXiv:2012.07421.

[513] Branwen, Gwern. 2011. "The neural net tank urban legend." https://gwern.net/tank.

[514] Kaufman, Shachar, Rosset, Saharon, Perlich, Claudia, and Stitelman, Ori. 2012. "Leakage in data mining: Formulation, detection, and avoidance." *ACM Transactions on Knowledge Discovery from Data (TKDD)*, 6(4):1–21.

[515] Marie, Benjamin, Fujita, Atsushi, and Rubino, Raphael. 2021. "Scientific credibility of machine translation research: A meta-evaluation of 769 papers." In *Proceedings of the 59th Annual Meeting of the Association for*

Computational Linguistics and the 11th International Joint Conference on Natural Language Processing, 7297–7306. Association for Computational Linguistics.

[516] Bouthillier, Xavier, Delaunay, Pierre, Bronzi, Mirko, Trofimov, Assya, Nichyporuk, Brennan, Szeto, Justin, Mohammadi Sepahvand, Nazanin, Raff, Edward, Madan, Kanika, Voleti, Vikram, et al. 2021. "Accounting for variance in machine learning benchmarks." *Proceedings of Machine Learning and Systems*, 3.

[517] Saitta, Lorenza, and Neri, Filippo. 1998. "Learning in the 'real world'." *Machine Learning*, 30(2):133–163.

[518] Salzberg, Steven L. 1999. "On comparing classifiers: A critique of current research and methods." *Data Mining and Knowledge Discovery*, 1(1):1–12.

[519] Radin, Joanna. 2017. " 'Digital natives': How medical and indigenous histories matter for big data." *Osiris*, 32(1):43–64.

[520] Kapoor, Sayash, and Narayanan, Arvind. 2022. "Leakage and the reproducibility crisis in ML-based science." *arXiv preprint* arXiv:2207.07048.

[521] Prabhu, Vinay Uday, and Birhane, Abeba. 2020. "Large image datasets: A pyrrhic win for computer vision?" *arXiv preprint* arXiv:2006.16923.

[522] Yang, Kaiyu, Qinami, Klint, Fei-Fei, Li, Deng, Jia, and Russakovsky, Olga. 2020. "Towards fairer datasets: Filtering and balancing the distribution of the people subtree in the ImageNet hierarchy." In *Proceedings of the Conference on Fairness, Accountability, and Transparency*, 547–558. ACM.

[523] Bender, Emily M., Gebru, Timnit, McMillan-Major, Angelina, and Shmitchell, Shmargaret. 2021. "On the dangers of stochastic parrots: Can language models be too big?" In *Proceedings of the Conference on Fairness, Accountability, and Transparency*, 610–623. ACM.

[524] Paullada, Amandalynne, Raji, Inioluwa Deborah, Bender, Emily M., Denton, Emily, and Hanna, Alex. 2020. "Data and its (dis)contents: A survey of dataset development and use in machine learning research." *arXiv preprint* arXiv:2012.05345.

[525] Gebru, Timnit, Morgenstern, Jamie, Vecchione, Briana, Vaughan, Jennifer Wortman, Wallach, Hanna, Daumé, Hal, III, and Crawford, Kate. 2018. "Datasheets for datasets." *arXiv preprint* arXiv:1803.09010.

[526] Jo, Eun Seo, and Gebru, Timnit. 2020. "Lessons from archives: Strategies for collecting sociocultural data in machine learning." In *Proceedings of the Conference on Fairness, Accountability, and Transparency*, 306–316. ACM.

[527] Wang, Angelina, Narayanan, Arvind, and Russakovsky, Olga. 2020. "REVISE: A tool for measuring and mitigating bias in visual datasets." In *European Conference on Computer Vision*, 733–751. Springer.

[528] Gonen, Hila, and Goldberg, Yoav. 2019. "Lipstick on a pig: Debiasing methods cover up systematic gender biases in word embeddings but do not remove them." *arXiv preprint* arXiv:1903.03862.

[529] Kuczmarski, James. 2018. "Reducing gender bias in Google Translate." *Google: The Keyword* (blog), December 6, 2018. https://blog.google/products/translate/reducing-gender-bias-google-translate/.

[530] Johnson, Melvin. 2020. "A scalable approach to reducing gender bias in Google Translate." *Google Research* (blog), April 22, 2020. https://ai.googleblog.com/2020/04/a-scalable-approach-to-reducing-gender.html.

[531] Steed, Ryan, and Caliskan, Aylin. 2021. "Image representations learned with unsupervised pre-training contain human-like biases." In *Proceedings of the Conference on Fairness, Accountability, and Transparency*, 701–713. ACM.

[532] Huang, Gary B., Ramesh, Manu, Berg, Tamara, and Learned-Miller, Erik. 2007. "Labeled faces in the wild: A database for studying face recognition in unconstrained environments." Technical report 07–49, University of Massachusetts, Amherst.

[533] Kumar, Neeraj, Berg, Alexander, Belhumeur, Peter N., and Nayar, Shree. 2011. "Describable visual attributes for face verification and image search." *IEEE Transactions on Pattern Analysis and Machine Intelligence*, 33(10):1962–1977.

[534] Jacobs, Abigail Z., and Wallach, Hanna. 2021. "Measurement and fairness." In *Proceedings of the Conference on Fairness, Accountability, and Transparency*, 375–385. ACM.

[535] Messick, Samuel. 1998. "Test validity: A matter of consequence." *Social Indicators Research*, 45(1):35–44.

[536] Dawes, Robyn. M, Faust, David, and Meehl, Paul E. 1989. "Clinical versus actuarial judgment." *Science*, 243(4899):1668–1674.

[537] Chaaban, Ibrahim, and Scheessele, Michael R. 2007. "Human performance on the USPS database." Report, Indiana University South Bend.

[538] He, Kaiming, Zhang, Xiangyu, Ren, Shaoqing, and Sun, Jian. 2015. "Delving deep into rectifiers: Surpassing human-level performance on imagenet classification." In *International Conference on Computer Vision*, 1026–1034.

[539] Shankar, Vaishaal, Roelofs, Rebecca, Mania, Horia, Fang, Alex, Recht, Benjamin, and Schmidt, Ludwig. 2020. "Evaluating machine accuracy on ImageNet." In *Proceedings of the 37th International Conference on Machine Learning*, 8634–8644. PMLR.

[540] Herlocker, Jonathan L., Konstan, Joseph A., Terveen, Loren G., and Riedl, John T. 2004. "Evaluating collaborative filtering recommender systems." *ACM Transactions on Information Systems (TOIS)*, 22(1): 5–53.

[541] Amatriain, Xavier, and Basilico, Justin. 2012. "Netflix recommendations: Beyond the 5 stars (part 1)." *Netflix Technology Blog*, April 6, 2012.

[542] Blodgett, Su Lin, Lopez, Gilsinia, Olteanu, Alexandra, Sim, Robert, and Wallach, Hanna. 2021. "Stereotyping Norwegian salmon: An inventory of pitfalls in fairness benchmark datasets." In *Proceedings of the 59th Annual Meeting of the Association for Computational Linguistics and the 11th International Joint Conference on Natural Language Processing*. Vol. 1: (long papers), 1004–1015. Association for Computational Linguistics.

[543] Bao, Michelle, Zhou, Angela, Zottola, Samantha, Brubach, Brian, Desmarais, Sarah, Horowitz, Aaron, Lum, Kristian, and Venkatasubramanian, Suresh. 2021. "It's COMPASlicated: The messy relationship between RAI datasets and algorithmic fairness benchmarks." *arXiv preprint* arXiv:2106.05498.

[544] Reichman, Nancy E., Teitler, Julien O., Garfinkel, Irwin, and McLanahan, Sara S. 2001. "Fragile families: Sample and design." *Children and Youth Services Review*, 23(4–5):303–326.

[545] Dwork, Cynthia, Feldman, Vitaly, Hardt, Moritz, Pitassi, Toniann, Reingold, Omer, and Roth, Aaron. 2015. "The reusable holdout: Preserving validity in adaptive data analysis." *Science*, 349(6248):636–638.

[546] Boyd, Danah, and Crawford, Kate. 2012. "Critical questions for big data: Provocations for a cultural, technological, and scholarly phenomenon." *Information, Communication & Society*, 15(5):662–679.

[547] Tufekci, Zeynep. 2014. "Big questions for social media big data: Representativeness, validity and other methodological pitfalls." In *Proceedings of the Eighth International AAAI Conference on Weblogs and Social Media*, 505–514. Association for the Advancement of Artificial Intelligence.

[548] Tufekci, Zeynep. 2014. "Engineering the public: Big data, surveillance and computational politics." *First Monday*, 19(7). https://firstmonday.org/article/view/4901/4097.

[549] Couldry, Nick, and Mejias, Ulises A. 2019. "Data colonialism: Rethinking big data's relation to the contemporary subject." *Television & New Media*, 20(4):336–349.

[550] Olteanu, Alexandra, Castillo, Carlos, Diaz, Fernando, and Kıcıman, Emre. 2019. "Social data: Biases, methodological pitfalls, and ethical boundaries." *Frontiers in Big Data*, 2:13.

[551] Crawford, Kate. 2021. *The atlas of AI*. Yale University Press.

[552] Liberman, Mark. 2015. "Reproducible research and the common task method." *Simmons Foundation Lecture*. https://www.simonsfoundation.org/event/reproducible-research-and-the-common-task-method/.

[553] Paullada, Amandalynne, Raji, Inioluwa Deborah, Bender, Emily M., Denton, Emily, and Hanna, Alex. 2021. "Data and its (dis)contents: A survey of dataset development and use in machine learning research." *Patterns*, 2(11):100336.

[554] Fabris, Alessandro, Messina, Stefano, Silvello, Gianmaria, and Susto, Gian Antonio. 2022. "Algorithmic fairness datasets: The story so far." *arXiv preprint* arXiv:2202.01711.

[555] Denton, Emily, Hanna, Alex, Amironesei, Razvan, Smart, Andrew, and Nicole, Hilary. 2021. "On the genealogy of machine learning datasets: A critical history of imagenet." *Big Data & Society*, 8(2): 20539517211035955.

[556] Raji, Inioluwa Deborah, Bender, Emily M., Paullada, Amandalynne, Denton, Emily, and Hanna, Alex. 2021. "AI and the everything in the whole wide world benchmark." *arXiv preprint* arXiv:2111.15366.

[557] Hand, David J. 2010. *Measurement theory and practice: The world through quantification*. Wiley.

[558] Hand, David J. 2016. *Measurement: A very short introduction*. Oxford University Press.

[559] Bandalos, Deborah L. 2018. *Measurement theory and applications for the social sciences*. Guilford.

[560] Liao, Thomas, Taori, Rohan, Raji, Inioluwa Deborah, and Schmidt, Ludwig. 2021. "Are we learning yet? A meta review of evaluation failures across machine learning." In *35th Conference on Neural Information Processing Systems Datasets and Benchmarks Track* (Round 2).

Index